STUDENT'S
SOLUTIONS MANUAL

Aimee L. Calhoun
Monroe Community College

Richard C. Stewart
Monroe Community College

A SURVEY OF MATHEMATICS
WITH APPLICATIONS

SEVENTH EDITION AND
EXPANDED SEVENTH EDITION

Allen R. Angel
Monroe Community College

Christine D. Abbott
Monroe Community College

Dennis C. Runde
Manatee Community College

PEARSON

Addison
Wesley

Boston   San Francisco   New York
London   Toronto   Sydney   Tokyo   Singapore   Madrid
Mexico City   Munich   Paris   Cape Town   Hong Kong   Montreal

ISBN   0-321-20597-9

6 7 8 9 10 BRG 07 06

PEARSON

Addison
Wesley

Table of Contents

ACKNOWLEDGMENTS

We would like to thank Allen Angel, Christine Abbott, and Dennis Runde, the authors of *A Survey of Mathematics with Applications*, for their support and encouragement; Joe Vetere from Addison Wesley for his computer expertise; Lauren Morse from Addison Wesley for her assistance; and Jane Cummings from Monroe Community College for her technical support.

Aimee L. Calhoun
Richard C. Stewart

To my wonderful husband, Justin, for his love and support throughout the process of co-authoring this manual;
my children, Melanie and Jacob, for being my inspiration;
my incredible parents and the rest of my loving family for always believing in me;
and to the loving memories of my grandparents.

Aimee L. Calhoun

I am grateful to my wife, Christy, who enthusiastically supported my efforts to contribute to this manual, and to my daughters, Sarah and Sheila, for their encouragement to serve as an educator.

Richard C. Stewart

CHAPTER ONE

CRITICAL THINKING SKILLS

Exercise Set 1.1

1. a) 1, 2, 3, 4, 5, …
   b) Counting numbers

3. A **conjecture** is a belief based on specific observations that has not been proven or disproven.

5. **Deductive reasoning** is the process of reasoning to a specific conclusion from a general statement.

7. Inductive reasoning

9. Inductive reasoning, because a general conclusion was made from observation of specific cases.

11. $1 \quad 5(1+4) \quad 10(4+6) \quad 10(6+4) \quad 5(4+1) \quad 1$

13. $5 \times 9 = 45$

15.

17.

19. 15, 18, 21 (Add 3 to previous number.)

21. $-1, 1, -1$ (Alternate -1 and 1.)

23. $\dfrac{1}{81}, \dfrac{1}{243}, \dfrac{1}{729}$ (Multiply previous number by $\dfrac{1}{3}$.)

25. 36, 49, 64 (The numbers in the sequence are the squares of the counting numbers.)

27. 34, 55, 89 (Each number in the sequence is the sum of the previous two numbers.)

29. There are three letters in the pattern. $39 \times 3 = 117$, so the 117th entry is the second R in the pattern. Therefore, the 118th entry is Y.

31. a) 36, 49, 64

    b) Square the numbers 6, 7, 8, 9 and 10.

    c) $8 \times 8 = 64 \qquad 9 \times 9 = 81$

    72 is not a square number since it falls between the two square numbers 64 and 81.

33. Blue: 1, 5, 7, 10, 12   Purple: 2, 4, 6, 9, 11
    Yellow: 3, 8

35. a) ≈ 58 million   b) ≈ 45 million
    c) We are using observation of specific cases to make a prediction.

37.

P	B	P	B
B	P	B	P
P	B	P	B
B	P	B	P

39. a) You should obtain the original number.

    b) You should obtain the original number.

    c) Conjecture: The result is always the original number.

    d) $n, 4n, 4n+8, \dfrac{4n+8}{4} = \dfrac{4n}{4} + \dfrac{8}{4} = n+2, n+2-2 = n$

41. a) You should obtain the number 5.

    b) You should obtain the number 5.

    c) Conjecture: No matter what number is chosen, the result is always the number 5.

    d) $n, n+1, n+(n+1) = 2n+1, 2n+1+9 = 2n+10, \dfrac{2n+10}{2} = \dfrac{2n}{2} + \dfrac{10}{2} = n+5, n+5-n = 5$

43. $999 \times 999 = 998,001$ is one counterexample.

45. Two is a counting number. The sum of 2 and 3 is 5. Five divided by two is $\dfrac{5}{2}$, which is not an even number.

47. One and two are counting numbers. The difference of 1 and 2 is $1-2 = -1$, which is not a counting number.

49. a) The sum of the measures of the interior angles should be $180°$.

    b) Yes, the sum of the measures of the interior angles should be $180°$.

    c) Conjecture: The sum of the measures of the interior angles of a triangle is $180°$.

51. 129, the numbers in positions are found as follows: $\begin{matrix} a & b \\ c & a+b+c \end{matrix}$

53. Counterexample has 14 letters, the upper left E get used twice.

54. c

Exercise Set 1.2

(Note: Answers in this section will vary depending on how you round your numbers. The answers may differ from the answers in the back of the textbook. However, your answers should be something near the answers given. All answers are approximate.)

1. $431 + 327.2 + 73.5 + 20.4 + 315.9 \approx 430 + 330 + 70 + 20 + 320 = 1170$

3. $297,700 \times 4087 \approx 300,000 \times 4000 = 1,200,000,000$

5. $\dfrac{405}{0.049} \approx \dfrac{400}{0.05} = 8000$

7. $0.049 \times 1989 \approx 0.05 \times 2000 = 100$

9. $51,608 \times 6981 \approx 52,000 \times 7000 = 364,000,000$

11. $592 \times 2070 \times 992.62$
    $\approx 600 \times 2000 \times 1000 = 1,200,000,000$

13. $52 \times \$0.37 \approx 50 \times \$0.40 = \$20$

15. $1521 + 1897 + 2324 + 2817$
    $\approx 1500 + 1900 + 2300 + 2800 = 8500$ mi

17. $\$2.29 + \$12.16 + \$4.97 + \$6.69 + \$49.76 + \$0.47$
    $+\$3.49 + \$5.65 \approx \$2 + \$12 + \$5 + \$7 + \$50 + \$0.50$
    $+\$3.50 + \$5.70 = \$85.70$

19. $\dfrac{\$44,569}{5} \approx \dfrac{\$45,000}{5} = \$9000$

21. $9 \times 5.12 \approx 9 \times 5 = 45$ lb

23. $\dfrac{23,663}{12} \approx \dfrac{24,000}{12} = 2000$ mi

25. $12(\$29.17 + \$39.95)$
    $\approx 12(\$30 + \$40) = 12(\$70) = \$840$

27. 15% of $\$38.60 \approx 15\%$ of $\$40 = 0.15 \times \$40 = \$6$

29. 100 Mexican pesos $= 100 \times 0.092$ U.S. dollars
    $\approx 100 \times 0.09$ U.S. dollars $= 9$ U.S. dollars
    $\$50 - \$9 = \$41$

31. ≈ 375 miles

33. a) $30.98\% \times 105$ million $\approx 31\% \times 105$ million
$= 0.31 \times 105$ million $= 32.55$ million ≈ 32.6 million
b) $18.41\% \times 3141$
$\approx 18\% \times 3100 = 0.18 \times 3100 = 558$ counties
c) The counties that use punch cards could be the largest counties with the most voters.

35. a) 4 million
b) 98 million
c) 98 million $- 37$ million $= 61$ million
d) 19 million $+ 78$ million $+ 82$ million $+$
61 million $+ 37$ million $= 277$ million

37. a) 83%
b) $65\% - 45\% = 20\%$
c) 83% of 110,567
$\approx 0.83 \times 110,567 = 91,770.61 \approx 91,771$ sq mi
d) No, since we are not given the area of each state.

39. 25

41. ≈ 90 berries

43. 150°

45. 10%

47. 9 square units

49. 150 feet

51.-59. Answers will vary.

60. There are 118 ridges around the edge.

61. There are 336 dimples on a regulation golf ball.

62. Answers will vary.

63. Answers will vary. The U.S. government categorized the middle class as $32,000 - $50,000 in 2001.

Exercise Set 1.3

1. $\dfrac{1 \text{ in.}}{50 \text{ mi}} = \dfrac{3.75 \text{ in.}}{x \text{ mi}}$
$1x = 50(3.75)$
$x = 187.5$ mi

3. $\dfrac{3 \text{ ft}}{1.2 \text{ ft}} = \dfrac{48.4 \text{ ft}}{x \text{ ft}}$
$3x = 1.2(48.4)$
$\dfrac{3x}{3} = \dfrac{58.08}{3}$
$x = \dfrac{58.08}{3} = 19.36$ ft

5. 11.5% of $4222 = 0.115($4222$) = 485.53
$4222 + $485.53 = $4707.53

7. $\dfrac{20,000 \text{ miles}}{20 \text{ miles per gallon}} = 1000$ gallons
Hawaii: $1000($2.02$) = 2020
South Carolina: $1000($1.22$) = 1220
$2020 - $1220 = $800

9. Denise parks her car for eight hours per day.
$5\left[$2.50 + $1.00(7 \text{ hours per day})\right]$
$= 5[$2.50 + $7.00] = 5($9.50$) = 47.50
Savings: $47.50 - $35.00 = $12.50

11. $120 + $80(15) = $120 + $1200 = 1320
Savings: $1320 - $1250 = $70

13. 20 year mortgage: $752.40(12)(20) = $180,576$
30 year mortgage: $660.60(12)(30) = $237,816$
Savings: $237,816 - $180,576 = $57,240

15. a) $\dfrac{86.5}{34} \approx 2.54; \dfrac{91.5}{36} \approx 2.54; \dfrac{96.5}{38} \approx 2.54;$
$\dfrac{101.5}{40} \approx 2.54; \dfrac{106.5}{42} \approx 2.54\ldots$
So, $48(2.54) \approx 122$.
b) Answers will vary. A close approximation can be obtained by multiplying the U.S. sizes by 2.54.

17. a) $\dfrac{460}{50} = 9.2$ min

    b) $\dfrac{1550}{25} = 62$ min

    c) $\dfrac{1400}{35} = 40$ min

    d) $\dfrac{1550}{25} + \dfrac{2200}{25} = \dfrac{3750}{25} = 150$ min

19. a) 11% of 273,300,000

    $= 0.11(273,300,000) = 30,063,000$

    b) 10% of 970,000 $= 0.10(970,000) = 97,000$

    c) 3% of 970,000 $= 0.03(970,000) = 29,100$

21. By mail: $(\$52.80 + \$5.60 + \$8.56) \times 4$

    $= \$66.96 \times 4 = \$267.84$

    Tire store: $\$324 + 0.08 \times \$324$

    $= \$324 + \$25.92 = \$349.92$

    Savings: $\$349.92 - \$267.84 = \$82.08$

23. a) $\$620(0.12) = \$74.40$

    b) $\$1200(0.22) = \$264$

    c) The store lost $\$1200 - \$1000 = \$200$ on
    the purchase.
    Store's profit: $\$264 - \$200 = \$64$

25. Let $x =$ the amount above $12,000$

    $\$4950 - \$1200 = \$3390$

    $\dfrac{0.15x}{0.15} = \dfrac{\$3390}{0.15}$

    $x = \$22,600$

    $\$12,000 + \$22.600 = \$34,600$

27. $7(2) + 5(1) + 4(29) + 3(201) + 2(1408) + 1(10,352)$

    $= 14 + 5 + 116 + 603 + 2816 + 10,352$

    $= 13,906$ violations

29. a) Yes, divide the total emissions by the emissions per capita.

    b) $\dfrac{6503.8}{24.3} \approx 267.646$ million ≈ 267.65 million

    c) $\dfrac{4964.8}{4.0} = 1241.2$ million or 1.2412 billion

31. Value after first year: $\$1000 + 0.10(\$1000)$

    $= \$1000 + \$100 = \$1100$

    Value after second year: $\$1100 - 0.10(\$1100)$

    $= \$1100 - \$110 = \$990$

    $990 is less than the intial investment of $1000.

33. a) $\dfrac{\$200}{\$41} \approx 4.87804878$ The maximum number of 10 packs is 4.

    $\$200 - (4 \times \$41) = \$200 - \$164 = \$36$, $\dfrac{\$36}{\$17} = 2.117647059$ Deirdre can also buy two 4 packs.

10 packs	4 packs	Number of rolls	Cost
4	2	$4(10) + 2(4) = 48$	$4(\$41) + 2(\$17) = \$198$
3	4	46	$191
2	6	44	$184
1	9	46	$194
0	11	44	$187

Maximum number of rolls of film is 48.

b) The cost is $198 when she purchases four 10 packs and two 4 packs.

31.   ≈ 375 miles

33.   a) 30.98%×105 million ≈ 31%×105 million

   = 0.31×105 million = 32.55 million ≈ 32.6 million

   b) 18.41%×3141

   ≈ 18%×3100 = 0.18×3100 = 558 counties

   c) The counties that use punch cards could be the largest counties with the most voters.

35.   a) 4 million

   b) 98 million

   c) 98 million − 37 million = 61 million

   d) 19 million + 78 million + 82 million + 61 million + 37 million = 277 million

37.   a) 83%

   b) 65% − 45% = 20%

   c) 83% of 110,567

   ≈ 0.83×110,567 = 91,770.61 ≈ 91,771 sq mi

   d) No, since we are not given the area of each state.

39.   25

41.   ≈ 90 berries

43.   150°

45.   10%

47.   9 square units

49.   150 feet

51.-59. Answers will vary.

60.   There are 118 ridges around the edge.

61.   There are 336 dimples on a regulation golf ball.

62.   Answers will vary.

63.   Answers will vary. The U.S. government categorized the middle class as $32,000 - $50,000 in 2001.

Exercise Set 1.3

1. $$\frac{1 \text{ in.}}{50 \text{ mi}} = \frac{3.75 \text{ in.}}{x \text{ mi}}$$
$$1x = 50(3.75)$$
$$x = 187.5 \text{ mi}$$

3. $$\frac{3 \text{ ft}}{1.2 \text{ ft}} = \frac{48.4 \text{ ft}}{x \text{ ft}}$$
$$3x = 1.2(48.4)$$
$$\frac{3x}{3} = \frac{58.08}{3}$$
$$x = \frac{58.08}{3} = 19.36 \text{ ft}$$

5.   11.5% of $4222 = 0.115($4222$) = 485.53
$4222 + $485.53 = $4707.53

7. $$\frac{20,000 \text{ miles}}{20 \text{ miles per gallon}} = 1000 \text{ gallons}$$

   Hawaii: 1000($2.02) = $2020

   South Carolina: 1000($1.22) = $1220

   $2020 − $1220 = $800

9.   Denise parks her car for eight hours per day.
$5[$2.50 + $1.00(7 \text{ hours per day})]$
$= 5[$2.50 + $7.00] = 5($9.50) = 47.50
   Savings: $47.50 − $35.00 = $12.50

11. $120 + $80(15) = $120 + $1200 = $1320
   Savings: $1320 − $1250 = $70

13.   20 year mortgage: $752.40(12)(20) = $180,576
   30 year mortgage: $660.60(12)(30) = $237,816
   Savings: $237,816 − $180,576 = $57,240

15.   a) $\frac{86.5}{34} \approx 2.54; \frac{91.5}{36} \approx 2.54; \frac{96.5}{38} \approx 2.54;$
$\frac{101.5}{40} \approx 2.54; \frac{106.5}{42} \approx 2.54\ldots$
   So, 48(2.54) ≈ 122.

   b) Answers will vary. A close approximation can be obtained by multiplying the U.S. sizes by 2.54.

17. a) $\dfrac{460}{50} = 9.2$ min

   b) $\dfrac{1550}{25} = 62$ min

   c) $\dfrac{1400}{35} = 40$ min

   d) $\dfrac{1550}{25} + \dfrac{2200}{25} = \dfrac{3750}{25} = 150$ min

19. a) 11% of 273,300,000

   $= 0.11(273,300,000) = 30,063,000$

   b) 10% of 970,000 $= 0.10(970,000) = 97,000$

   c) 3% of 970,000 $= 0.03(970,000) = 29,100$

21. By mail: $(\$52.80 + \$5.60 + \$8.56) \times 4$

   $= \$66.96 \times 4 = \$267.84$

   Tire store: $\$324 + 0.08 \times \$324$

   $= \$324 + \$25.92 = \$349.92$

   Savings: $\$349.92 - \$267.84 = \$82.08$

23. a) $\$620(0.12) = \$74.40$

   b) $\$1200(0.22) = \$264$

   c) The store lost $\$1200 - \$1000 = \$200$ on the purchase.

   Store's profit: $\$264 - \$200 = \$64$

25. Let $x =$ the amount above $12,000

   $\$4950 - \$1200 = \$3390$

   $\dfrac{0.15x}{0.15} = \dfrac{\$3390}{0.15}$

   $x = \$22,600$

   $\$12,000 + \$22.600 = \$34,600$

27. $7(2) + 5(1) + 4(29) + 3(201) + 2(1408) + 1(10,352)$

   $= 14 + 5 + 116 + 603 + 2816 + 10,352$

   $= 13,906$ violations

29. a) Yes, divide the total emissions by the emissions per capita.

   b) $\dfrac{6503.8}{24.3} \approx 267.646$ million ≈ 267.65 million

   c) $\dfrac{4964.8}{4.0} = 1241.2$ million or 1.2412 billion

31. Value after first year: $\$1000 + 0.10(\$1000)$

   $= \$1000 + \$100 = \$1100$

   Value after second year: $\$1100 - 0.10(\$1100)$

   $= \$1100 - \$110 = \$990$

   $990 is less than the intial investment of $1000.

33. a) $\dfrac{\$200}{\$41} \approx 4.87804878$ The maximum number of 10 packs is 4.

   $\$200 - (4 \times \$41) = \$200 - \$164 = \$36$, $\dfrac{\$36}{\$17} = 2.117647059$ Deirdre can also buy two 4 packs.

10 packs	4 packs	Number of rolls	Cost
4	2	$4(10) + 2(4) = 48$	$4(\$41) + 2(\$17) = \$198$
3	4	46	$191
2	6	44	$184
1	9	46	$194
0	11	44	$187

   Maximum number of rolls of film is 48.

   b) The cost is $198 when she purchases four 10 packs and two 4 packs.

35. a) water/milk: $3(1) = 3$ cups   salt: $3\left(\dfrac{1}{8}\right) = \dfrac{3}{8}$ tsp

    cream: $3(3) = 9$ tbsp $= \dfrac{9}{16}$ cup (because 16 tbsp = 1 cup)

    b) water/milk: $\dfrac{2+3.75}{2} = \dfrac{5.75}{2} = 2.875$ cups $= 2\dfrac{7}{8}$ cups

    salt: $\dfrac{0.25+0.5}{2} = \dfrac{0.75}{2} = 0.375$ tsp $= \dfrac{3}{8}$ tsp   cream: $\dfrac{0.5+0.75}{2} = \dfrac{1.25}{2} = 0.625$ cups $= \dfrac{5}{8}$ cup

    $$= \dfrac{5}{8}(16 \text{ tbsp}) = 10 \text{ tbsp}$$

    c) water/milk: $3\dfrac{3}{4} - 1 = \dfrac{15}{4} - \dfrac{4}{4} = \dfrac{11}{4} = 2\dfrac{3}{4}$ cups

    salt: $\dfrac{1}{2} - \dfrac{1}{8} = \dfrac{4}{8} - \dfrac{1}{8} = \dfrac{3}{8}$ tsp   cream: $\dfrac{3}{4} - \dfrac{3}{16} = \dfrac{12}{16} - \dfrac{3}{16} = \dfrac{9}{16}$ cup = 9 tbsp

    d) Differences exist in water/milk because the amount for 4 servings is not twice that for 2 servings. Differences also exist in Cream of Wheat because $\dfrac{1}{2}$ cup is not twice 3 tbsp.

37. 1 ft^2 would be 12 in. by 12 in.

    Thus, $1 \text{ ft}^2 = 12 \text{ in.} \times 12 \text{ in.} = 144 \text{ in.}^2$

39. Area of original rectangle = lw

    Area of new rectangle = $(2l)(2w) = 4lw$

    Thus, if the length and width of a rectangle are doubled, the area is 4 times as large.

41. 1 and 9

    $1 \times 9 = 9$

    $1 + 9 = 10$

43. Left side: $1(-6) = -6$ Right side: $1(2) = 2$

    $2(-2) = -4$ $1(3) = 3$

    $-6 + -4 = -10$ $1(6) = 6$

                                     $2 + 3 + 6 = 11$

    Place it at -1 so the left side would total $-10 + -1 = -11$.

45.

Birds	Lizards	Number of Heads	Number of Feet
8	14	22	$8(2) + 14(4) = 72$
9	13	22	$9(2) + 13(4) = 70$
10	12	22	$10(2) + 12(4) = 68$

    Therefore, there are 10 birds and 12 lizards.

    $\dots + (2 \times 2) + (1 \times 1)$

    $\dots 30$

49.

51.

8	6	16
18	10	2
4	14	12

53. $6+10+8+4=28; 3+7+5+1=16;$

$10+14+12+8=44$

The sum of the four corner entries is

4 times the number in the center of the middle row.

55. $63, 36, 99$. Multiply the number in the center of the middle   row by 9.

57. $3 \times 2 \times 1 = 6$ ways

59.

	7	
3	1	4
5	8	6
	2	

Other answers are possible, but 1 and 8 must appear in the center.

61.

1	2	3	4	5
2	3	4	5	1
3	4	5	1	2
4	5	1	2	3
5	1	2	3	4

Other answers are possible.

63.   Mary is the skier.

65.   Areas of the colored regions are:

1×1, 1×1, 2×2, 3×3, 5×5, 8×8, 13×13,

21×21; $1 + 1 + 4 + 9 + 25 + 64 + 169 + 441$

$= 714$ square units

66.   1 giraffe = 2 frogs, 1 giraffe = 3 lions, 3 lions = 2 frogs. Therefore, $\dfrac{3}{3}$ lion = $\dfrac{2}{3}$ frog.

Therefore, 1 lion = $\dfrac{2}{3}$ frog. 1 lion = 2 ostriches. Therefore, $\dfrac{2}{3}$ frog = 2 ostriches.

$\dfrac{2}{3}\left(\dfrac{3}{2}\right)$ frog = $2\left(\dfrac{3}{2}\right)$ ostriches. Therefore, 1 frog = 3 ostriches.

Review Exercises

1.   23, 28, 33 (Add 5 to previous number.)

2.   25, 36, 49 (next three perfect squares)

3.   −48, 96, −192 (Multiply previous number by –2.)

4.   25, 32, 40 (19 + 6 = 25, 25 + 7 = 32, 32 + 8 = 40)

5.   15, 9, 2 $(20 - 5 = 15, 15 - 6 = 9, 9 - 7 = 2)$

6. $\dfrac{3}{8}, \dfrac{3}{16}, \dfrac{3}{32}$ (Multiply previous number by $\dfrac{1}{2}$.)

7.

8.

9. c

10.   a) The original number and the final number are the same.

b) The original number and the final number are the same.

c) Conjecture: The final number is the same as the original number.

d) $n, 2n, 2n+10, \dfrac{2n+10}{2} = \dfrac{2n}{2} + \dfrac{10}{2} = n+5, n+5-5 = n$

11. This process will always result in an answer of 3. $n, n+5, 6(n+5) = 6n+30, 6n+30-12$

$$= 6n+18, \frac{6n+18}{2} = \frac{6n}{2} + \frac{18}{2} = 3n+9, \frac{3n+9}{3} = \frac{3n}{3} + \frac{9}{3} = n+3, n+3-n = 3$$

12. $1^2 + 2^2 = 5, 5$ is an odd number.

(Note: Answers for Ex. 13 - 25 will vary depending on how you round your numbers. The answers may differ from the answers in the back of the textbook. However, your answers should be something near the answers given. All answers are approximate.)

13. $210,302 \times 1992 \approx 210,000 \times 2000 = 420,000,000$

14. $346.2 + 96.402 + 1.04 + 897 + 821$
$\approx 350 + 100 + 0 + 900 + 800 = 2150$

15. 21% of $1012 \approx 20\%$ of 1000
$= 0.20 \times 1000 = 200$

16. Answers will vary.

17. $82 \times \$1.09 \approx 80 \times \$1.10 = \$88$

18. 6% of $\$202 \approx 6\%$ of $200 = 0.06 \times 200 = \$12$

19. $\dfrac{1.1 \text{ mi}}{22 \text{ min}} \approx \dfrac{1 \text{ mi}}{20 \text{ min}} = \dfrac{3 \text{ mi}}{60 \text{ min}} = 3 \text{ mph}$

20. $\$2.49 + \$0.79 + \$1.89 + \$0.10 + \$2.19 + \$6.75$
$\approx \$2 + \$1 + \$2 + \$0 + \$2 + \$7 = \$14.00$

21. $5 \text{ in.} = \dfrac{20}{4} \text{ in.} = 20\left(\dfrac{1}{4}\right) \text{ in.} = 20(0.1) \text{ mi} = 2 \text{ mi}$

22. 70%

23. 5%

24. 13 square units

25. Length $= 1.75$ in., $1.75(12.5) = 21.875 \approx 22$ ft

Height $= 0.625$ in., $0.625(12.5) = 7.8125 \approx 8$ ft

26. $\$2.00 + 7(\$1.50) = \$2.00 + \$10.50 = \$12.50$

Change: $\$20.00 - \$12.50 = \$7.50$

27. $4(\$2.69) = \$10.76$ for four six-packs

Savings: $\$10.76 - \$9.60 = \$1.16$

28. Akala's: $2 \text{ hr} = 120 \text{ min}, \dfrac{120}{15} = 8, 8 \times \$15 = \$120$

Berkman's: $2 \text{ hr} = 120 \text{ min}, \dfrac{120}{30} = 4,$

$4 \times \$25 = \$100$

Berkman's is the better deal by
$\$120 - \$100 = \$20.00.$

29. To produce the 52 Oscars he found:
$52 \times \$327 = \$17,004$
He was awarded
$\$50,000 - \$17,004 = \$32,996$ more.

30. $\$1.50 + \left[\left(10 - \dfrac{1}{5}\right)(5)\right]\$0.30$

$= \$1.50 + \left[\left(\dfrac{50}{5} - \dfrac{1}{5}\right)(5)\right]\$0.30$

$= \$1.50 + \left[\dfrac{49}{5}(5)\right]\$0.30$

$= \$1.50 + 49 \times \$0.30 = \$1.50 + \$14.70 = \$16.20$

31. 10% of $\$530 = 0.10 \times \$530 = \$53$
$\$53 \times 7 = \$371$
Savings: $\$371 - \$60 = \$311$

32. $\dfrac{1.5 \text{ mg}}{10 \text{ lb}} = \dfrac{x \text{ mg}}{47 \text{ lb}}$

$10x = 47(1.5)$

$\dfrac{10x}{10} = \dfrac{70.5}{10}$

$x = 7.05 \text{ mg}$

33. $\$3800 - 0.30(\$3800) = \$3800 - \$1140$

$= \$2660$ take-home

28% of $\$2660 = 0.28 \times \$2660 = \$744.80$

34. 9 A.M. Eastern is 6 A.M. Pacific,
from 6 A.M. Pacific to 1:35 P.M. Pacific
is 7 hr 35 min , 7 hr 35 min $-$ 50 min stop
$= 6$ hr 45 min

35. $3 \text{ P.M.} - 4 \text{ hr} = 11 \text{ A.M.}$

    July 26, 11:00 A.M.

36. a) $1 \text{ in.} \times 1 \text{ in.} = 2.54 \text{ cm} \times 2.54 \text{ cm}$

    $= 6.4516 \text{ cm}^2 \approx 6.45 \text{ cm}^2$

    b) $1 \text{ in.} \times 1 \text{ in.} \times 1 \text{ in.}$

    $= 2.54 \text{ cm} \times 2.54 \text{ cm} \times 2.54 \text{ cm}$

    $= 16.387064 \text{ cm}^3 \approx 16.39 \text{ cm}^3$

    c) $\dfrac{1 \text{ in.}}{2.54 \text{ cm}} = \dfrac{x \text{ in.}}{1 \text{ cm}}$

    $2.54x = 1(1)$

    $\dfrac{2.54x}{2.54} = \dfrac{1}{2.54}$

    $x = 0.393700787 \approx 0.39 \text{ in.}$

37. Each figure has an additional two dots. To get the hundredth figure, 97 more figures must be drawn, $97(2) = 194$ dots added to the third figure. Thus, $194 + 7 = 201$.

38.

21	7	8	18
10	16	15	13
14	12	11	17
9	19	20	6

39.

23	25	15
13	21	29
27	17	19

40. 59 min 59 sec Since it doubles every second, the jar was half full 1 second earlier than 1 hour.

41. 6

42. Nothing. Each friend paid $9 for a total of $27; $25 to the hotel, $2 to the clerk.

    $25 for the room + $3 for each friend + $2 for the clerk = $30

43. Let $x =$ the total weight of the four women

    $\dfrac{x}{4} = 130, \quad x = 520, \quad \dfrac{520 + 180}{5} = \dfrac{700}{5} = 140 \text{ lb}$

44. Yes; 3 quarters and 4 dimes, or 1 half dollar, 1 quarter and 4 dimes, or 1 quarter and 9 dimes. Other answers are possible.

45. $6 \text{ cm} \times 6 \text{ cm} \times 6 \text{ cm} = 216 \text{ cm}^3$

46. Place six coins in each pan with one coin off to the side. If it balances, the heavier coin is the one on the side. If the pan does not balance, take the six coins on the heavier side and split them into two groups of three. Select the three heavier coins and weigh two coins. If the pan balances, it is the third coin. If the pan does not balance, you can identify the heavier coin.

47. $\dfrac{n(n+1)}{2} = \dfrac{500(501)}{2} = \dfrac{250,500}{2} = 125,250$

48. 16 blue: 4 green → 8 blue, 2 yellow → 5 blue, 2 white → 3 blue

49. 90: 101, 111, 121, 131, 141, 151, 161, 171, 181, 191, …

50. The fifth figure will be an octagon with sides of equal length. Inside the octagon will be a seven sided figure with each side of equal length. The figure will have one antenna.

51. 61: The sixth figure will have 6 rows of 6 tiles and 5 rows of 5 tiles ($6 \times 6 + 5 \times 5 = 36 + 25 = 61$).

52. Some possible answers are given below. There are other possibilities.

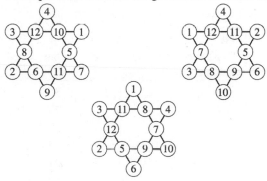

53. a) 2

b) There are 3 choices for the first spot. Once that person is standing, there are 2 choices for the second spot and 1 for the third. Thus, $3 \times 2 \times 1 = 6$.

c) $4 \times 3 \times 2 \times 1 = 24$

d) $5 \times 4 \times 3 \times 2 \times 1 = 120$

e) $n(n-1)(n-2)\cdots 1$, (or $n!$), where $n =$ the number of people in line

Chapter Test

1. 18, 21, 24   (Add 3 to previous number.)

2. $\dfrac{1}{81}, \dfrac{1}{243}, \dfrac{1}{729}$ (Multiply previous number by $\dfrac{1}{3}$.)

3. a) The result is the original number plus 1.

b) The result is the original number plus 1.

c) Conjecture: The result will always be the original number plus 1.

d) $n, 5n, 5n+10, \dfrac{5n+10}{5} = \dfrac{5n}{5} + \dfrac{10}{5} = n+2, n+2-1 = n+1$

(Note: Answers for #3 - #6 will vary depending on how you round your numbers. The answers may differ from the answers in the back of the textbook. However, your answers should be something near the answers given. All answers are approximate.)

4. $0.06 \times 98,000 \approx 0.06 \times 100,000 = 6000$

5. $\dfrac{102,000}{0.00302} \approx \dfrac{100,000}{0.003} \approx 33,000,000$

6. 7 square units

7. a) $\dfrac{130 \text{ lb}}{63 \text{ in.}} \approx 2.0635$

$\dfrac{2.0635}{63 \text{ in.}} = 0.032754$

$0.032754 \times 703 \approx 23.03$

b) He is in the at risk range.

8. $122.13 - \$9.63 = \$112.50$

$\dfrac{\$112.50}{\$0.72} = 156.25 \text{ therms}$

$156.25 \text{ therms} + \text{ first 3 therms} = 159.25 \text{ therms}$

9. $\dfrac{\$15}{\$2.59} \approx 5.79$

The maximum number of 6 packs is 5.

$\$15.00 - (5 \times \$2.59) = \$15.00 - \$12.95 = \$2.05$

$\dfrac{\$2.05}{\$0.80} = 2.5625$

Thus, two individual cans can be purchased.

6 packs	Indiv. cans	Number of cans
5	2	32
4	5	29
3	9	27
2	12	24
1	15	21
0	18	18

The maximum number of cans is 32.

10. 1 cut yields 2 equal pieces. Cut each of these 2 equal pieces to get 4 equal pieces.

$3 \text{ cuts} \rightarrow 3(2.5 \text{ min}) = 7.5 \text{ min}$

11. 2.5 in. by 1.875 in.

$\approx 2.5 \times 15.8 \text{ by } 1.875 \times 15.8 = 39.5 \text{ in. by } 29.625 \text{ in.}$

$\approx 39.5 \text{ in. by } 29.6 \text{ in.}$

(The actual dimensions are 100.5 cm by 76.5 cm.)

12. $12.75 \times 40 = \$510$

$\$12.75 \times 1.5 \times 10 = \$191.25$

$\$510 + \$191.25 = \$701.25$

$\$701.25 - \$652.25 = \$49.00$

13.

40	15	20
5	25	45
30	35	10

14. Mary drove the first 15 miles at 60 mph which took $\dfrac{15}{60} = \dfrac{1}{4}$ hr, and the second 15 miles at 30 mph which took $\dfrac{15}{30} = \dfrac{1}{2}$ hr for a total time of $\dfrac{3}{4}$ hr. If she drove the entire 30 miles at 45 mph, the trip would take $\dfrac{30}{45} = \dfrac{2}{3}$ hr (40 min) which is less than $\dfrac{3}{4}$ hr (45 min).

15. $2 \times 6 \times 8 \times 9 \times 13 = 11,232$; 11 does not divide 11,232.

16. 243 jelly beans; $260 - 17 = 243, 234 + 9 = 243, 274 - 31 = 243$

17. a) $3 \times \$3.99 = \$11.97$

b) $9(\$1.75 \times 0.75) = 11.8125 \approx \$11.81$

c) $\$11.97 - \$11.81 = \$0.16$ Using the coupon is least expensive by $0.16.

18. 8: $\$ \rightarrow$ on $* \rightarrow$ off

$\$\$\$\$, \$\$\$*, \$\$*\$, \$*\$\$, *\$\$\$, *\$*\$, *\$\$*, \$*\$*$

CHAPTER TWO

SETS

Exercise Set 2.1

1. A **set** is a collection of objects.

3. Description: the set of counting numbers less than 7

   Roster form: $\{1, 2, 3, 4, 5, 6\}$

   Set-builder notation: $\{x | x \in N \text{ and } x < 7\}$

5. An **infinite** set is a set that is not finite.

7. Two sets are **equivalent** if they contain the same number of elements.

9. A set that contains no elements is called the **empty set** or **null set**.

11. Set A and set B can be placed in **one-to-one correspondence** if every element of set A can be matched with exactly one element of set B and every element of set B can be matched with exactly one element of set A.

13. Not well defined, "large" is interpreted differently by different people.

15. Well defined, the contents can be clearly determined.

17. Well defined, the contents can be clearly determined.

19. Infinite, the number of elements in the set is not a natural number.

21. Infinite, the number of elements in the set is not a natural number.

23. Infinite, the number of elements in the set is not a natural number.

25. $\{$Atlantic, Pacific, Arctic, Indian$\}$

27. $\{11, 12, 13, 14, \ldots, 177\}$

29. $B = \{2, 4, 6, 8, \ldots\}$

31. $\{\ \}$ or $\varnothing$

33. $E = \{6, 7, 8, 9, \ldots, 71\}$

35. $\{$Sony DSC-S50, Sony DSC-S70, Sony Mavica FD-90$\}$

37. $\{$Sony Mavica FD-73, Olympus D-360L, Sony DSC-S50, Kodak DC215, H-P Photo Smart C315$\}$

39. $\{2002, 2003, 2004, 2005, 2006, 2007, 2008\}$

41. $\{2002, 2005, 2006, 2007, 2008\}$

43. $B = \{x | x \in N \text{ and } 3 < x < 11\}$ or

    $B = \{x | x \in N \text{ and } 4 \leq x \leq 10\}$

45. $C = \{x | x \in N \text{ and } x \text{ is a multiple of } 3\}$

47. $E = \{x | x \in N \text{ and } x \text{ is odd}\}$

49. $C = \{x | x \text{ is February}\}$

51. Set A is the set of natural numbers less than or equal to 7.

53. Set V is the set of vowels in the English alphabet.

55. Set C is the set of companies that make calculators.

57. Set B is the set of members of the Beatles.

59.   {St. Louis}

61.   { } or ∅

63.   {1999, 2000, 2001, 2002}

65.   {1999, 2001, 2002}

67.   False; {b} is a set, and not an element of the set.

69.   False; h is not an element of the set.

71.   False; 3 is an element of the set.

73.   True; *Titanic* is an element of the set.

75. $n(A) = 4$

77. $n(C) = 0$

79.   Both; A and B contain exactly the same elements.

81.   Neither; the sets have a different number of elements.

83.   Equivalent; both sets contain the same number of elements, 3.

85.   a) Set A is the set of natural numbers greater than 2. Set B is the set of all numbers greater than 2.

   b) Set A contains only natural numbers. Set B contains other types of numbers, including fractions and decimal numbers.

   c) $A = \{3, 4, 5, 6, \ldots\}$

   d) No

87.   Cardinal; 12 tells how many.

89.   Ordinal; sixteenth tells Lincoln's relative position.

91.   Answers will vary.

93.   Answers will vary.

95.

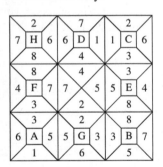

Exercise Set 2.2

1.   Set A is a **subset** of set B, symbolized by $A \subseteq B$, if and only if all the elements of set A are also elements of set B.

3.   If $A \subseteq B$, then every element of set A is also an element of set B. If $A \subset B$, then every element of set A is also an element of set B and set $A \neq$ set B.

5. $2^n - 1$, where n is the number of elements in the set.

7.   False; gold is an element of the set, not a subset.

9.   True; the empty set is a subset of every set.

11.   True; 5 is not an element of {2, 4, 6}.

13.   False; the set {∅} contains the element ∅.

15.   True; { } and ∅ each represent the empty set.

17.   False; the set {0} contains the element 0.

19.   False; {swimming} is a subset, not an element.

21.   True; the empty set is a subset of every set, including itself.

23.   False; no set is a proper subset of itself.

25. $B \subseteq A, B \subset A$

27. $B \subseteq A, B \subset A$

29. $B \subseteq A, B \subset A$

29. $B \subseteq A, B \subset A$

31. $A = B, A \subseteq B, B \subseteq A$

33.   { } is the only subset.

35.   { },{pen},{pencil},{pen, pencil}

37.   a) { },{a},{b},{c},{d},{a,b},{a,c},{a,d},
  {b,c},{b,d},{c,d},{a,b,c},{a,b,d},
  {a,c,d},{b,c,d},{a,b,c,d}

  b) All the sets in part a) are proper subsets of
  A except {a,b,c,d}.

39.   False; A could be equal to B.

41.   True; every set is a subset of itself.

43.   True; $\varnothing$ is a proper subset of every set except itself.

45.   True; every set is a subset of the universal set.

47.   True; $\varnothing$ is a proper subset of every set except itself and $U \neq \varnothing$.

49.   True; $\varnothing$ is a subset of every set.

51.   The number of different variations of the house is equal to the number of subsets of
  {deck, jacuzzi, security system, hardwood flooring} which is $2^4 = 2\times2\times2\times2 = 16$.

53.   The number of different variations is equal to the number of subsets of
  {call waiting, call forwarding, caller identification, three way calling, voice mail, fax line},

  which is $2^6 = 2\times2\times2\times2\times2\times2 = 64$.

55. $E = F$ since they are both subsets of each other.

57.   a) Yes, because a is a member of set D.

  b) No, c is an element of set D.

  c) Yes, each element of {a,b} is an element of set D.

59.   A one element set has one proper subset, namely the empty set. A one element set has two subsets, namely itself
  and the empty set. One is one-half of two. Thus, the set must have one element.

60. Yes   61. Yes   62. No

Section 2.3

1.

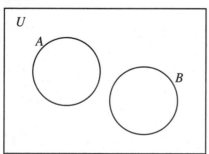

3.

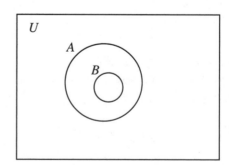

5.

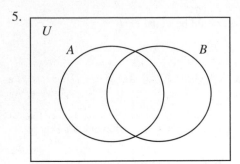

7. Combine the elements from set *A* and set *B* into one set. List any element that is contained in both sets only once.

9. Take the elements common to both set *A* and set *B* .

11. a) *Or* is generally interpreted to mean *union*.

    b) *And* is generally interpreted to mean *intersection*.

13. Region II, the intersection of the two sets.

15.

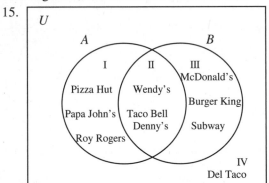

17.

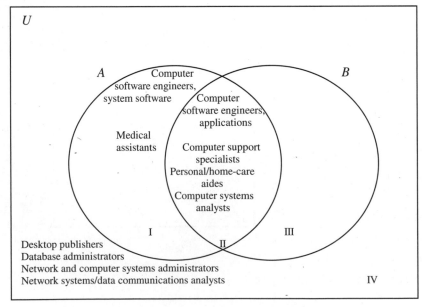

19. The set of U.S. colleges and universities that are not in the state of North Dakota

21. The set of insurance companies in the U.S. that do not offer life insurance

23. The set of insurance companies in the U.S. that offer life insurance or car insurance

25. The set of insurance companies in the U.S. that offer life insurance and do not offer car insurance

27. The set of U.S. corporations whose headquarters are in New York State and whose chief executive officer is a woman

29. The set of U.S. corporations whose chief executive officer is not a woman and who employ at least 100 people

31. The set of U.S. corporations whose headquarters are in New York State or whose chief executive officer is a woman or that employ at least 100 people

33. $A = \{a, b, c, h, t, w\}$

35. $A \cap B = \{a, b, c, h, t, w\} \cap \{a, f, g, h, r\} = \{a, h\}$

37. $A \cup B = \{a, b, c, h, t, w\} \cup \{a, f, g, h, r\} = \{a, b, c, f, g, h, r, t, w\}$

39. $A' \cap B' = \{a, b, c, h, t, w\}' \cap \{a, f, g, h, r\}' = \{f, g, r, p, m, z\} \cap \{c, w, b, t, p, m, z\} = \{p, m, z\}$

41. $A = \{L, \Delta, @, *, \$\}$

43. $U = \{L, \Delta, @, *, \$, R, \square, \alpha, \infty, \Sigma, Z\}$

45. $A \cap B = \{L, \Delta, @, *, \$\} \cap \{*, \$, R, \square, \alpha\} = \{*, \$\}$

47. $A' \cap B = \{L, \Delta, @, *, \$\}' \cap \{*, \$, R, \square, \alpha\} = \{R, \square, \alpha, \infty, \Sigma, Z\} \cap \{*, \$, R, \square, \alpha\} = \{R, \square, \alpha\}$

49. $A \cup B = \{1, 2, 4, 5, 8\} \cup \{2, 3, 4, 6\} = \{1, 2, 3, 4, 5, 6, 8\}$

51. $B' = \{2, 3, 4, 6\}' = \{1, 5, 7, 8\}$

53. $(A \cup B)'$　From #49, $A \cup B = \{1, 2, 3, 4, 5, 6, 8\}$. $(A \cup B)' = \{1, 2, 3, 4, 5, 6, 8\}' = \{7\}$

55. $(A \cup B)' \cap B$　From #53, $(A \cup B)' = \{7\}$. $(A \cup B)' \cap B = \{7\} \cap \{2, 3, 4, 6\} = \{ \ \}$

57. $(B \cup A)' \cap (B' \cup A')$　From #53, $(A \cup B)' = (B \cup A)' = \{7\}$.

$(B \cup A)' \cap (B' \cup A') = \{7\} \cap \left(\{2, 3, 4, 6\}' \cup \{1, 2, 4, 5, 8\}'\right) = \{7\} \cap \left(\{1, 5, 7, 8\} \cup \{3, 6, 7\}\right)$

$= \{7\} \cap \{1, 3, 5, 6, 7, 8\} = \{7\}$

59. $B' = \{b, c, d, f, g\}' = \{a, e, h, i, j, k\}$

61. $A \cap C = \{a, c, d, f, g, i\} \cap \{a, b, f, i, j\} = \{a, f, i\}$

63. $(A \cap C)'$　From #61, $A \cap C = \{a, f, i\}$. $(A \cap C)' = \{a, f, i\}' = \{b, c, d, e, g, h, j, k\}$

65. $A \cup (C \cap B)' = \{a, c, d, f, g, i\} \cup \left(\{a, b, f, i, j\} \cap \{b, c, d, f, g\}\right)' = \{a, c, d, f, g, i\} \cup \{b, f\}'$

$= \{a, c, d, f, g, i\} \cup \{a, c, d, e, g, h, i, j, k\} = \{a, c, d, e, f, g, h, i, j, k\}$

67. $(A' \cup C) \cup (A \cap B) = \left(\{a, c, d, f, g, i\}' \cup \{a, b, f, i, j\}\right) \cup \left(\{a, c, d, f, g, i\} \cap \{b, c, d, f, g\}\right)$

$= \left(\{b, e, h, j, k\} \cup \{a, b, f, i, j\}\right) \cup \{c, d, f, g\} = \{a, b, e, f, h, i, j, k\} \cup \{c, d, f, g\}$

$= \{a, b, c, d, e, f, g, h, i, j, k\}, \text{ or } U$

For exercises 69-81: $U = \{1, 2, 3, 4, 5, 6, 7, 8, 9\}$, $A = \{1, 3, 5, 7, 9\}$, $B = \{2, 4, 6, 8\}$, $C = \{1, 2, 3, 4, 5\}$

69. $A \cap B = \{1,3,5,7,9\} \cap \{2,4,6,8\} = \{\ \}$

71. $A' \cup B = \{1,3,5,7,9\}' \cup \{2,4,6,8\} = \{2,4,6,8\} \cup \{2,4,6,8\} = \{2,4,6,8\}$, or B

73. $A \cap C' = \{1,3,5,7,9\} \cap \{1,2,3,4,5\}' = \{1,3,5,7,9\} \cap \{6,7,8,9\} = \{7,9\}$

75. $(B \cap C)' = (\{2,4,6,8\} \cap \{1,2,3,4,5\})' = \{2,4\}' = \{1,3,5,6,7,8,9\}$

77. $(C \cap B) \cup A$ From #75, $C \cap B = \{2,4\}$. $(C \cap B) \cup A = \{2,4\} \cup \{1,3,5,7,9\} = \{1,2,3,4,5,7,9\}$

79. $(A' \cup C) \cap B = \left(\{1,3,5,7,9\}' \cup \{1,2,3,4,5\}\right) \cap \{2,4,6,8\} = (\{2,4,6,8\} \cup \{1,2,3,4,5\}) \cap \{2,4,6,8\}$

    $= \{1,2,3,4,5,6,8\} \cap \{2,4,6,8\} = \{2,4,6,8\}$, or B

81. $(A' \cup B') \cap C = \left(\{1,3,5,7,9\}' \cup \{2,4,6,8\}'\right) \cap \{1,2,3,4,5\}$

    $= (\{2,4,6,8\} \cup \{1,3,5,7,9\}) \cap \{1,2,3,4,5\} = \{1,2,3,4,5,6,7,8,9\} \cap \{1,2,3,4,5\} = \{1,2,3,4,5\}$, or C

83. A set and its complement will always be disjoint since the complement of a set is all of the elements in the universal set that are not in the set. Therefore, a set and its complement will have no elements in common.

    For example, if $U = \{1,2,3\}$, $A = \{1,2\}$, and $A' = \{3\}$, then $A \cap A' = \{\ \}$.

85. Let $A = \{$visitors who visited the Hollywood Bowl$\}$ and $B = \{$visitors who visited Disneyland$\}$.

    $n(A \cup B) = n(A) + n(B) - n(A \cap B) = 27 + 38 - 16 = 49$

87. a) $A \cup B = \{a,b,c,d\} \cup \{b,d,e,f,g,h\} = \{a,b,c,d,e,f,g,h\}$, $n(A \cup B) = 8$,

    $A \cap B = \{a,b,c,d\} \cap \{b,d,e,f,g,h\} = \{b,d\}$, $n(A \cap B) = 2$.

    $n(A) + n(B) - n(A \cap B) = 4 + 6 - 2 = 8$

    Therefore, $n(A \cup B) = n(A) + n(B) - n(A \cap B)$.

    b) Answers will vary.

    c) Elements in the intersection of A and B are counted twice in $n(A) + n(B)$.

    $A' \cap B'$ or $(A \cup B)'$ defines Region IV.

89. $A \cup B = \{1,2,3,4,...\} \cup \{4,8,12,16,...\} = \{1,2,3,4,...\}$, or A

91. $B \cap C = \{4,8,12,16,...\} \cap \{2,4,6,8,...\} = \{4,8,12,16,...\}$, or B

93. $A \cap C = \{1,2,3,4,...\} \cap \{2,4,6,8,...\} = \{2,4,6,8,...\}$, or C

95. $B' \cap C = \{4,8,12,16,...\}' \cap \{2,4,6,8,...\} = \{0,1,2,3,5,6,7,9,10,11,13,14,15,...\} \cap \{2,4,6,8,...\}$

    $= \{2,6,10,14,18,...\}$

97. $(A \cap C) \cap B'$ From #93, $A \cap C = C$. $(A \cap C) \cap B' = C \cap B'$.

    From #95, $B' \cap C = C \cap B' = \{2,6,10,14,18,...\}$

99. $A \cup A' = U$

101. $A \cup \varnothing = A$

103. $A' \cup U = U$

105. $A \cup U = U$

107. If $A \cap B = B$, then $B \subseteq A$.

109. If $A \cap B = \varnothing$, then A and B are disjoint sets.

111. If $A \cap B = A$, then $A \subseteq B$.

113. $A - B = \{b,c,e,f,g,h\} - \{a,b,c,g,i\} = \{e,f,h\}$

115. $A' - B = \{b,c,e,f,g,h\}' - \{a,b,c,g,i\}$

  $= \{a,d,i,j,k\} - \{a,b,c,g,i\} = \{d,j,k\}$

117. $A - B = \{2,4,5,7,9,11,13\} - \{1,2,4,5,6,7,8,9,11\}$

  $= \{13\}$

119. $(A - B)'$ From #117, $A - B = \{13\}$.

  $(A - B)' = \{13\}'$

  $= \{1,2,3,4,5,6,7,8,9,10,11,12,14,15\}$

121. $(B - A)'$ From #118, $B - A = \{1,6,8\}$.

  $(B - A)' = \{1,6,8\}'$

  $= \{2,3,4,5,7,9,10,11,12,13,14,15\}$

123. Complement

124.
```
N O I E L A
T C T S E T
S E U B P U
R D S R I N
E I A G O N
T N I R A M
```

Exercise Set 2.4

1. 8

3. Regions II, IV, VI

  then region II contains $10 - 6 = 4$ elements.

5. $B \cap C$ is represented by regions V and VI. If $B \cap C$ contains 12 elements and region V contains 4 elements, then region VI contains $12 - 4 = 8$ elements.

7. a) Yes

  $A \cup B = \{1,4,5\} \cup \{1,4,5\} = \{1,4,5\}$

  $A \cap B = \{1,4,5\} \cap \{1,4,5\} = \{1,4,5\}$

  b) No, one specific case cannot be used as proof.

  c)

	$A \cup B$		$A \cap B$
Set	Regions	Set	Regions
A	I, II	A	I, II
B	II, III	B	II, III
$A \cup B$	I, II, III	$A \cap B$	II

  Since the two statements are not represented by the same regions, $A \cup B \neq A \cap B$ for all sets A and B.

9.
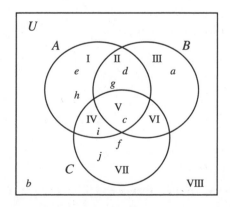

11.
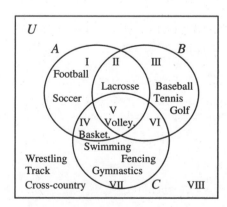

13.

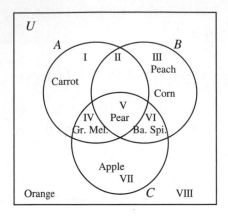

15.

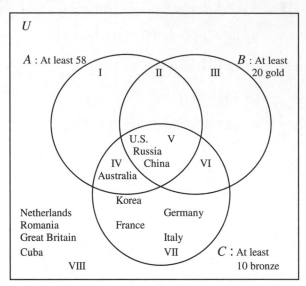

17. Harvard, V

19. Boston College, VIII

21. Northwestern, VI

23. Washington D.C., IV

25. Denver, II

27. Rochester, NY, VII

29. VI

31. III

33. III

35. V

37. II

39. VII

41. I

43. VIII

45. VI

47. $A = \{1, 2, 3, 4, 5, 6\}$

49. $B = \{3, 4, 5, 7, 8, 9, 12\}$

51. $A \cap B = \{3, 4, 5\}$

53. $(B \cap C)' = \{1, 2, 3, 6, 9, 10, 11, 12\}$

55. $A \cup B = \{1, 2, 3, 4, 5, 6, 7, 8, 9, 12\}$

57. $(A \cup C)' = \{9, 11, 12\}$

59. $A' = \{7, 8, 9, 10, 11, 12\}$

61. $(A \cup B)'$ $A' \cap B'$

Set	Regions	Set	Regions
A	I, II	A	I, II
B	II, III	A'	III, IV
$A \cup B$	I, II, III	B	II, III
$(A \cup B)'$	IV	B'	I, IV
		$A' \cap B'$	IV

Both statements are represented by the same region, IV,
of the Venn diagram. Therefore, $(A \cup B)' = A' \cap B'$
for all sets A and B.

63. $A' \cup B'$ $A \cap B$

Set	Regions	Set	Regions
A	I, II	A	I, II
A'	III, IV	B	II, III
B	II, III	$A \cap B$	II
B'	I, IV		
$A' \cup B'$	I, III, IV		

Since the two statements are not represented by the same
regions, $A' \cup B' \neq A \cap B$ for all sets A and B.

65. $A' \cup B'$ $(A \cup B)'$

Set	Regions	Set	Regions
A	I, II	A	I, II
A'	III, IV	B	II, III
B	II, III	$A \cup B$	I, II, III
B'	I, IV	$(A \cup B)'$	IV
$A' \cup B'$	I, III, IV		

Since the two statements are not represented by the same

regions, $A' \cup B' \neq (A \cup B)'$ for all sets A and B.

67. $(A' \cap B)'$ $A \cup B'$

Set	Regions	Set	Regions
A	I, II	A	I, II
A'	III, IV	B	II, III
B	II, III	B'	I, IV
$A' \cap B$	III	$A \cup B'$	I, II, IV
$(A' \cap B)'$	I, II, IV		

Both statements are represented by the same regions,
I, II, IV, of the Venn diagram. Therefore,

$(A' \cap B)' = A \cup B'$ for all sets A and B.

69. $A \cap (B \cup C)$ $(A \cap B) \cup C$

Set	Regions	Set	Regions
B	II, III, V, VI	A	I, II, IV, V
C	IV , V, VI, VII	B	II, III, V, VI
$B \cup C$	II, III, IV, V, VI, VII	$A \cap B$	II, V
A	I, II, IV, V	C	IV, V, VI, VII
$A \cap (B \cup C)$	II, IV, V	$(A \cap B) \cup C$	II, IV, V, VI, VII

Since the two statements are not represented by the same regions, $A \cap (B \cup C) \neq (A \cap B) \cup C$

for all sets A, B, and C.

71. $A \cap (B \cup C)$ $(B \cup C) \cap A$

Set	Regions	Set	Regions
B	II, III, V, VI	B	II, III, V, VI
C	IV , V, VI, VII	C	IV, V, VI, VII
$B \cup C$	II, III, IV, V, VI, VII	$B \cup C$	II, III, IV, V, VI, VII
A	I, II, IV, V	A	I, II, IV, V
$A \cap (B \cup C)$	II, IV, V	$(B \cup C) \cap A$	II, IV, V

Both statements are represented by the same regions, II, IV, V, of the Venn diagram.

Therefore, $A \cap (B \cup C) = (B \cup C) \cap A$ for all sets A, B, and C.

73. $A\cap(B\cup C)$ $(A\cap B)\cup(A\cap C)$

Set	Regions	Set	Regions
B	II, III, V, VI	A	I, II, IV, V
C	IV, V, VI, VII	B	II, III, V, VI
$B\cup C$	II, III, IV, V, VI, VII	$A\cap B$	II, V
A	I, II, IV, V	C	IV, V, VI, VII
$A\cap(B\cup C)$	II, IV, V	$A\cap C$	IV, V
		$(A\cap B)\cup(A\cap C)$	II, IV, V

Both statements are represented by the same regions, II, IV, V, of the Venn diagram.

Therefore, $A\cap(B\cup C)=(A\cap B)\cup(A\cap C)$ for all sets $A, B,$ and C.

75. $A\cap(B\cup C)'$ $A\cap(B'\cap C')$

Set	Regions	Set	Regions
B	II, III, V, VI	B	II, III, V, VI
C	IV, V, VI, VII	B'	I, IV, VII, VIII
$B\cup C$	II, III, IV, V, VI, VII	C	IV, V, VI, VII
$(B\cup C)'$	I, VIII	C'	I, II, III, VIII
A	I, II, IV, V	$B'\cap C'$	I, VIII
$A\cap(B\cup C)'$	I	A	I, II, IV, V
		$A\cap(B'\cap C')$	I

Both statements are represented by the same region, I, of the Venn diagram.

Therefore, $A\cap(B\cup C)' = A\cap(B'\cap C')$ for all sets $A, B,$ and C.

77. $(A\cup B)'\cap C$ $(A'\cup C)\cap(B'\cup C)$

Set	Regions	Set	Regions
A	I, II, IV, V	A	I, II, IV, V
B	II, III, V, VI	A'	III, VI, VII, VIII
$A\cup B$	I, II, III, IV, V, VI	C	IV, V, VI, VII
$(A\cup B)'$	VII, VIII	$A'\cup C$	III, IV, V, VI, VII, VIII
C	IV, V, VI, VII	B	II, III, V, VI
$(A\cup B)'\cap C$	VII	B'	I, IV, VII, VIII
		$B'\cup C$	I, IV, V, VI, VII, VIII
		$(A'\cup C)\cap(B'\cup C)$	IV, V, VI, VII, VIII

Since the two statements are not represented by the same regions, $(A\cup B)'\cap C \neq (A'\cup C)\cap(B'\cup C)$

for all sets $A, B,$ and C.

79. $(A \cup B)'$

81. $(A \cup B) \cap C'$

83. a) $(A \cup B) \cap C = (\{1,2,3,4\} \cup \{3,6,7\}) \cap \{6,7,9\} = \{1,2,3,4,6,7\} \cap \{6,7,9\} = \{6,7\}$

$(A \cap C) \cup (B \cap C) = (\{1,2,3,4\} \cap \{6,7,9\}) \cup (\{3,6,7\} \cap \{6,7,9\}) = \varnothing \cup \{6,7\} = \{6,7\}$

Therefore, for the specific sets, $(A \cup B) \cap C = (A \cap C) \cup (B \cap C)$.

b) Answers will vary.

c) $(A \cup B) \cap C$ $(A \cap C) \cup (B \cap C)$

Set	Regions	Set	Regions
A	I, II, IV, V	A	I, II, IV, V
B	II, III, V, VI	C	IV, V, VI, VII
$A \cup B$	I, II, III, IV, V, VI	$A \cap C$	IV, V
C	IV, V, VI, VII	B	II, III, V, VI
$(A \cup B) \cap C$	IV, V, VI	$B \cap C$	V, VI
		$(A \cap C) \cup (B \cap C)$	IV, V, VI

Both statements are represented by the same regions, IV, V, VI, of the Venn diagram.

Therefore, $(A \cup B) \cap C = (A \cap C) \cup (B \cap C)$ for all sets $A, B,$ and C.

85.

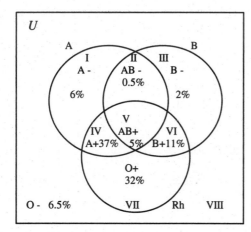

87. a) A : Office Building Construction Projects, B : Plumbing Projects, C : Budget Greater Than \$300,000

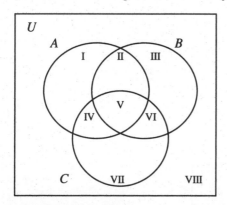

b) Region V; $A \cap B \cap C$ c) Region VI; $A' \cap B \cap C$ d) Region I; $A \cap B' \cap C'$

89. a)

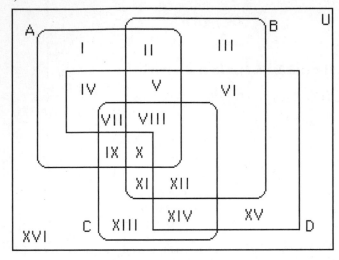

b)

Region	Set	Region	Set
I	$A \cap B' \cap C' \cap D'$	IX	$A \cap B' \cap C \cap D'$
II	$A \cap B \cap C' \cap D'$	X	$A \cap B \cap C \cap D'$
III	$A' \cap B \cap C' \cap D'$	XI	$A' \cap B \cap C \cap D'$
IV	$A \cap B' \cap C' \cap D$	XII	$A' \cap B \cap C \cap D$
V	$A \cap B \cap C' \cap D$	XIII	$A' \cap B' \cap C \cap D'$
VI	$A' \cap B \cap C' \cap D$	XIV	$A' \cap B' \cap C \cap D$
VII	$A \cap B' \cap C \cap D$	XV	$A' \cap B' \cap C' \cap D$
VIII	$A \cap B \cap C \cap D$	XVI	$A' \cap B' \cap C' \cap D'$

90.

```
S U B  T R H M D F R T S
  T S J O N B U I E L E M
W E E S N E R N I T V O
  P R S A K C P A R N E P
O R P O N R R O Y I M T
  I R J I E E S O C L U S
F M U N G R T R N E P P
```

Exercise Set 2.5

1. a) 33, Region I
   b) 29, Region III
   c) 27, Region IV

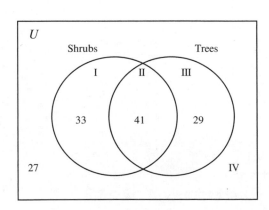

3. a) 17, Region I
    b) 12, Region III
    c) 59, the sum of the numbers in Regions I, II, III

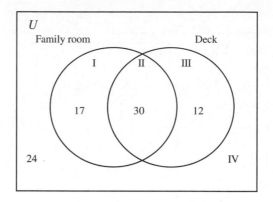

5. a) 27, Region VIII
    b) 80, Region VII
    c) 340, the sum of the numbers in Regions I, III, VII
    d) 55, the sum of the numbers in Regions II, IV, VI
    e) 337, the sum of the numbers in Regions I, II, III, IV, V, VI

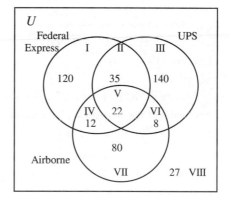

7. a) 22, Region I
    b) 11, Region II
    c) 64, the sum of the numbers in Regions I, II, III, IV, V, VI
    d) 50, the sum of the numbers in Regions I, II, III
    e) 23, the sum of the numbers in Regions II, IV, VI

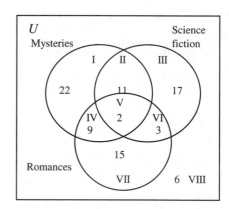

9. a) 17, Region I
    b) 27, Region VII
    c) 2, Region II
    d) 31, the sum of the numbers in Regions I, II, III
    e) 2, Region VIII

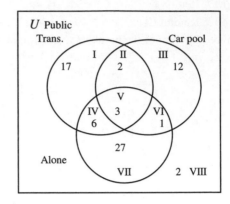

10. a) 496, the sum of the numbers in
    Regions I, II, III, IV, V, VI, VII, VIII
    b) 132, Region IV
    c) 29, Region III
    d) 328, the sum of the numbers in Regions II, IV, VI
    e) 470, the sum of the numbers in Regions I, II, III, IV, V, VI, VII

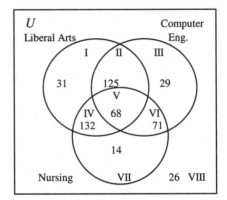

11. a) 10, the sum of the numbers in Regions III and VI
    b) 15, the sum of the numbers in Regions I, II, III, IV, V, VI
    c) 0, Region II
    d) 6, Region VIII

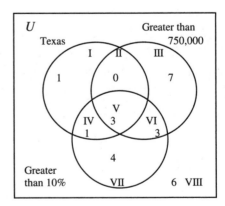

13. The Venn diagram shows the number of cars driven by women is 37, the sum of the numbers in Regions II, IV, V. This exceeds the 35 women the agent claims to have surveyed.

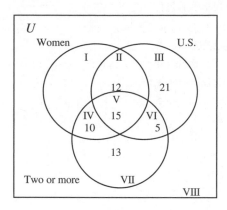

15. a) 410, the sum of the numbers in Regions I through VII
    b) 35, Region V
    c) 90, Region VIII
    d) 50, the sum of the numbers in Regions II, IV, VI
    The number of farmers growing wheat only, Region I, is 125. The number growing corn only, Region III, is 110. The number growing oats only, Region VII, is 90. 60 farmers grew wheat and corn, Regions II and V. 200 farmers grew wheat. Therefore, the sum of the numbers in Regions I, II, IV, V must equal 200.

    $125 + 60 +$ number in Region IV $= 200$.

    Thus, the number in Region IV is 15.

16. a) 10   b) 10   c) 6

Exercise Set 2.6

1. An **infinite set** is a set that can be placed in a one-to-one correspondence with a proper subset of itself.

3. $\{7, 8, 9, \ 10, 11, \ldots, n + 6, \ldots\}$
    $\downarrow \downarrow \downarrow \downarrow \downarrow \qquad \downarrow$
    $\{8, 9, 10, 11, 12, \ldots, n + 7, \ldots\}$

5. $\{3, 5, \ 7, \ 9, 11, \ldots, 2n + 1, \ldots\}$
    $\downarrow \downarrow \downarrow \downarrow \downarrow \qquad \downarrow$
    $\{5, 7, \ 9, 11, 13, \ldots, 2n + 3, \ldots\}$

7. $\{4, \ 7, \ 10, 13, 16, \ldots, 3n + 1, \ldots\}$
    $\downarrow \downarrow \ \downarrow \ \downarrow \ \downarrow \qquad \downarrow$
    $\{7, 10, 13, 16, 19, \ldots, 3n + 4, \ldots\}$

9. $\{6, \ \ 11, 16, 21, 26, \ldots, 5n+1, \ldots\}$
    $\downarrow \ \downarrow \ \downarrow \ \downarrow \ \downarrow \qquad \downarrow$
    $\{11, 16, 21, 26, 31, \ldots, 5n+6, \ldots\}$

11. $\left\{\dfrac{1}{2}, \dfrac{1}{4}, \dfrac{1}{6}, \dfrac{1}{8}, \ldots, \ \dfrac{1}{2n}, \ldots\right\}$
    $\quad \downarrow \downarrow \downarrow \downarrow \qquad \downarrow$
    $\left\{\dfrac{1}{4}, \dfrac{1}{6}, \dfrac{1}{8}, \dfrac{1}{10}, \ldots, \dfrac{1}{2n+2}, \ldots\right\}$

13. $\{1,\ 2,\ \ 3,\ 4,\ ...,\ \ n,\ ...\}$
$\quad\downarrow\downarrow\ \downarrow\ \downarrow\ \quad\ \downarrow$
$\{6, 12, 18, 24, ..., 6n, ...\}$

15. $\{1, 2, 3, 4,\ \ 5, ...,\ \ n, ...\}$
$\quad\downarrow\downarrow\downarrow\downarrow\ \downarrow\quad\ \downarrow$
$\{4, 6, 8, 10, 12, ..., 2n + 2, ...\}$

17. $\{1, 2, 3, 4,\ \ 5, ...,\ \ \ n, ...\}$
$\quad\downarrow\downarrow\downarrow\downarrow\ \downarrow\quad\ \ \downarrow$
$\{2, 5, 8, 11, 14, ..., 3n - 1, ...\}$

19. $\{1,\ 2,\ 3,\ 4,\ \ 5, ...,\quad n\ , ...\}$
$\quad\downarrow\ \downarrow\ \downarrow\ \downarrow\ \downarrow\qquad\ \downarrow$
$\{5, 8, 11, 14, 17, ..., 3n + 2, ...\}$

21. $\{\ \ 1, 2,\ 3, 4, 5, ...,\ \ n, ...\}$
$\quad\ \ \downarrow\ \downarrow\ \downarrow\ \downarrow\ \downarrow\qquad\ \downarrow$
$$\left\{\frac{1}{3},\frac{1}{4},\frac{1}{5},\frac{1}{6},\frac{1}{7},...,\frac{1}{n+2},...\right\}$$

23. $\{1,\ 2, 3,\ 4,\ \ 5, ...,\ \ n, ...\}$
$\quad\downarrow\ \downarrow\ \downarrow\ \downarrow\ \downarrow\quad\ \downarrow$
$\{1, 4,\ 9, 16, 25, ..., n^2, ...\}$

25. $\{1, 2,\ \ 3,\ 4,\ \ 5, ...,\ \ \ n, ...\}$
$\quad\downarrow\ \downarrow\ \downarrow\ \downarrow\ \downarrow\qquad\ \downarrow$
$\{3, 9, 27, 81, 243, ..., 3^n, ...\}$

27. $=$ 28. $=$

29. $=$ 30. $=$

31. $=$ 32. a) Answers will vary.

b) No

Review Exercises

1. True   2. False; the word *best* makes the statement not well defined.

3. True   4. False; no set is a proper subset of itself.

5. False; the elements 6, 12, 18, 24, ... are members of both sets.   6. True

7. False; both sets do not contain exactly the same elements.   8. True

9. True   10. True

11. True   12. True

13. True   14. True

15. $A = \{7, 9, 11, 13, 15\}$ 16. $B = \{$California, Oregon, Idaho, Utah, Arizona$\}$

17. $C = \{1, 2, 3, 4, ..., 296\}$ 18. $D = \{9, 10, 11, 12, ..., 96\}$

19. $A = \{x \mid x \in N \text{ and } 52 < x < 100\}$ 20. $B = \{x \mid x \in N \text{ and } x > 63\}$

21. $C = \{x \mid x \in N \text{ and } x < 3\}$ 22. $D = \{x \mid x \in N \text{ and } 23 \le x \le 41\}$

23. A is the set of capital letters in the English alphabet from E through M, inclusive.

24. B is the set of U.S. coins with a value of less than one dollar.

25. C is the set of the last three lowercase letters in the English alphabet.

26. D is the set of numbers greater than or equal to 3 and less than 9.

27. $A \cap B = \{1,3,5,6\} \cap \{5,6,9,10\} = \{5,6\}$

28. $A \cup B' = \{1,3,5,6\} \cup \{5,6,9,10\}' = \{1,3,5,6\} \cup \{1,2,3,4,7,8\} = \{1,2,3,4,5,6,7,8\}$

29. $A' \cap B = \{1,3,5,6\}' \cap \{5,6,9,10\} = \{2,4,7,8,9,10\} \cap \{5,6,9,10\} = \{9,10\}$

30. $(A \cup B)' \cup C = (\{1,3,5,6\} \cup \{5,6,9,10\})' \cup \{1,6,10\} = \{1,3,5,6,9,10\}' \cup \{1,6,10\}$

   $= \{2,4,7,8\} \cup \{1,6,10\} = \{1,2,4,6,7,8,10\}$

31. $2^4 = 2 \times 2 \times 2 \times 2 = 16$

32. $2^4 - 1 = (2 \times 2 \times 2 \times 2) - 1 = 16 - 1 = 15$

33.

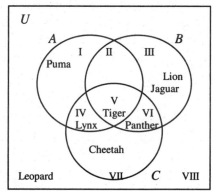

34. $A \cup B = \{b, e, g, k, c, d, f, a\}$

35. $A \cap B' = \{b, d\}$

36. $A \cup B \cup C = \{b, e, g, k, c, d, f, a, i\}$

37. $A \cap B \cap C = \{f\}$

38. $(A \cup B) \cap C = \{d, f, a\}$

39. $(A \cap B) \cup C = \{e, g, f, d, a, i\}$

40. $(A' \cup B')'$ $\qquad$ $A \cap B$

Set	Regions	Set	Regions
A	I, II	A	I, II
A'	III, IV	B	II, III
B	II, III	$A \cap B$	II
B'	I, IV		
$A' \cup B'$	I, III, IV		
$(A' \cup B')'$	II		

Both statements are represented by the same region, II, of the Venn diagram. Therefore, $(A' \cup B')' = A \cap B$ for all sets A and B.

41. $(A\cup B')\cup(A\cup C')$ $A\cup(B\cap C)'$

Set	Regions
A	I, II, IV, V
B	II, III, V, VI
B'	I, IV, VII, VIII
$A\cup B'$	I, II, IV, V, VII, VIII
C	IV, V, VI, VII
C'	I, II, III, VIII
$A\cup C'$	I, II, III, IV, V, VIII
$(A\cup B')\cup(A\cup C')$	I, II, III, IV, V, VII, VIII

Set	Regions
B	II, III, V, VI
C	IV, V, VI, VII
$B\cap C$	V, VI
$(B\cap C)'$	I, II, III, IV, VII, VIII
A	I, II, IV, V
$A\cup(B\cap C)'$	I, II, III, IV, V, VII, VIII

Both statements are represented by the same regions, I, II, III, IV, V, VII, VIII, of the Venn diagram.

Therefore, $(A\cup B')\cup(A\cup C') = A\cup(B\cap C)'$ for all sets $A, B,$ and C.

42. II

43. V

44. VIII

45. IV

46. IV

47. VII

48. The company paid $450 since the sum of the numbers in Regions I through IV is 450.

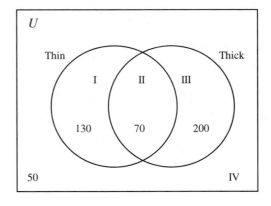

49. a) 315, the sum of the numbers in Regions I through VIII
    b) 10, Region III
    c) 30, Region II
    d) 110, the sum of the numbers in Regions III, VI, VII

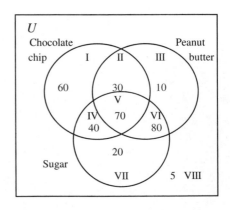

50. a) 38, Region I
    b) 298, the sum of the numbers in Regions I, III, VII
    c) 28, Region VI
    d) 236, the sum of the numbers in Regions I, IV, VII
    e) 106, the sum of the numbers in Regions II, IV, VI

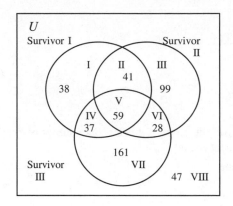

51. $\{2, 4, 6,\ 8,\ 10,\ \ldots, 2n, \ldots\}$
    ↓ ↓ ↓ ↓ ↓   ↓
    $\{4, 6,\ 8, 10, 12,\ \ldots, 2n + 2, \ldots\}$

52. $\{3, 5, 7,\ 9,\ 11,\ \ldots, 2n + 1, \ldots\}$
    ↓ ↓ ↓ ↓ ↓   ↓
    $\{5, 7, 9, 11, 13,\ \ldots, 2n + 3, \ldots\}$

53. $\{1, 2,\ \ 3,\ 4,\ \ 5,\ \ldots,\ \ \ n, \ldots\}$
    ↓ ↓ ↓ ↓ ↓   ↓
    $\{5, 8,\ \ 11, 14, 17,\ \ldots, 3n + 2, \ldots\}$

54. $\{1, 2,\ \ 3,\ 4,\ \ \ 5,\ \ldots,\ \ \ n, \ldots\}$
    ↓ ↓ ↓ ↓ ↓   ↓
    $\{4, 9,\ 14, 19, 24,\ \ldots, 5n - 1, \ldots\}$

Chapter Test

1. True

2. False; the sets do not contain exactly the same elements.

3. True

4. False; the second set has no subset that contains the element 7.

5. False; the empty set is a proper subset of every set except itself.

6. False; the set has $2^3 = 2 \times 2 \times 2 = 8$ subsets.

7. True

8. False; for any set A, $A \cup A' = U$, not $\{\ \}$.

9. True

10. $A = \{1, 2, 3, 4, 5, 6, 7, 8\}$

11. Set A is the set of natural numbers less than 9.

12. $A \cap B = \{3, 5, 7, 9\} \cap \{7, 9, 11, 13\} = \{7, 9\}$

13. $A \cup C' = \{3, 5, 7, 9\} \cup \{3, 11, 15\}' = \{3, 5, 7, 9\} \cup \{5, 7, 9, 13\} = \{3, 5, 7, 9, 13\}$

14. $A \cap (B \cap C)' = \{3, 5, 7, 9\} \cap (\{7, 9, 11, 13\} \cap \{3, 11, 15\})' = \{3, 5, 7, 9\} \cap \{11\}' = \{3, 5, 7, 9\} \cap \{3, 5, 7, 9, 13, 15\}$
    $= \{3, 5, 7, 9\}$, or A.

15. $n(A \cap B') = n\left(\{3, 5, 7, 9\} \cap \{7, 9, 11, 13\}'\right) = n\left(\{3, 5, 7, 9\} \cap \{3, 5, 15\}\right) = n\left(\{3, 5\}\right) = 2$

16.

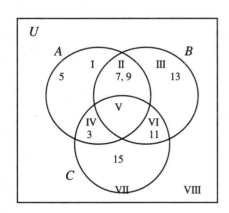

17. $A \cap (B \cup C')$ $(A \cap B) \cup (A \cap C')$

Set	Regions	Set	Regions
B	II, III, V, VI	A	I, II, IV, V
C	IV, V, VI, VII	B	II, III, V, VI
C'	I, II, III, VIII	$A \cap B$	II, V
$B \cup C'$	I, II, III, V, VI, VIII	C	IV, V, VI, VII
A	I, II, IV, V	C'	I, II, III, VIII
$A \cap (B \cup C')$	I, II, V	$A \cap C'$	I, II
		$(A \cap B) \cup (A \cap C')$	I, II, V

Both statements are represented by the same regions, I, II, V, of the Venn diagram.

Therefore, $A \cap (B \cup C') = (A \cap B) \cup (A \cap C')$ for all sets $A, B,$ and C.

18.

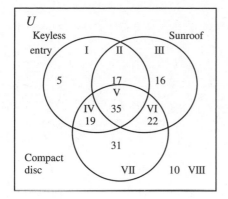

a) 52, the sum of the numbers in
   Regions I, III, VII
b) 10, Region VIII
c) 93, the sum of the numbers in
   Regions II, IV, V, VI
d) 17, Region II
e) 38, the sum of the numbers in
   Regions I, II, III
f) 31, Region VII

19. $\{7, 8,\ 9,\ 10, 11, \ldots, n + 6, \ldots\}$
    $\downarrow \downarrow \downarrow\ \ \downarrow\ \ \downarrow\qquad\downarrow$
    $\{8, 9,\ 10, 11, 12, \ldots, n + 7, \ldots\}$

20. $\{1, 2, 3, 4,\ 5, \ldots,\ \ n, \ldots\}$
    $\downarrow \downarrow \downarrow \downarrow\ \ \downarrow\qquad\downarrow$
    $\{1, 3, 5,\ 7, 9, \ldots, 2n - 1, \ldots\}$

CHAPTER THREE

LOGIC

Exercise Set 3.1

1.a. A simple statement is a sentence that conveys one idea and can be identified as either true or false.

  b. Statements consisting of two or more simple statements are called compound statements

.

3. a) Some are
    b) All are
    c) Some are not
    d) None are

5. a) →
    b) ∨
    c) ∧
    d) ~
    e) ↔

7. When a compound statement contains more than one connective a comma can be used to indicate which simple statements are to be grouped together. When writing a statement symbolically, the simple statements on the same side of the comma are to be grouped together within parentheses.

9. compound; conjunction, ∧

11. compound; biconditional ↔

13. compound; disjunction, ∨

15. simple statement

17. compound; negation, ~

19. compound; conjunction, ∧

21. compound; negation, ~

23. No picnic tables are portable.

25. Some chicken do not fly.

27. All turtles have claws.

29. Some bicycles have three wheels.

31. All pine trees produce pine cones.

33. No pedestrians are in the crosswalk.

35. ~ p

37. ~ q ∨ ~ p

39. ~ p → ~ q

41. ~ q → ~ p

43. ~ p ∧ ~ q

45. ~ (q → ~ p)

47. Firemen do not work hard.

49. Firemen wear red suspenders or firemen work hard.

51. Firemen do not work hard if and only if firemen do not wear red suspenders.

53. It is false that firemen wear red suspenders or firemen work hard.

55. Firemen do not work hard and firemen do not wear red suspenders.

57. $(p \lor \sim q) \to r$

59. $(p \land q) \lor r$

61. $p \to (q \lor \sim r)$

63. $(r \leftrightarrow q) \land p$

65. $q \to (p \leftrightarrow r)$

67. The water is 70° or the sun is shining, and we do not go swimming.

69. The water is not 70° , and the sun is shining or we go swimming.

71. If we do not go swimming, then the sun is shining and the water is 70°.

73. If the sun is shining then we go swimming, and the water is 70°.

75. The sun is shinning if and only if the water is 70°, and we go swimming.

77. Not permissible. In the list of choices, the connective "or" is the exclusive or, thus one can order either the soup or the salad but not both items.

79. Not permissible. Potatoes and pasta cannot be ordered together.

81. a) $(\sim p) \to q$ b) conditional   83. a) $(\sim q) \wedge (\sim r)$ b) conjunction

85. a) $(p \vee q) \to r$ b) conditional   87. a) $r \to (p \vee q)$ b) conditional

89. a) $(\sim p) \leftrightarrow (\sim q \to r)$ b) biconditional   91. $(r \wedge \sim q) \to (q \wedge \sim p)$ b) conditional

93. a) $\sim [(p \wedge q) \leftrightarrow (p \vee r)]$ b) negation   95. a) r: retired; c: started concrete business;
$$r \wedge \sim c$$
b) conjunction

97. a) b: below speed limit; p: pulled over; $\sim(b \to \sim p)$ 99. a) f: food has fiber; v: food has vitamins   h: be

b) negation   healthy; $(f \vee v) \to h$ b). conditional

101. a) c: may take course; f: fail previous   103. a) c: classroom is empty; w: is the

exam; p: passed placement test;   weekend   s: is 7:00 a.m.; $(c \leftrightarrow w) \vee s$

$c \leftrightarrow (\sim f \vee p)$ b) biconditional   b) disjunction

105. $[(\sim q) \to (r \vee p)] \leftrightarrow [(\sim r) \wedge q]$,
biconditional

107. a) The conjunction and disjunction have   107. c) If we evaluate the truth table for $p \vee q \wedge r$

the same dominance.   using the order $(p \vee q) \wedge r$ we get a different

b) Answers will vary.   solution than if we used the order $p \vee (q \wedge r)$.
Therefore, unless we are told where the
parentheses belong, we do not know which
solution is correct.

Exercise Set 3.2

1. a) $2^2 = 2 \times 2 = 4$ distinct cases

b)

	p	q
Case 1:	T	T
Case 2:	T	F
Case 3:	F	T
Case 4:	F	F

3. a)

p	q	p	$\vee$	q
T	T	T	T	T
T	F	T	T	F
F	T	F	T	T
F	F	F	F	F
		1	3	2

b) Only in Case 4, in which both simple
statements are false.

5.

p	p	$\vee$	$\sim p$
T	T	T	F
F	F	T	T

7.

p	q	p	$\wedge$	$\sim q$
T	T	T	F	F
T	F	T	T	T
F	T	F	F	F
F	F	F	F	T
		1	3	2

9.

p	q	~ (p ∨ ~ q)
T	T	F T T F
T	F	F T T T
F	T	T F F F
F	F	F F T T
		4 1 3 2

11.

p	q	~(p ∧ ~ q)
T	T	T T F F
T	F	F T T T
F	T	T F F F
F	F	T F F T
		4 1 3 2

13.

p	q	r	~q ∨ (p ∧ r)
T	T	T	F T T T T
T	T	F	F F T F F
T	F	T	T T T T T
T	F	F	T T T F F
F	T	T	F F F F T
F	T	F	F F F F F
F	F	T	T T F F T
F	F	F	T T F F F
			1 5 2 4 3

15.

p	q	r	r ∨ (p ∧ ~ q)
T	T	T	T T T F F
T	T	F	F F T F F
T	F	T	T T T T T
T	F	F	F T T T T
F	T	T	T T F F F
F	T	F	F F F F F
F	F	T	T T F F T
F	F	F	F F F F T
			1 5 2 4 3

17.

p	q	r	~q ∧ (r ∨ ~ p)
T	T	T	F F T T F
T	T	F	F F F F F
T	F	T	T T T T F
T	F	F	T F F F F
F	T	T	F F T T T
F	T	F	F F F T T
F	F	T	T T T T T
F	F	F	T T F T T
			1 5 2 4 3

19.

p	q	r	(~ q ∧ r) ∨ p
T	T	T	F F T T T
T	T	F	F F F T T
T	F	T	T T T T T
T	F	F	T F F T T
F	T	T	F F T F F
F	T	F	F F F F F
F	F	T	T T T T F
F	F	F	T F F F F
			1 3 2 5 4

21. p: Meetings are dull.

q: Teaching is fun.

In symbolic form the statement is p ∧ q.

p	q	p ∧ q
T	T	T
T	F	F
F	T	F
F	F	F
		1

23. p: Bob will get a haircut.

q: Bob will shave his beard.

In symbolic form the statement is p ∧ ~ q.

p	q	p	∧	~ q
T	T	T	F	F
T	F	T	T	T
F	T	F	F	F
F	F	F	F	T
		1	3	2

25. p: Jasper Adams is the tutor.

q: Mark Russo is a secretary.

In symbolic form the statement is ~ (p ∧ q).

p	q	~ (p ∧ q)
T	T	F T T T
T	F	T T F F
F	T	T F F T
F	F	T F F F
		4 1 3 2

27. p: The copier is out of toner.

q: The lens is dirty.

r : The corona wires are broken.

The statement is p ∨ (q ∨ r).

p	q	r	p ∨ (q ∨ r)
T	T	T	T T T
T	T	F	T T T
T	F	T	T T T
T	F	F	T T F
F	T	T	F T T
F	T	F	F T T
F	F	T	F T T
F	F	F	F F F
			2 3 1

29. p: Congress must act on the bill.
   q: The President signs the bill.
   In symbolic form, the statement is
   p ∧ (q ∨ ~ q).

p	q	p ∧(q∨ ~q)
T	T	TT TT F
T	F	TT FT T
F	T	FF TT F
F	F	FF FT T
		15 24 3

31. (a) ~ p ∨ (q ∨ r)
       F ∨ (F ∧ T)
       F ∨   F
           F
   Therefore the statement is false.
   (b) ~ p ∨ (q ∧ r)
       T ∨ (T ∧ T)
       T ∨   T
           T
   Therefore the statement is true.

33. (a) (~ q ∧ ~p) ∨ ~ r
       (T ∧ F) ∨ F
          F   ∨ F
             F
   Therefore the statement is false.
   (b) (~ q ∧ ~p) ∨ ~ r
       (F ∧ T) ∨ F
          F   ∨ F
             F
   Therefore the statement is false.

35. (a) (p ∧ ~q) ∨ r
       (T ∧ T) ∨ T
          T   ∨ T
             T
   Therefore the statement is true.
   (b) (p ∧ ~q) ∨ r
       (F ∧ F) ∨ T
          F   ∨ T
             T
   Therefore the statement is true.

37. (a) (~ r ∧ p) ∨ q
       (F ∧ T) ∨ F
          F   ∨ F
             F
   Therefore the statement is false.
   (b) (~ r ∧ p) ∨ q
       (F ∧ F) ∨ T
          F   ∨ T
             T
   Therefore the statement is true.

39. (a) (~q ∨ ~p) ∧ r
       (T ∨ F) ∧ T
          T   ∧ T
             T
   Therefore the statement is true.
   (b) (~q ∨ ~p) ∧ r
       (F ∨ T) ∧ T
          T   ∧ T
             T
   Therefore the statement is true.

41. (a) (~p ∨ ~q) ∨ (~r ∨ q)
       (F ∨ T) ∨ (F ∨ F)
          T   ∨   F
             T
   Therefore the statement is true.
   (b) (~p ∨ ~q) ∨ (~r ∨ q)
       (T ∨ F) ∨ (F ∨ T)
          T   ∨   T
             T
   Therefore the statement is true.

43. 3 + 5 = 4 + 47 or 10 − 9 = 9 − 10
       8 = 8   ∨   1 ≠ -1
         T   ∨   F
                  T
   Therefore the statement is true.

45. E: Elvis was a singer.
    C: Chickens can swim.

    E ∨ C
    T ∨ F
        T

    Therefore the statement is true.

47. U2: U2 is a rock band.
    DW: Denzel Washington is an actor.
    JS: Jerry Seinfeld is a comedian.

    (U2 ∧ DW) ∧ ~JS
    (T ∧ T) ∧   F
        T   ∧ F
                F

    Therefore the statement is false.

49. CR: Cal Ripken played football.
    GB: Bush was prime minister of England.
    CP: Colon Powell was in the Army.

    (CR ∨ GB) ∧ CP
    (F ∨ F) ∧   T
        F   ∧ T
            F

    Therefore the statement is false.

51. p: 30 pounds of cheese was consumed by
        the average American in 1909.
    q: The average American consumed 154
        pounds of sweetners in 2001.

    p ∧ ~ q
    F ∧ ~ T
    F ∧ F
    False

53. p: In 1909, average American ate
        approximately the same amount of
        fish and poultry.
    q: Between 1909 and 2001, average
        American consumed more poultry.

    p ∧ q
    T ∧ T
    True

55. p: 30% of Americans get 6 hours of sleep.
    q: 9% get 5 hours of sleep.

    ~ (p ∧ q)
    ~ (F ∧ T)
       ~F
    True

57. p: 13% of Americans get ≤ 5 hrs. of sleep.
    q: 32% of Americans get ≥ 6 hrs. of sleep.
    r: 30% of Americans get ≥ 8 hrs. of sleep.

    (p ∨ q) ∧ r
    (T ∨ F) ∧ F
        T   ∧ F
    False

59. p ∧ ~q

61. p ∨ ~q

63. (r ∨ q) ∧ p

65. q ∨ (p ∧ ~r)

67. (a) Mr. Duncan qualifies for the loan.
       Mrs. Tuttle qualifies for the loan.
    (b) The Rusineks do not qualify
        because their gross income is too low.

69. (a) Wing Park qualifies for the special
        fare.
    (b) The other 4 do not qualify:
        Gina V. returns after 04/01;
        Kara S. returns on Monday;
        Christos G. does not stay
        at least one Saturday; and
        Alex C. returns on Monday.

71.

p	q	r	[(q ∧ ~ r) ∧ (~ p ∨ ~ q)] ∨ (p ∨ ~ r)				
T	T	T	F	F	F	T	T
T	T	F	T	F	F	T	T
T	F	T	F	F	T	T	T
T	F	F	F	F	T	T	T
F	T	T	F	F	T	F	F
F	T	F	T	T	T	T	T
F	F	T	F	F	T	F	F
F	F	F	F	F	T	T	T
			1	3	2	5	4

73. Yes

p	q	r	(p ∧ ~q) ∨ r			(q ∧ ~r) ∨ p		
T	T	T	F	T	T	F	T	T
T	T	F	F	F	F	T	T	T
T	F	T	T	T	T	F	T	T
T	F	F	T	T	F	F	T	T
F	T	T	F	T	T	F	F	F
F	T	F	F	F	F	T	T	F
F	F	T	F	T	T	F	F	F
F	F	F	F	F	F	F	F	F

Exercise Set 3.3

1.a)

p	q	p	→	q
T	T	T	T	T
T	F	T	F	F
F	T	F	T	T
F	F	F	T	F
		1	3	2

b) The conditional statement is false only in the case
   when antecedent is true and the consequent is false,
   otherwise it is true.

3.a) Substitute the truth values for the simple statement. Then evaluate the compound statement for that specific case.

 b) $[(p \leftrightarrow q) \vee (\sim r \rightarrow q)] \rightarrow \sim r$
 $[(T \leftrightarrow T) \vee (\sim T \rightarrow T)] \rightarrow \sim T$
 $[\quad T \quad \vee (\sim T \rightarrow T)] \rightarrow T$
 $[\quad T \quad \vee \quad T \quad] \rightarrow T$
 $\qquad T \qquad\qquad \rightarrow T$
 $\qquad\qquad\qquad\qquad T$

In this specific case the statement is true.

5. A self-contradiction is a compound statement that is false in every case.

7.

p	q	$\sim q \rightarrow \sim p$		
T	T	F	T	F
T	F	T	F	F
F	T	F	T	T
F	F	T	T	T
		1	3	2

9.

p	q	$\sim (q \rightarrow p)$	
T	T	F	T
T	F	F	T
F	T	T	F
F	F	F	T
		2	1

11.

p	q	$\sim q$	$\leftrightarrow$	p
T	T	F	F	T
T	F	T	T	T
F	T	F	T	F
F	F	T	F	F
		1	3	2

13.

p	q	p	$\leftrightarrow$	$(q \vee p)$
T	T	T	T	T
T	F	T	T	T
F	T	F	F	T
F	F	F	T	F
		1	3	2

15.

p	q	$q \rightarrow (p \rightarrow \sim q)$				
T	T	T	F	T	F	F
T	F	F	T	T	T	T
F	T	T	T	F	T	F
F	F	F	T	F	T	T
		4	5	1	3	2

17.

p	q	r	r	∧	(~q	→	p)
T	T	T	T	T	F	T	T
T	T	F	F	F	F	T	T
T	F	T	T	T	T	T	T
T	F	F	F	F	T	T	T
F	T	T	T	T	F	T	F
F	T	F	F	F	F	T	F
F	F	T	T	T	T	F	F
F	F	F	F	F	T	F	F
			4	5	1	3	2

19.

p	q	r	(q	↔	p)	∧	~ r
T	T	T		T		F	F
T	T	F		T		T	T
T	F	T		F		F	F
T	F	F		F		F	T
F	T	T		F		F	F
F	T	F		F		F	T
F	F	T		T		F	F
F	F	F		T		T	T
				1		3	2

21.

p	q	r	(q	∨	~ r)	↔	~ p
T	T	T	T	T	F	F	F
T	T	F	T	T	T	F	F
T	F	T	F	F	F	T	F
T	F	F	F	T	T	F	F
F	T	T	T	T	F	T	T
F	T	F	T	T	T	T	T
F	F	T	F	F	F	F	T
F	F	F	F	T	T	T	T
			1	3	2	5	4

23.

p	q	r	(~ r	∨	~q)	→	p
T	T	T	F	F	F	T	T
T	T	F	T	T	F	T	T
T	F	T	F	T	T	T	T
T	F	F	T	T	T	T	T
F	T	T	F	F	F	T	F
F	T	F	T	T	F	F	F
F	F	T	F	T	T	F	F
F	F	F	T	T	T	F	F
			1	3	2	5	4

25.

p	q	r	(p	→	q)	↔	(~q	→	~r)
T	T	T		T		T	F	T	F
T	T	F		T		T	F	T	T
T	F	T		F		T	T	F	F
T	F	F		F		F	T	T	T
F	T	T		T		T	F	T	F
F	T	F		T		T	F	T	T
F	F	T		T		F	T	F	F
F	F	F		T		T	T	T	T
				1		5	2	4	3

27.

p	q	r	p	→	(q	∧	r)
T	T	T	T	T		T	
T	T	F	T	F		F	
T	F	T	T	F		F	
T	F	F	T	F		F	
F	T	T	F	T		T	
F	T	F	F	T		F	
F	F	T	F	T		F	
F	F	F	F	T		F	
			1	3		2	

29.

p	q	r	(p	↔	~q)	∨	r
T	T	T		F		T	T
T	T	F		F		F	F
T	F	T		T		T	T
T	F	F		T		T	F
F	T	T		T		T	T
F	T	F		T		T	F
F	F	T		F		T	T
F	F	F		F		F	F
				1		3	2

31.

p	q	r	(~ p	→	q)	∨	r
T	T	T		T		T	T
T	T	F		T		T	F
T	F	T		T		T	T
T	F	F		T		T	F
F	T	T		T		T	T
F	T	F		T		T	F
F	F	T		F		T	T
F	F	F		F		F	F
				1		3	2

33.

p	q	p	→	~ q
T	T	T	F	F
T	F	T	T	T
F	T	F	T	F
F	F	F	T	T
		1	3	2

neither

35.

p	q	p	∧	(q ∧ ~ p)
T	T	T	F	F
T	F	T	F	F
F	T	F	F	T
F	F	F	F	F
		1	3	2

self-contradiction

37.

p	q	(~ q	→	p)	∨	~ q
T	T		T		T	F
T	F		T		T	T
F	T		T		T	F
F	F		F		T	T
			1		3	2

tautology

39.

p	q	p	→	(p ∧ q)
T	T	T	T	T
T	F	T	F	F
F	T	F	T	F
F	F	F	T	F
		1	3	2

not an implication

41.

p	q	(q ∧ p)	→	(p ∧ q)
T	T	T	T	T
T	F	F	T	F
F	T	F	T	F
F	F	F	T	F
		1	3	2

an implication

43.

p	q	[(p → q)	∧	(q → p)]	→	(p ↔ q)
T	T	T	T	T	T	T
T	F	F	F	T	T	F
T	T	T	F	F	T	F
T	F	T	T	T	T	T
		1	3	2	5	4

an implication

45. $\quad p \to (\sim q \land r)$
$\quad\ \ T \to (\ T \land T)$
$\quad\ \ T \to \quad\ T$
$\quad\qquad T$

47. $\quad (q \land \sim p) \leftrightarrow \sim r$
$\quad\ \ F \land F\ \leftrightarrow F$
$\quad\qquad F\quad \leftrightarrow F$
$\quad\qquad\quad T$

49. $\quad (\sim p \land \sim q) \lor \sim r$
$\quad\ \ (F \land\ T) \lor F$
$\quad\qquad F\quad \lor F$
$\quad\qquad\quad F$

51. $\quad (p \land r) \leftrightarrow (p \lor \sim q)$
$\quad\ \ (T \land T) \leftrightarrow (T \lor T)$
$\quad\qquad T\ \leftrightarrow\quad T$
$\quad\qquad\quad T$

53. $\quad (\sim p \leftrightarrow r) \lor (\sim q \leftrightarrow r)$
$\quad\ \ (F \leftrightarrow T) \lor (T \leftrightarrow T)$
$\quad\qquad T\quad \lor\quad T$
$\quad\qquad\quad T$

55. $\quad \sim [(p \lor q) \leftrightarrow (p \to \sim r)]$
$\quad\ \ \sim [(T \lor F) \leftrightarrow (T \to F)]$
$\quad\ \ \sim [T\quad \leftrightarrow\quad F]$
$\quad\qquad \sim F$
$\quad\qquad\quad T$

57. If $10 + 5 = 15$, then $56 \div 7 = 8$.
$\qquad T \to T$
$\qquad\quad T$

59. A triangle has four sides or a square has three sides, and a rectangle has four sides.
$\qquad (F \lor F) \land T$
$\qquad\ F\quad \land T$
$\qquad\qquad F$

61. Dell makes computers, if and only if Gateway makes computers or Canon makes printers.
$\qquad T \leftrightarrow (T \lor T)$
$\qquad T \leftrightarrow\quad T$
$\qquad\qquad T$

63. Valentine's Day is in February or President's Day is in March, and Thanksgiving Day is in November.
$\qquad (T \lor F) \land T$
$\qquad\ T \land T$
$\qquad\quad T$

65. Io has a diameter of 1000–3161 miles, or Thebe may have water, and Io may have atmosphere.
$\qquad (T \lor F) \land T$
$\qquad\ T \land T$
$\qquad\quad T$

67. Phoebe has a larger diameter than Rhea if and only if Callisto may have water ice, and Calypso has a diameter of 6–49 miles.
$\qquad (F \leftrightarrow T) \land T$
$\qquad\ F\quad \land T$
$\qquad\qquad F$

69. The most common cosmetic surgery procedure for females is liposuction or the most common procedure for males is eyelid surger, and 20% of male cosmetic surgery is for nose reshaping.
$\qquad (T \lor F) \land F$
$\qquad\quad T\quad \land F$
$\qquad\qquad F$

For 71–75, p: Muhundan spoke at the teachers' conference.
$\qquad q$: Muhundan received the outstanding teacher award
$\qquad$ Assume p and q are true.

71. $\quad p \to q$
$\quad\ T \to T$
$\quad\quad T$

73. $\quad \sim p \to q$
$\quad\ F \to T$
$\quad\quad T$

75. $\quad q \to p$
$\quad\ T \to T$
$\quad\quad T$

77. No, the statement only states what will occur if your sister gets straight A's. If your sister does not get straight A's, your parents may still get her a computer.

79.

p	q	r	[(p ∨ q)	→	~ r]	↔	(p ∧ ~ q)
T	T	T	T	F	F	T	F
T	T	F	T	T	T	F	F
T	F	T	T	F	F	F	T
T	F	F	T	T	T	T	T
F	T	T	T	F	F	T	F
F	T	F	T	T	T	F	F
F	F	T	F	T	F	F	F
F	F	F	F	T	T	F	F
			1	4	2	5	3

81. The statement may be expressed as $(p \rightarrow q) \vee (\sim p \rightarrow q)$, where p: It is a head and q: I win.

p	q	(p → q)	∨	(~ p → q)
T	T	T	T	T
T	F	F	T	T
F	T	T	T	T
F	F	F	T	F
		1	3	2

The statement is a tautology.

83.
Tiger	Boots	Sam	Sue
Blue	Yellow	Red	Green
Nine Lives	Whiskas	Friskies	Meow Mix

84. The birth order is Mary, Annie, Katie; Katie and Annie are telling the truth.

Exercise Set 3.4

1. Two statements are equivalent if both statements have exactly the same truth values in the answer column of the truth table.

3. The two statements must be equivalent. A biconditional is a tautology only when the statements on each side of the biconditional are equivalent.

5. a) $q \rightarrow p$ b) $\sim p \rightarrow \sim q$ c) $\sim q \rightarrow \sim p$

7. $\sim p \vee q$

9. $\sim p \vee \sim q \Leftrightarrow \sim (p \wedge q)$ by the first of De Morgan's Laws.

11. Using DeMorgan's Laws on the statement $\sim(p \wedge q)$, we get $\sim(p \wedge q) \leftrightarrow \sim p \vee \sim q$ and $\sim p \vee \sim q$ is not equivalent to $\sim p \wedge \sim q$.

13. Yes, $\sim(p \vee \sim q) \leftrightarrow \sim p \wedge \sim(\sim q) \leftrightarrow \sim p \wedge q$

15. Yes, $(\sim p \wedge \sim q) \rightarrow r$

17. Yes, $\sim(p \rightarrow \sim q) \Leftrightarrow \sim(\sim p \vee \sim q) \Leftrightarrow p \wedge q$

19.

p	q	p → q	~ p ∨ q
T	T	T	F T T
T	F	F	F F F
F	T	T	T T T
F	F	T	T T F
		1	1 3 2

The statements are equivalent.

21.

p	q	r	(p ∧ q) ∧ r	p ∧ (q ∧ r)
T	T	T	T TT	T T T
T	T	F	T FF	T F F
T	F	T	F FT	T F F
T	F	F	F FF	T F F
F	T	T	F FT	F F T
F	T	F	F FF	F F F
F	F	T	F FT	F F F
F	F	F	F FF	F F F
			1 3 2	2 3 1

The statements are equivalent.

23.

p	q	r	(p ∨ q) ∨ r	p ∨ (q ∨ r)
T	T	T	T TT	T T T
T	T	F	T TF	T T T
T	F	T	T TT	T T T
T	F	F	T TF	T T F
F	T	T	T TT	F T T
F	T	F	T TF	F T T
F	F	T	F TT	F T T
F	F	F	F FF	F F F
			1 3 2	2 3 1

The statements are equivalent.

25.

p	q	r	p ∧ (q ∨ r)	(p ∧ q) ∨ r
T	T	T	TT TTT	TTTTT
T	T	F	TT TTF	TTTTT
T	F	T	TT FTT	TFF TT
T	F	F	TF FFF	TFF FF
F	T	T	FF TTT	FFT TT
F	T	F	FF TTF	FFT FF
F	F	T	FF FTT	FFF TT
F	F	F	FF FFF	FFF FF
			1 5 2 4 3	1 3 2 5 4

The statements are not equivalent.

27.

p	q	r	(p → q) ∧ (q → r)	(p → q) → r
T	T	T	T T T	T TT
T	T	F	T F F	T FF
T	F	T	F F T	F TT
T	F	F	F F T	F TF
F	T	T	T T T	T TT
F	T	F	T F F	T FF
F	F	T	T T T	T TT
F	F	F	T T T	T FF
			1 3 2	1 3 2

The statements are not equivalent.

29.

p	q	(p → q) ∧ (q → p)	p ↔ q
T	T	T T T	T
T	F	F F T	F
F	T	T F F	F
F	F	T T T	T
		1 3 2	1

The statements are equivalent.

31. p: The Mississippi River runs through Ohio.

q: The Ohio River runs through Mississippi.

In symbolic form, the statement is ~ (p ∨ q).

Applying DeMorgan's Laws we get: ~ p ∧ ~ q.

The Mississippi River does not run through Ohio and the Ohio River does not run through Miss.

33. p: The snowmobile was an Arctic Cat.

q: The snowmobile was a Ski-Do.

In symbolic form, the statement is ~ p ∧ ~ q.

Applying DeMorgan's Laws we get: ~ (p ∨ q).

It is false that the snowmobile was an Arctic Cat or a Ski-Do.

35. p: The hotel has a weight room.

q: The conference center has an auditorium.

In symbolic form, the statement is ~ p ∨ ~ q.

Applying DeMorgan's Laws we get: ~ (p ∧ q).

It is false that the hotel has a weight room and the conference center has an auditorium.

37. p: We go to Cozumel.

q: We will go snorkeling.

r: We will go to Senior Frogs.

In symbolic form, the statement is

p → (q ∨ ~ r). Applying DeMorgan's Laws we get: p → ~ (~ q ∧ r). If we go to Cozumel, then it is false that we will not go snorkeling and we will go to Senior Frogs.

39. p: You drink a glass of orange juice.
    q: You'll get a full day's supply of folic acid.
    In symbolic form, the statement is p → q.
    Since p → q ⇔ ~ p ∨q, an equivalent
    Statement is: You do not drink a glass of OJ
    or you will get a full day's supply of folic acid.

41. p: Bob the Tomato visited the nursing home.
    q: Bob the Tomato visited the Cub Scout
       meeting.
    In symbolic form, the statement is p ∨~ q.
    Since ~p → ~q ⇔ ~ p ∨q, an equivalent
    Statement is: If Bob the Tomato did not visit
    the nursing home, then he did not visit the
    Cub Scout meeting.

43. p: The plumbers meet in Kansas City.
    q: The *Rainmakers* will provide the
       entertainment.
    In symbolic form, the statement is ~ (p → q).
    ~ (p → q) ⇔ ~ (~ p ∨ q) ⇔ p ∧ ~ q . The
    plumbers meet in KC and the *Rainmakers* will
    not provide the entertainment.

45. p: It is cloudy.
    q: The front is coming through.
    In symbolic form, the statement is (p → q) ∧
    (q → p). (p → q) ∧(q → p) ⇔ p ↔ q. It is
    cloudy if and only if the front is coming
    through.

47. p: The chemistry teacher teaches mathematics.
    q: There is a shortage of math. teachers.
    In symbolic form, the statement is p ↔ q.
    (p → q) ∧(q → p) ⇔ p ↔ q.
    If the chemistry teacher teaches math., then
    there is a shortage of math. teachers and if
    there is a shortage of math. teachers, then the
    chemistry teacher teaches math.

49. Converse: If I finish the book in 1 week, then it is interesting.
    Inverse: If the book is not interesting, then I will not finish it
       in 1 week.
    Contrapositive: If I do not finish the book in
       one week, then it is not interesting.

51. Converse: If you can watch TV, then you
       finish your HW.
    Inverse: If you do not finish your HW, then
       you cannot watch TV.
    Contrapositive: If you cannot watch TV, then
       you did not finish your HW.

53. Converse: If I scream, then that annoying paper clip (Clippie)
       shows up on my screen.
    Inverse: If Clippie does not show up on my
       screen, then I will not scream.
    Contrapositive: If I do not scream, then
       Clippie does not show up on my screen.

55. Converse: If we go down to the marina and
       take out a sailboat, then the sun is shining.
    Inverse: If the sun is not shining, then we do
       not go down to the marina or we do not take
       out a sailboat.
    Contrapositive: If we do not go down to the
       marina or we do not take out a sailboat, then
       the sun is not shining.

57. If a natural number is divisible by 10, then it is divisible by 5. True

59. If a natural number is not divisible by 6, then it is not divisible by 3. False

61. If two lines are not parallel, then the two lines intersect in at least one point. True

63. If the polygon is a quadrilateral, then the sum of the interior angles is 360 degrees. True

65. p: Maria has retired.

q: Maria is still working.

In symbolic form, the statements are:

a) $\sim p \vee q$, b) $q \rightarrow \sim p$, c) $p \rightarrow \sim q$

Statement (c) is the contrapositive of statement. (b). Therefore, statements (b) and (c) are equivalent.

p	q	$\sim p \vee q$	$q \rightarrow \sim p$
T	T	F TT	TF F
T	F	F FF	FT F
F	T	T TT	TT T
F	F	T TF	FT T
		1 32	13 2

Since the truth tables for (a) and (b) are different we conclude that only statements (b) and (c) are equivalent.

67. p: The car is reliable.

q: The car is noisy.

In symbolic form, the statements are: a) $\sim p \wedge q$, b) $\sim p \rightarrow \sim q$, c) $\sim (p \vee \sim q)$. If we use DeMorgan's Laws on statement (a), we get statement (c). Therefore, statements (a) and (c) are equivalent. If we look at the truth tables for statements (a), (b), and (c), we see that only statements (a) and (c) are equivalent.

		a)	b)	c)
p	q	$\sim p \wedge q$	$\sim p \rightarrow \sim q$	$\sim (p \vee \sim q)$
T	T	F F T	F T F	F TT F
T	F	F F F	F T T	F TT T
F	T	T T T	T F F	T FF F
F	F	T F F	T T T	F FT T
		1 3 2	1 3 2	4 1 3 2

69. p: Today is Sunday.

q: The library is open.

In symbolic form, the statements are: a) $\sim p \vee q$, b) $p \rightarrow \sim q$, c) $q \rightarrow \sim p$. Looking at the truth table for all three statements, we can determine that only statements (b) and (c) are equivalent.

		a)	b)	c)
p	q	$\sim p \vee q$	$p \rightarrow \sim q$	$q \rightarrow \sim p$
T	T	F TT	TF F	T F F
T	F	F FF	TT T	F T F
F	T	T TT	FT F	T T T
F	F	T TF	FT T	F T T
		1 32	13 2	13 2

71. p: The grass grows.

q: The trees are blooming.

In symbolic form, the statements are: a) $p \wedge q$, b) $q \rightarrow \sim p$, c) $\sim q \vee \sim p$. Using the fact that $p \rightarrow q \Leftrightarrow \sim p \vee q$, on statement (b) we get $\sim q \vee \sim p$. Therefore, statements (b) and (c) are equivalent. Looking at the truth table for statements (a) and (b) we can conclude that only statements (b) and (c) are equivalent.

p	q	$p \wedge q$	$q \rightarrow \sim p$
T	T	T	TF F
T	F	F	FT F
F	T	F	TT T
F	F	F	FT T
		1	13 2

73. p: You drink milk.
    q: Your cholesterol count will be lower.
    In symbolic form, the statements are:
    a) ~ (~ p → q), b) q ↔ p, and c) ~ (p → ~ q) .

P	q	~ (~ p → q)	q ↔ p	~ (p → ~ q)
T	T	F FT T	T T T	T T F F
T	F	F FT F	F F T	F T T T
F	T	F TT T	T F F	F F T F
F	F	T TF F	F T F	F F T T
		4 1 3 2	5 7 6	11 8 10 9

Therefore, none of the statements are equivalent.

75. p: The pay is good.
    q: Today is Monday.
    r : I will take the job.
    Looking at the truth tables for statements (a), (b), and (c), we can determine that none of these statements are equivalent.

p	q	r	a) (p ∧ q) → r	b) ~ r → ~(p ∨ q)	c) (p ∧ q) ∨ r
T	T	T	T	T T F T F T	T T T
T	T	F	T	F F T F F T	T T F
T	F	T	F	T T F T F T	F T T
T	F	F	F	T F T F F T	F F F
F	T	T	F	T T F T F T	F T T
F	T	F	F	T F T F F T	F F F
F	F	T	F	T T F T T F	F T T
F	F	F	F	T F T T T F	F F F
			1	3 2 1 4 3 2	1 3 2

77. p: The package was sent by Federal Express.
    q: The package was sent by United Parcel Service.
    r : The package arrived on time.
    Using the fact that p → q ⇔ ~ p ∨ q to rewrite statement (c), we get p ∨ (~ q ∧ r). Therefore, statements (a) and (c) are equivalent. Looking at the truth table for statements (a) and (b), we can conclude that only statements (a) and (c) are equivalent.

p	q	r	a) p ∨ (~ q ∧ r)	b) r ⇔ (p ∨ ~ q)
T	T	T	T T F F T	T T T T F
T	T	F	T T F F F	F F T T F
T	F	T	T T T T T	T T T T T
T	F	F	T T T F F	F F T T T
F	T	T	F F F F T	T F F F F
F	T	F	F F F F F	F T F F F
F	F	T	F T T T T	T T F T T
F	F	F	F F T F F	F F F T T
			1 5 2 4 3	1 5 2 4 3

79. p: The car needs oil.
    q: The car needs gas.
    r : The car is new.
    In symbolic form, the statements are: a) p ∧ (q ∨ r),
    b) p ∧ ~ (~ q ∧ ~ r), and c) p → (q ∨ ~ r). If we use DeMorgan's Laws on the disjunction in statement (a), we obtain p ∧ ~ (~ q ∧ ~ r). Therefore, statements (a) and (b) are equivalent. If we compare the truth tables for (a) and (c) we see that they are not equivalent. Therefore, only statements (a) and (b) are equivalent.

p	q	r	p ∧ (q ∨ r)	p → (q ∨ ~ r)
T	T	T	T T T	T T T T F
T	T	F	T T T	T T T T T
T	F	T	T T T	T F F F F
T	F	F	T F F	T T F T T
F	T	T	F F T	F T T T F
F	T	F	F F T	F T T T T
F	F	T	F F T	F T F F F
F	F	F	F F F	F T F T T
			1 3 2	1 5 2 4 3

81. True. If $p \rightarrow q$ is false, it must be of the form $T \rightarrow F$. Therefore, the converse must be of the form $F \rightarrow T$, which is true.

83. False. A conditional statement and its contrapositive always have the same truth values.

85. If we use DeMorgan's Laws to rewrite ~ p ∨q, we get ~ (p ∧~ q). Since ~ p ∨q ⇔ ~ (p ∧~ q) and p → q ⇔ ~ p ∨q, we can conclude that p → q ⇔ ~ (p ∧~ q). Other answers are possible.

87. Research problem -- Answers will vary.

89. (a) conditional; (b) biconditional; (c) inverse; (d) converse; (e) contrapositive

Exercise Set 3.5

1. An argument is valid when its conclusion necessarily follows from the given set of premises.

3. Yes. It is not necessary for the premises or the conclusion to be true statements for the argument to be valid.

5. Yes. If the conclusion follows from the set of premises, then the argument is valid, even if the premises are false.

7. a) $p \rightarrow q$ b) Pizza is served on time or is free.
    ~ p   The pizza was not served on time.
    ∴ q   The pizza is free.

9. a) p → q   b) If the sky is clear, then it will be hot.
    q → r   If it is hot, then will wear shorts.
    ∴ p → r   If sky is clear, then wear shorts.

11. a) p → q   b) If you wash my car, then I pay you $5.
    ~ p   You did not wash my car.
    ∴ ~ q   I will not give you $5.

13. This argument is the law of detachment and therefore it is valid.

15.

p	q	[(p ∧ ~ q) ∧ q] → ~ p
T	T	T F F F T T F
T	F	T T T F F T F
F	T	F F F F T T T
F	F	F F T F F T T
		1 3 2 5 4 7 6

The argument is valid.

17.

p	q	[~ p ∧ (p ∨ q)] → ~ q
T	T	F F T T F
T	F	F F T T T
F	T	T T T F F
F	F	T F F T T
		1 3 2 5 4

The argument is invalid.

19. This argument is the fallacy of the inverse. Therefore it is invalid.

21.

p	q	[(~ p → q) ∧ ~ q] → ~ p
T	T	F T T F F T F
T	F	F T F T T F F
F	T	T T T F F T T
F	F	T F F F T T T
		1 3 2 5 4 7 6

The argument is invalid.

23. This argument is the law of syllogism and therefore it is valid.

25.

p	q	r	[(p ↔ q) ∧ (q ∧ r)] → (p ∨ r)
T	T	T	T T T T T
T	T	F	T F F T T
T	F	T	F F F T T
T	F	F	F F F T T
F	T	T	F F T T T
F	T	F	F F F T F
F	F	T	T F F T T
F	F	F	T F F T F
			1 3 2 5 4

The argument is valid.

27.

p	q	r	[(r ↔ p) ∧ (~p ∧ q)] → (p ∧ r)
T	T	T	T F FFT T T
T	T	F	F F FFT T F
T	F	T	T F FFF T T
T	F	F	F F FFF T F
F	T	T	F F TTT T F
F	T	F	T T TTT F F
F	F	T	F F TFF T F
F	F	F	T F TFF T F
			1 5 2 4 3 7 6

The argument is invalid.

29.

p	q	r	[(p→ q) ∧ (q ∨ r) ∧ (r ∨ p)] → p
T	T	T	T T T T T T T
T	T	F	T T T T T T T
T	F	T	F F T F T T T
T	F	F	F F F F T T T
F	T	T	T T T T T F F
F	T	F	T T T F F T F
F	F	T	T T T T T F F
F	F	F	T F F F F T F
			1 3 2 5 4 7 6

The argument is invalid.

31.

p	q	r	[(p → q) ∧ (r → ~ p) ∧ (p ∨ r)] → (q ∨ ~ p)
T	T	T	T F TF F F T T TTF
T	T	F	T T FT F T T T TTF
T	F	T	F F TF F F T T FFF
T	F	F	F F FT F F T T FFF
F	T	T	T T TT T T T T TTT
F	T	F	T T FT T F F T TTT
F	F	T	T T TT T T T T FTT
F	F	F	T T FT T F F T FTT
			1 5 24 3 7 6 11 9 10 8

The argument is valid.

33. p: Will Smith wins an Academy Award.
 q: Will Smith retires from acting.

p	q	$[(p \to q) \wedge \sim p] \to \sim q$
T	T	T F F T F
T	F	F F F T T
F	T	T T T F F
F	F	T T T T T
		1 3 2 5 4

The argument is invalid (fallacy of the inverse.).

35. p: The baby is a boy.
 q: The baby will be named Alexander Martin.

p	q	$[(p \to q) \wedge q] \to p$
T	T	T T T T
T	F	F F F T T
F	T	T T T F F
F	F	T F F T F
		1 3 2 5 4

The argument is valid (law of detachment).

37. p: Monkeys can fly.
 q: Scarecrows can dance.
 $[(p \to q) \wedge \sim q] \to \sim p$

 This argument is valid because of the Law of Contraposition.

39. p: The orange was left on the tree for one year.
 q: The orange is ripe.

p	q	$[(p \to q) \wedge q] \to p$
T	T	T T T T T
T	F	F F F T T
F	T	T T T F F
F	F	T F F T F
		1 3 2 5 4

This is the Fallacy of the Converse; thus

41. p: The X-Games will be in San Diego.
 q: The X-Games will be in Corpus Christi.

P	q	$[(p \vee q) \wedge \sim p] \to q$
T	T	T F F T T
T	F	T F F T F
F	T	T F T T T
F	F	F T F T F
		1 3 2 5 4

The argument is valid.

43. p: It is cold.
 q: The graduation will be held indoors.
 r: The fireworks will be postponed.
 $[(p \to q) \wedge (q \to r)] \to (p \to r)$

This argument is valid because of the Law of Syllogism.

45. f: The canteen is full
 w: We can go for a walk.
 t: We will get thirsty.

f	w	t	$[(f \to w) \wedge (w \wedge \sim t)] \to (w \to \sim f)$
T	T	T	T F T F F T T F F
T	T	F	T T T T T F T F F
T	F	T	F F F F F T F T F
T	F	F	F F F F T T F T F
F	T	T	T F T F F T T T T
F	T	F	T T T T T T T T T
F	F	T	T F F F F T F T T
F	F	F	T F F F T T F T T
			1 5 2 4 3 9 6 8 7

The argument is invalid.

47. s: It is snowing.
    g: I am going skiing.
    c: I will wear a coat.

s	g	c	[(s ∧ g) ∧ (g → c)] → (s → c)
T	T	T	T T T T T
T	T	F	T F F T F
T	F	T	F F T T T
T	F	F	F F T T F
F	T	T	F F T T T
F	T	F	F F F T T
F	F	T	F F T T T
F	F	F	F F T T T
			1 3 2 5 4

The argument is valid.

49. h: The house has electric heat.
    b: The Flynns will buy the house.
    p: The price is less than $100,000.

h	b	p	[(h → b) ∧ (~p → ~b)] → (h → p)
T	T	T	T T F T F T T
T	T	F	T F T F F T F
T	F	T	F F F T T T T
T	F	F	F F T T T T F
F	T	T	T T F T F T T
F	T	F	T F T F F T T
F	F	T	T T F T T T T
F	F	F	T T T T T T T
			1 5 2 4 3 7 6

The argument is valid.

51. p: The prescription is called in to Walgreen's.
    q. You pick up the prescription at 4:00 p.m.

    p → q

    ~q

∴ ~p

The argument is the law of contraposition and is valid.

53. t: The television is on.
    p: The plug is plugged in.

t	p	[(t ∨ ~ p) ∧ (p)] → t
T	T	T F T T T
T	F	T T T T T
F	T	F F F T F
F	F	T T F T F
		2 1 3 5 4

The argument is valid.

55. t: The test was easy.
    g: I received a good grade.

t	g	[(t ∧ g) ∧ (~ t ∨ ~ g)] → ~ t
T	T	T F F F F T F
T	F	F F F T T T F
F	T	F F T T F T T
F	F	F F T T T T T
		1 5 2 4 3 7 6

The argument is valid.

57. c: The baby is crying.
    h: The baby is hungry.

c	h	[(c ∧ ~ h) ∧ (h → c)] → h
T	T	T F F F T T T
T	F	T T T T T F F
F	T	F F F F F T T
F	F	F F T F T T F
		1 3 2 5 4 7 6

The argument is invalid.

59. f: The football team wins the game.   f → d
    d: Dave played quarterback.   d → ~ s
    s: The team is in second place.   ∴ f → s

    Using the law of syllogism f → ~s, so this argument is invalid.

61. Your face will break out. (law of detachment)

63. p: A tick is an insect.

 q: A tick is an arachnid.

p	q	[(p ∨ q)	∧	~ p]	→	q
T	T	T	F	F	T	T
T	F	T	F	F	T	F
F	T	T	T	T	T	T
F	F	F	F	T	T	F
		1	3	2	5	4

 The argument is valid.

70. p: I think.

 q: I am.

 p → q

 ~q

 ∴ ~p

 By the Fallacy of the Inverse, the argument is invalid.

65. You did not close the deal. (law of contraposition)

67. If you do not pay off your credit card bill, then the bank makes money. (law of syllogism)

69. No. An argument is <u>invalid</u> only when the conjunction of the premises is true and the conclusion is false.

Exercise Set 3.6

1. The argument is invalid.
3. The conclusion necessarily follows from the set of premises.
5. Yes. If the conjunction of the premises is false in all cases, then the argument is valid regardless of the truth value of the conclusion.

7.

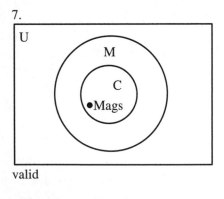

valid

9.

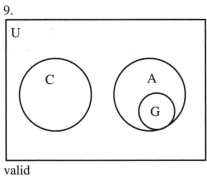

valid

11.

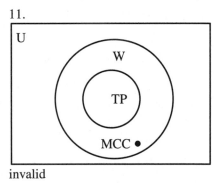

invalid

13.

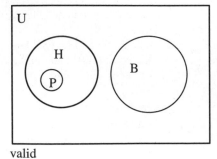

valid

15.

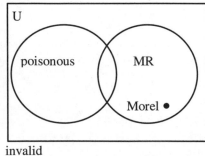

invalid

17.

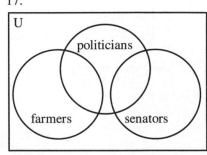

invalid

19.

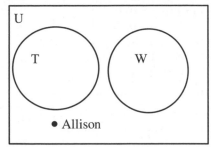

invalid

21.

physics

math

people

valid

23.

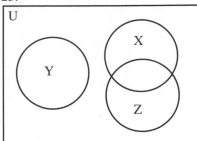

invalid

25.

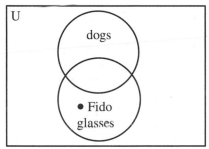

invalid

27.

P

F

G

S

valid

29.

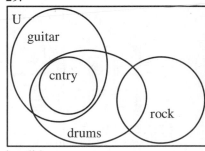

invalid

31. $[(p \rightarrow q) \wedge (p \vee q)] \rightarrow \sim p$ can be expressed as a set statement by $[(P' \cup Q) \cap (P \cup Q)] \subseteq P'$. If this statement is true, then the argument is valid; otherwise, the argument is invalid.

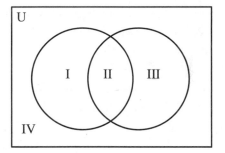

Set	Regions
$P' \cup Q$	II, III, IV
$P \cup Q$	I, II, III
$(P' \cup Q) \cap (P \cup Q)$	II, III
P'	III, IV

Since $(P' \cup Q) \cap (P \cup Q)$ is not a subset of P', the argument is invalid.

Review Exercises

1. No rock bands play ballads.
2. All bananas are ripe.
3. Some chickens have lips.
4. Some panthers are not endangered.
5. Some pens do not use ink.
6. Some rabbits wear glasses.
7. The coffee is Maxwell House or the coffee is hot.
8. The coffee is not hot and the coffee is strong.
9. If the coffee is hot, then the coffee is strong and it is not Maxwell House.
10. The coffee is Maxwell House if and only if the coffee is not strong.
12. The coffee is not Maxwell House, if and only if the coffee is strong and the coffee is not hot.
11. The coffee is Maxwell House or the coffee is not hot, and the coffee is not strong.

13. $r \wedge q$
14. $p \rightarrow r$
15. $(r \rightarrow q) \vee \sim p$
16. $(q \leftrightarrow p) \wedge \sim r$
17. $(r \wedge q) \vee \sim p$
18. $\sim (r \wedge q)$

19.

p	q	(p ∨ q) ∧ ~ p
T	T	T F F
T	F	T F F
F	T	T T T
F	F	F F T
		1 3 2

20.

p	q	q ↔ (p ∨ ~ q)
T	T	T T T T F
T	F	F F T T T
F	T	T F F F F
F	F	F F F T T
		1 5 2 4 3

21.

p	q	r	(p ∨ q) ↔ (p ∨ r)
T	T	T	T T T
T	T	F	T T T
T	F	T	T T T
T	F	F	T T T
F	T	T	T T T
F	T	F	T F F
F	F	T	F F T
F	F	F	F T F
			1 3 2

22.

p	q	r	p ∧ (~ q ∨ r)
T	T	T	T T F T T
T	T	F	T F F F F
T	F	T	T T T T T
T	F	F	T T T T F
F	T	T	F F F T T
F	T	F	F F F F F
F	F	T	F F T T T
F	F	F	F F T T F
			4 5 1 3 2

23.

p	q	r	P → (q ∧ ~ r)
T	T	T	T F T F F
T	T	F	T T T T T
T	F	T	T F F F F
T	F	F	T F F F T
F	T	T	F T T F F
F	T	F	F T T T T
F	F	T	F T F F F
F	F	F	F T F F T
			4 5 1 3 2

24.

p	q	r	(p ∧ q) → ~ r
T	T	T	T F F
T	T	F	T T T
T	F	T	F T F
T	F	F	F T T
F	T	T	F T F
F	T	F	F T T
F	F	T	F T F
F	F	F	F T T
			1 3 2

25. p: 7 is odd.   p → q
    q: 11 is even.   T → F
                              F

26. p: The St. Louis arch is in St. Louis.   p ∨ q
    q: Abraham Lincoln is buried in   T ∨ F
       in Grant's Tomb.   T

27. p: Oregon borders the Pacific Ocean.
    q: California borders the Atlantic Ocean.
    r: Minnesota is south of Texas.
    (p ∨ q) → r
    (T ∨ F) → F
       T → F
          F

28. p: 15 − 7 = 22   (p ∨ q) ∧ r
    q: 4 + 9 = 13   (F ∨ T) ∧ T
    r : 9 − 8 = 1   T   ∧ T
                              T

29. p: 32% of OR's electricity - coal   (p ↔ q) ∨ r
    q: 54% of OR's electricity - hydro   (T ↔ T) ∨ F
    r: 38% of OR's electricity - nuclear   T   ∨ F
                                                  T

30. p: 3% of OR's electricity - gas/oil   p → (q ∧ r)
    q: 45% of OR's electricity - coal   F → (F ∧ T)
    r: 3% of OR's electricity - nuclear   F →   F
                                               T

31. (p → ~ r) ∨ (p ∧ q)
   (T → T) ∨ (T ∧ F)
     T   ∨   F
         T

32. (p ∨ q) ↔ (~ r ∧ p)
   (T ∨ F) ↔ (T ∧T)
     T ↔ T
        T

33. ~ r ↔ [(p ∨ q) ↔ ~ p]
   T ↔ [(T ∨ F) ↔ F]
   T ↔ [T ↔ F]
   T ↔   F
     F

34. ~ [(q ∧ r) → (~ p ∨ r)]
   ~ [(F ∧ F) → (F ∨ F)]
   ~ [F →   F]
     ~ T
      F

35.

p	q	~ p ∨ ~ q	~ p ↔ q
T	T	F F F	F F T
T	F	F T T	F T F
F	T	T T F	T T T
F	F	T T T	T F F
		1 3 2	1 3 2

The statements are not equivalent.

36. Using the fact that (p → q) ⇔ (~ p ∨ q), we can conclude that ~ p → ~ q ⇔ p ∨ ~ q.

37.

p	q	r	~ p ∨ (q ∧ r)	(~ p ∨ q) ∧ (~ p ∨ r)
T	T	T	F T T T	F T T T F T T
T	T	F	F F F F	F T T F F F F
T	F	T	F F F F	F F F F F T T
T	F	F	F F F F	F F F F F F F
F	T	T	T T T T	T T T T T T T
F	T	F	T T F T	T T T T T T F
F	F	T	T T F T	T T F T T T T
F	F	F	T T F T	T T F T T T F
			2 3 1	1 3 2 7 4 6 5

The statements are equivalent.

38.

p	q	(~ q → p) ∧ p	~ (~ p ↔ q) ∨ p
T	T	F T T T T	T F F T T T
T	F	T T T T T	F F T F T T
F	T	F T F F F	F T T T F F
F	F	T F F F F	T T F T F F
		1 3 2 5 4	4 1 3 2 6 5

The statements are not equivalent.

39. p: Johnny Cash is in the Rock and Roll (R&R) Hall of Fame.

q: India Arie recorded *Acoustic Soul*.
In symbolic form, the statement is p ∧ q. Using DeMorgan's Laws, we get p ∧ q ⇔ ~ (~ p ∨ ~ q). It is false that Johnny Cash is not in the R&R Hall of Fame or India Arie did not sing *Acoustic Soul*.

40. p: Her foot fell asleep.

q: She has injured her ankle.
In symbolic form, the statement is p ∨ q. Using the fact that p → q ⇔ ~ p ∨ q, we can rewrite the given statement as ~ p → q. If her foot did not fall asleep, then she has injured her ankle.

41. p: Altec Lansing only produces speakers.

    q: Harmon Kardon only produces stereo receivers.
    The symbolic form is ~ (p ∨ q).
    Using DeMorgan's Laws, we get ~ (p ∨ q) ⇔ ~ p ∧ ~ q.
    Altec Lansing does not produce only speakers and
    Harmon Karson does not produce only stereo receivers.

42. p: Travis Tritt won an Academy Award.

    q: Randy Jackson does commercials for
      Milk Bone dog biscuits.
    The symbolic form is ~ p ∧ ~ q.
    Using DeMorgan's Laws, we get
    ~ p ∧ ~ q ⇔ ~ (p ∨ q). It is false
    that Travis Tritt won an Academy
    Award or Randy Jackson does
    commercials for Milk Bone dog
    biscuits.

43. p: The temperature is above 32 degrees Fahrenheit.
    q: We will go ice fishing at O'Leary's Lake.
    The symbolic form is ~ p → q.
    Using DeMorgan's Laws, we get ~ p → q ⇔ p ∨ q.
    The temperature is above 32 degrees Fahrenheit or
    we will go ice fishing at O'Leary's Lake.

44. Converse: If you enjoy life, then you
    hear a beautiful songbird today.
    Inverse: If you do not hear a beautiful
    songbird today, then you do not
    enjoy life.
    Contrapositive: If you do not enjoy
    life, then you do not hear a beautiful
    songbird today.

45. Converse: If the quilt has a uniform design, then
    you followed the correct pattern.
    Inverse: If you do not follow the correct pattern,
    then the quilt will not have a uniform design.
    Contrapositive: If the quilt does not have a
    uniform design, then you did not follow the
    correct pattern.

46. Converse: If Maureen Gerald is helping
    at school, then she is not in
    attendance.
    Inverse: If Maureen Gerald is in attendance,
    then she is not helping at school.
    Contrapositive: If Maureen Gerald is not
    helping at school, then she is in attendance.

47. Converse: If we do not buy a desk at Miller's
    Furniture, then the desk is made by Winner's Only
    and is in the Rose catalog.
    Inverse: If we will buy a desk at Miller's furniture,
    then the desk is not made by Winner's Only or it
    is not in the Rose catalog.
    Contrapositive: If the desk is not made by
    Winner's Only or is not in the Rose catalog,
    then we will buy a desk at Miller's Furniture.

48. Converse: If I let you attend the prom,
    then you get straight A's on your report
    card.
    Inverse: If you do not get straight A's on
    your report card, then I will not let you
    attend the prom.
    Contrapositive: If I do not let you attend
    the prom, then you do not get straight
    A's on your report card.

49. p: The temperature is over 80^0.

q: The air conditioner will come on.

In symbolic form, the statements are: a) $p \rightarrow q$, b) $\sim p \vee q$, and c) $\sim (p \wedge \sim q)$. Using the fact that $p \rightarrow q$ is equivalent to $\sim p \vee q$, statements (a) and (b) are equivalent. Using DeMorgan's Laws on statement (b) we get $\sim (p \wedge \sim q)$.

Therefore all 3 statements are equivalent.

51. p: $2 + 3 = 6$.

q: $3 + 1 = 5$.

In symbolic form, the statements are: a) $p \rightarrow q$, b) $p \leftrightarrow \sim q$, and c) $\sim q \rightarrow \sim p$.

Statement (c) is the contrapositive of statement (a). Therefore statements (a) and (c) are equivalent. For p and q false, (a) and (c) are true but (b) is false, so (b) is not equivalent to (a) and (c).

50. p: The screwdriver is on the workbench.

q: The screwdriver is on the counter.

In symbolic form, the statements are: a) $p \leftrightarrow \sim q$, b) $\sim q \rightarrow \sim p$, and c) $\sim (q \wedge \sim p)$. Looking at the truth tables for statements (a), (b), and (c) we can conclude that none of the statements are equivalent.

		a)	b)	c)
p	q	$p \leftrightarrow \sim q$	$\sim q \rightarrow \sim p$	$\sim (q \wedge \sim p)$
T	T	T F F	F T F	T T F F
T	F	T T T	T F F	T F F F
F	T	F T F	F T T	F T T T
F	F	F F T	T T T	T F F T
		1 3 2	1 3 2	4 1 3 2

52. p: The sale is on Tuesday.

q: I have money.

r : I will go to the sale.

In symbolic form the statements are: a) $(p \wedge q) \rightarrow r$, b) $r \rightarrow (p \wedge q)$, and c) $r \vee (p \wedge q)$. The truth table for statements (a), (b), and (c) shows that none of the statements are equivalent.

p	q	r	$(p \wedge q) \rightarrow r$	$r \rightarrow (p \wedge q)$	$r \vee (p \wedge q)$
T	T	T	T T T	T T T	T T T
T	T	F	T F F	F T T	F T T
T	F	T	F T T	T F F	T T F
T	F	F	F T F	F T F	F F F
F	T	T	F T T	T F F	T T F
F	T	F	F T F	F T F	F F F
F	F	T	F T T	T F F	T T F
F	F	F	F T F	F T F	F F F
			1 3 2	1 3 2	1 3 2

53.

p	q	$[(p \rightarrow q) \wedge \sim p] \rightarrow q$
T	T	T F F T T
T	F	F F F T F
F	T	T T T T T
F	F	T T T F F
		1 3 2 5 4

The argument is invalid.

54.

p	q	r	$[(p \wedge q) \wedge (q \rightarrow r)] \rightarrow (p \rightarrow r)$
T	T	T	T T T T T
T	T	F	T F F T F
T	F	T	F F T T T
T	F	F	F F T T F
F	T	T	F F T T T
F	T	F	F F F T T
F	F	T	F F T T T
F	F	F	F F T T T
			1 3 2 5 4

The argument is valid.

55. p: Nicole is in the hot tub.   p ∨ q
    q: Nicole is in the shower.   p
                                    ∴ ~ q

p	q	[(p ∨ q)	∧	p]	→	~ q
T	T	T	T	T	F	F
T	F	T	T	T	T	T
F	T	T	F	F	T	F
F	F	F	F	F	T	T
		1	3	2	5	4

The argument is invalid.

56. p: The car has a sound system.   p → q
    p: Rick will buy the car.   ~ r → ~ q
    r: The price is less than $18,000. ∴ p → r

p	q	r	[(p → q)	∧	(~ r → ~q)]	→	(p → r)
T	T	T	T	T	F T F	T	T
T	T	F	T	F	T F F	T	F
T	F	T	F	F	F T T	T	F
T	F	F	F	F	T T T	T	F
F	T	T	T	T	F T F	T	T
F	T	F	T	F	T F F	T	T
F	F	T	T	T	F T T	T	T
F	F	F	T	T	T T T	T	T
			1	5	2 4 3	7	6

The argument is valid.

57.

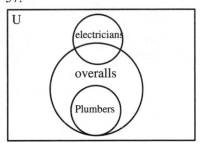

invalid

58.

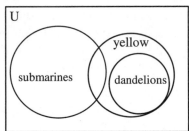

invalid

Chapter Test

1. (p ∧ r) ∨ ~ q

2. (r → q) ∨ ~ p

3. ~ (r ↔ ~ q)

4. Ann is the secretary, if and only if Dick is the vice president and Elaine is the president.

5. It is false that if Ann is the secretary, then Elaine is not the president.

6.

p	q	r	[~ (p → r)]	∧	q
T	T	T	F	T	F T
T	T	F	T	F	T T
T	F	T	F	T	F F
T	F	F	T	F	F F
F	T	T	F	T	F T
F	T	F	F	T	F T
F	F	T	F	T	F F
F	F	F	F	T	F F
			2	1	4 3

7.

p	q	r	(q ↔ ~ r)	∨	p
T	T	T	T F	F	T T
T	T	F	T T	T	T T
T	F	T	F T	F	T T
T	F	F	F F	T	T T
F	T	T	T F	F	F F
F	T	F	T T	T	T F
F	F	T	F T	F	T F
F	F	F	F F	T	F F
			1 3	2	5 4

8. p: $2 + 6 = 8$
   q: $7 - 12 = 5$

   $p \vee q$
   $T \vee F$
   T

9. p: A scissors can cut paper.
   q: A dime equals 2 nickels.
   r : Louisville is a city in Kentucky.

   $(p \vee q) \leftrightarrow r$
   $(T \vee T) \leftrightarrow T$
   $T \quad \leftrightarrow T$
   T

10. $(r \vee q) \leftrightarrow (p \wedge \sim q)$
    $(T \vee F) \leftrightarrow (T \wedge T)$
    $T \quad \leftrightarrow \quad T$
    T

11. $[\sim(r \rightarrow \sim p)] \wedge (q \rightarrow p)$
    $[\sim(T \rightarrow F\ )] \wedge (F \rightarrow T)$
    $[\ \sim(F)\ \ ] \wedge \quad T$
    $T \quad \wedge \quad T$
    T

12. By DeMorgan's Laws ,
    $\sim p \vee q \Leftrightarrow \sim (\sim (\sim p) \wedge \sim q)$.
    and this is equivalent to
    $\sim(p \wedge \sim q)$.

13. p: The bird is red.
    q: It is a cardinal.
    In symbolic form the statements
    are: a) $p \rightarrow q$, b) $\sim p \vee q$,
    and c) $\sim p \rightarrow \sim q$.
    Statement (c) is the inverse of
    statement (a) and thus they cannot
    be equivalent. Using the fact that
    $p \rightarrow q \Leftrightarrow \sim p \vee q$, to rewrite
    statement (a) we get $\sim p \vee q$.
    Therefore statements (a) and (b)
    are equivalent.

14. p: The test is today.   q: The concert is tonight.   In symbolic form the
    statements are: a) $\sim (p \vee q)$, b) $\sim p \wedge \sim q$, and $\sim p \rightarrow \sim q$.
    Applying DeMorgan's Law to statement (a) we get: $\sim p \wedge \sim q$.
    Therefore statements (a) and (b) are equivalent. When we compare
    the truth tables for statements (a), (b), and (c) we see that only
    statements (a) and (b) are equivalent.

p	q	$\sim$ (p $\vee$ q)		$\sim$ p $\wedge$ $\sim$ q			$\sim$ p $\rightarrow$ $\sim$ q		
T	T	F	T	F	F	F	F	T	F
T	F	F	T	F	F	T	F	T	T
F	T	F	T	T	F	F	T	F	F
F	F	T	F	T	T	T	T	T	T
		2	1	1	3	2	1	3	2

15. s: The soccer team won the game.
    f: Sue played fullback.
    p: The team is in second place.
    This argument is the law of
    syllogism and therefore it is
    valid. $s \rightarrow f$
                 $\underline{f \rightarrow p}$
                 $s \rightarrow p$
    This argument is the law of
    syllogism and therefore it is valid.

16.

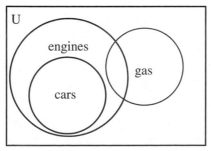

Fallacy

17. Some leopards are not spotted.

18. No Jacks-in-the-box are electronic.

19. Converse: If today is Saturday, then the
    garbage truck comes.
    Inverse: If the garbage truck does not
    come today, then today is not Saturday.
    Contrapositive: If today is not Saturday,
    then the garbage truck does not come.

20. Yes. An argument is valid when its conclusion necessarily follows from the given set of
    premises. It doesn't matter whether the conclusion is a true or false statement.

CHAPTER FOUR

SYSTEMS OF NUMERATION

Exercise Set 4.1

1. A **number** is a quantity, and it answers the question, "How many?" A **numeral** is a symbol used to represent the number.

3. A **system of numeration** consists of a set of numerals and a scheme or rule for combining the numerals to represent numbers.

5. The Hindu-Arabic numeration system

7. In a **multiplicative system**, there are numerals for each number less than the base and for powers of the base. Each numeral less than the base is multiplied by a numeral for the power of the base, and these products are added to obtain the number.

9. $100 + 10 + 10 + 10 + 10 + 1 + 1 = 142$

11. $1000 + 1000 + 100 + 100 + 100 + 100 + 10 + 10$
$+ 1 + 1 + 1 = 2423$

13. $100,000 + 100,000 + 100,000 + 10,000$
$+ 10,000 + 10,000 + 1000 + 1000 + 1000 + 1000$
$+ 100 + 100 + 10 + 1 + 1 + 1 + 1 = 334,214$

15.

17.

19.

21. $10 + (10 - 1) = 19$

23. $500 + (50 - 10) + 5 + 1 + 1 = 547$

25. $1000 + (500 - 100) + (100 - 10) + 1 + 1 = 1492$

27. $1000 + 1000 + (1000 - 100) + (50 - 10) + 5 + 1$
$= 2946$

29. $10(1000) + 1000 + 1000 + 500 + 100 + 50 + 10$
$+ 5 + 1 = 12,666$

31. $9(1000) + (500 - 100) + 50 + 10 + (5 - 1) = 9464$

33. LIX

35. CXXXIV

37. MMV

39. $\overline{\text{IV}}$DCCXCIII

41. $\overline{\text{IX}}$CMXCIX

43. $\overline{\text{XX}}$DCXLIV

45. $7(10) + 4 = 74$

47. $4(1000) + 8(10) + 1 = 4081$

49. $8(1000) + 5(100) + 5(10) = 8550$

51. $4(1000) + 3 = 4003$

53. 五十三

55. 三百七十八七千零五十六

57. 四千二百六十

59. （as shown）

61. $300 + 40 + 1 = 341$

63. $20(1000) + 2(1000) + 500 + 5 = 22,505$

65. $9(1000) + 600 + 7 = 9607$

67. $\nu\ \theta$

69. $\psi\ \kappa\ 2$

71. $\pi'\ \beta'\ \psi\ \delta$

73. Advantage: You can write some numbers more compactly.
    Disadvantage: There are more numerals to memorize.

75. Advantage: You can write some numbers more compactly.
    Disadvantage: There are more numerals to memorize.
    The Hindu-Arabic system has fewer symbols, more compact notation, the inclusion of zero, and the capability of expressing decimal numbers and fractions.

77. $1000 + (1000 - 100) + 10 + 10 + 10 + 5 + 1 = 1936$,

    𐤀999999999∩∩IIIIII,

    $\alpha'\ \pi\ \lambda\ 2$,

    一千九百三十六

79. $400 + 20 + 2 = 422$, 9999∩II, CDXXII,

    四百二十二

81. $\pi'\ Q'\ \theta'\ \pi\ Q\ \theta$

83. Turn the book upside down

84. MM

85. 1888, MDCCCLXXXVIII

Exercise Set 4.2

1. A base 10 place-value system

3. $40 \rightarrow$ four tens, $400 \rightarrow$ four hundreds

5. A true positional-value system requires a base and a set of symbols, including a symbol for zero and one for each counting number less than the base.

7. Write each digit times its corresponding positional value.

9. a) There may be confusion because numbers could be interpreted in different ways. For example, $\mathbf{\uparrow}$ could be interpreted to be either 1 or 60.

   b) $\mathbf{\uparrow\uparrow}$ $\mathbf{<\uparrow\uparrow\uparrow}$ for both numbers; $133 = 2(60) + 13(1)$ and $7980 = 133(60)$

11. $1, 20, 18 \times 20, 18 \times (20)^2, 18 \times (20)^3$

13. $(6 \times 10) + (3 \times 1)$

15. $(3 \times 100) + (5 \times 10) + (9 \times 1)$

17. $(8 \times 100) + (9 \times 10) + (7 \times 1)$

19. $(4 \times 1000) + (3 \times 100) + (8 \times 10) + (7 \times 1)$

21. $(1 \times 10,000) + (6 \times 1000) + (4 \times 100) + (0 \times 10) + (2 \times 1)$

23. $(3 \times 100,000) + (4 \times 10,000) + (6 \times 1000) + (8 \times 100) + (6 \times 10) + (1 \times 1)$

25. $(10 + 10 + 10 + 10 + 1 + 1)(1) = 42$

27. $(10 + 1 + 1 + 1)(60) + (1 + 1 + 1 + 1)(1) = 13(60) + 4(1) = 780 + 4 = 784$

29. $1(60^2) + (10 + 10 + 1)(60) + (10 - (1 + 1))(1) = 3600 + 21(60) + (10 - 2)(1) = 3600 + 1260 + 8 = 4868$

31. 88 is 1 group of 60 and 28 units remaining. $\mathbf{\uparrow}$ $\mathbf{<<<\uparrow\uparrow\uparrow}$

33. 295 is 4 groups of 60 and 55 units remaining. $\mathbf{\uparrow\uparrow\uparrow\uparrow}$ $\mathbf{<<<<<<\uparrow\uparrow\uparrow\uparrow\uparrow}$

35. 3685 is 1 group of 3600, 1 group of 60, and 25 units remaining. $\mathbf{\uparrow}$ $\mathbf{\uparrow}$ $\mathbf{<<\uparrow\uparrow\uparrow\uparrow\uparrow}$

37. $4(20) + 12(1) = 80 + 12 = 92$

39. $12(18 \times 20) + 0(20) + 1(1) = 4320 + 0 + 1 = 4321$

41. $11(18 \times 20) + 2(20) + 0(1) = 3960 + 40 + 0 = 4000$

43. $\overset{\bullet\bullet}{\equiv}$

45.
$$
\begin{array}{r}
14 \\
20 \overline{)297} \\
280 \\
\hline
17
\end{array}
$$
$$297 = 14(20) + 17(1)$$

47.

$$360 \overline{\smash{\big)}\ \begin{array}{r} 6 \\ 2163 \\ \underline{2160} \\ 3 \end{array}}$$

$$2163 = 6(360) + 0(20) + 3(1)$$

49. Advantages: In general, a place-value system is more compact; large and small numbers can be written more easily; there are fewer symbols to memorize.

Disadvantage: If many of the symbols in the numeral represent zero, then a place-value system may be less compact.

51. Hindu-Arabic: $10+10+10+1+1+1=33$

Mayan: $33 = 1(20) + 13(1)$

53.

$$\left(\triangle \times \square^2\right) + \left(\square \times \square\right) + \left(\lozenge \times 1\right)$$

55. a) No largest number; The positional values are $\ldots,(60)^3,(60)^2,60,1$.

b) $999,999 = 4(60)^3 + 37(60)^2 + 46(60) + 39(1)$

57. $2(60) + 23(1) = 120 + 23 = 143$

23

$143 + 23 = 166$

$166 = 2(60) + 46(1)$

59. $7(18 \times 20) + 6(20) + 15(1) = 2520 + 120 + 15 = 2655$

$6(18 \times 20) + 7(20) + 13(1) = 2160 + 140 + 13 = 2313$

$2655 + 2313 = 4968$

$4968 = 13(18 \times 20) + 14(20) + 8(1)$

61.

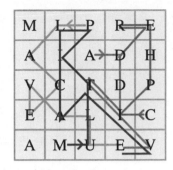

Exercise Set 4.3

1. Answers will vary.

$\underline{3.}\ 5_6 = 5(1) = 5$

5. $42_5 = 4(5) + 2(1) = 20 + 2 = 22$

7. $1011_2 = 1(2^3) + 0(2^2) + 1(2) + 1(1) = 8 + 0 + 2 + 1 = 11$

9. $84_{12} = 8(12) + 4(1) = 96 + 4 = 100$

11. $565_8 = 5(8^2) + 6(8) + 5(1) = 5(64) + 48 + 5 = 320 + 48 + 5 = 373$

13. $20432_5 = 2(5^4) + 0(5^3) + 4(5^2) + 3(5) + 2(1) = 2(625) + 0 + 4(25) + 15 + 2 = 1250 + 0 + 100 + 15 + 2 = 1367$

15. $4003_6 = 4(6^3) + 0(6^2) + 0(6) + 3(1) = 4(216) + 0 + 0 + 3 = 864 + 0 + 0 + 3 = 867$

17. $123_8 = 1(8^2) + 2(8) + 3(1) = 64 + 16 + 3 = 83$

19. $14705_8 = 1(8^4) + 4(8^3) + 7(8^2) + 0(8) + 5(1) = 4096 + 4(512) + 7(64) + 0 + 5 = 4096 + 2048 + 448 + 0 + 5 = 6597$

21. To convert 8 to base 2   ... 16 8 4 2 1

$$
\begin{array}{cccc}
\;\;1 & \;\;0 & \;\;0 & \;\;0 \\[-2pt]
8\,\overline{\big|\,8} & 4\,\overline{\big|\,0} & 2\,\overline{\big|\,0} & 1\,\overline{\big|\,0} \\[-2pt]
\underline{8} & \underline{0} & \underline{0} & \underline{0} \\[-2pt]
0 & 0 & 0 & 0 \quad 8 = 1000_2
\end{array}
$$

23. To convert 23 to base 2   ... 32 16 8 4 2 1

$$
\begin{array}{ccccc}
\;\;1 & \;\;0 & \;\;1 & \;\;1 & \;\;1 \\[-2pt]
16\,\overline{\big|\,23} & 8\,\overline{\big|\,7} & 4\,\overline{\big|\,7} & 2\,\overline{\big|\,3} & 1\,\overline{\big|\,1} \\[-2pt]
\underline{16} & \underline{0} & \underline{4} & \underline{2} & \underline{1} \\[-2pt]
7 & 7 & 3 & 1 & 0 \quad 23 = 10111_2
\end{array}
$$

25. To convert 635 to base 6   ... 1296 216 36 6 1

$$
\begin{array}{cccc}
\;\;2 & \;\;5 & \;\;3 & \;\;5 \\[-2pt]
216\,\overline{\big|\,635} & 36\,\overline{\big|\,203} & 6\,\overline{\big|\,23} & 1\,\overline{\big|\,5} \\[-2pt]
\underline{432} & \underline{180} & \underline{18} & \underline{5} \\[-2pt]
203 & 23 & 5 & 0 \quad 635 = 2535_6
\end{array}
$$

27. To convert 2061 to base 12   ... 20,736 1728 144 12 1

$$
\begin{array}{cccc}
\;\;1 & \;\;2 & \;\;3 & \;\;9 \\[-2pt]
1728\,\overline{\big|\,2061} & 144\,\overline{\big|\,333} & 12\,\overline{\big|\,45} & 1\,\overline{\big|\,9} \\[-2pt]
\underline{1728} & \underline{288} & \underline{36} & \underline{9} \\[-2pt]
333 & 45 & 9 & 0 \quad 2061 = 1239_{12}
\end{array}
$$

29. To convert 529 to base 8   ... 4096   512   64   8   1

$$512 \overline{\smash{)}\ 529} \quad 64 \overline{\smash{)}\ 17} \quad 8 \overline{\smash{)}\ 17} \quad 1 \overline{\smash{)}\ 1}$$

         1   0   2   1

        512   0   16   1

         17   17   1   0 $529 = 1021_8$

31. To convert 2867 to base 12   ... 20,736   1728   144   12   1

             1   7   10   11

$$1728 \overline{\smash{)}\ 2867} \quad 144 \overline{\smash{)}\ 1139} \quad 12 \overline{\smash{)}\ 131} \quad 1 \overline{\smash{)}\ 11}$$

         1728   1008   120   11

         1139   131   11   0 $2867 = 17TE_{12}$

33. To convert 1011 to base 2   ... 1024   512   256   128   64   32   16   8   4   2   1

        1   1   1   1   1

$$512 \overline{\smash{)}\ 1011} \quad 256 \overline{\smash{)}\ 499} \quad 128 \overline{\smash{)}\ 243} \quad 64 \overline{\smash{)}\ 115} \quad 32 \overline{\smash{)}\ 51}$$

       512   256   128   64   32

       499   243   115   51   19

        1   0   0   1   1

$$16 \overline{\smash{)}\ 19} \quad 8 \overline{\smash{)}\ 3} \quad 4 \overline{\smash{)}\ 3} \quad 2 \overline{\smash{)}\ 3} \quad 1 \overline{\smash{)}\ 1}$$

       16   0   0   2   1

        3   3   3   1   0 $1011 = 1111110011_2$

35. To convert 2307 to base 8   ... 4096   512   64   8   1

        4   4   0   3

$$512 \overline{\smash{)}\ 2307} \quad 64 \overline{\smash{)}\ 259} \quad 8 \overline{\smash{)}\ 3} \quad 1 \overline{\smash{)}\ 3}$$

       2048   256   0   3

       259   3   3   0 $2307 = 4403_8$

37. $735_{16} = 7(16^2) + 3(16) + 5(1) = 7(256) + 48 + 5 = 1792 + 48 + 5 = 1845$

39. $6D3B7_{16} = 6(16^4) + 13(16^3) + 3(16^2) + 11(16) + 7(1) = 6(65,536) + 13(4096) + 3(256) + 176 + 7$

     $= 393,216 + 53,248 + 768 + 176 + 7 = 447,415$

41. To convert 573 to base 16   ... 4096   256   16   1

        2   3   13 = D

$$256 \overline{\smash{)}\ 573} \quad 16 \overline{\smash{)}\ 61} \quad 1 \overline{\smash{)}\ 13}$$

       512   48   13

        61   13   0 $573 = 23D_{16}$

43. To convert 5478 to base 16   ... 65,536   4096   256   16   1

         1   5   6   6

$$4096 \overline{\smash{)}\ 5478} \quad 256 \overline{\smash{)}\ 1382} \quad 16 \overline{\smash{)}\ 102} \quad 1 \overline{\smash{)}\ 6}$$

       4096   1280   96   6

       1382   102   6   0 $5478 = 1566_{16}$

45. To convert 2005 to base 2   ... 2048   1024   512   256   128   64   32   16   8   4   2   1

        1   1   1   1   1

$$1024 \overline{\smash{)}\ 2005} \quad 512 \overline{\smash{)}\ 981} \quad 256 \overline{\smash{)}\ 469} \quad 128 \overline{\smash{)}\ 213} \quad 64 \overline{\smash{)}\ 85}$$

       1024   512   256   128   64

       981   469   213   85   21

        0   1   0   1   0   1

$$32 \overline{\smash{)}\ 21} \quad 16 \overline{\smash{)}\ 21} \quad 8 \overline{\smash{)}\ 5} \quad 4 \overline{\smash{)}\ 5} \quad 2 \overline{\smash{)}\ 1} \quad 1 \overline{\smash{)}\ 1}$$

        0   16   0   4   0   1

       21   5   5   1   1   0 $2005 = 11111010101_2$

47. To convert 2005 to base 5   ... 3125   625   125   25   5   1

$$\begin{array}{r} 3 \\ 625 \overline{)\ 2005} \\ \underline{1875} \\ 130 \end{array} \quad \begin{array}{r} 1 \\ 125 \overline{)\ 130} \\ \underline{125} \\ 5 \end{array} \quad \begin{array}{r} 0 \\ 25 \overline{)\ 5} \\ \underline{0} \\ 5 \end{array} \quad \begin{array}{r} 1 \\ 5 \overline{)\ 5} \\ \underline{5} \\ 0 \end{array} \quad \begin{array}{r} 0 \\ 1 \overline{)\ 0} \\ \underline{0} \\ 0 \end{array}$$

$$2005 = 31010_5$$

49. To convert 2005 to base 12   ... 20,736   1728   144   12   1

$$\begin{array}{r} 1 \\ 1728 \overline{)\ 2005} \\ \underline{1728} \\ 277 \end{array} \quad \begin{array}{r} 1 \\ 144 \overline{)\ 277} \\ \underline{144} \\ 133 \end{array} \quad \begin{array}{r} 11 = E \\ 12 \overline{)\ 133} \\ \underline{132} \\ 1 \end{array} \quad \begin{array}{r} 1 \\ 1 \overline{)\ 1} \\ \underline{1} \\ 0 \end{array}$$

$$2005 = 11E1_{12}$$

51. Incorrect; there is no 5 in base 5.

53. Correct

55. Incorrect; there is no 8 in base 7.

57. $2(5) + 3(1) = 10 + 3 = 13$

59. $2(5^2) + 4(5) + 3(1) = 2(25) + 20 + 3$

$= 50 + 20 + 3 = 73$

61. To convert   ... 25   5   1

$$\begin{array}{r} 3 = \ominus \\ 5 \overline{)\ 19} \\ \underline{15} \\ 4 \end{array} \quad \begin{array}{r} 4 = \oslash \\ 1 \overline{)\ 4} \\ \underline{4} \\ 0 \end{array}$$

$$19 = \ominus \oslash {}_5$$

63. To convert   ... 125   25   5   1

$$\begin{array}{r} 2 = \oslash \\ 25 \overline{)\ 74} \\ \underline{50} \\ 24 \end{array} \quad \begin{array}{r} 4 = \oslash \\ 5 \overline{)\ 24} \\ \underline{20} \\ 4 \end{array} \quad \begin{array}{r} 4 = \oslash \\ 1 \overline{)\ 4} \\ \underline{4} \\ 0 \end{array}$$

$$74 = \oslash \oslash \oslash {}_5$$

65. $1(4) + 3(1) = 4 + 3 = 7$ **67.** $2(4^2) + 1(4) + 0(1) = 2(16) + 4 + 0 = 32 + 4 + 0 = 36$

For #69-71, blue = 0 = b, red = 1 = r, gold = 2 = go, green = 3 = gr

69. To convert   ... 16   4   1

$$\begin{array}{r} 2 = \boxed{go} \\ 4 \overline{)\ 10} \\ \underline{8} \\ 2 \end{array} \quad \begin{array}{r} 2 = \boxed{go} \\ 1 \overline{)\ 2} \\ \underline{2} \\ 0 \end{array}$$

$$10 = \boxed{go}\ \boxed{go}\ {}_4$$

71. To convert   ... 64   16   4   1

$$\frac{3}{16\overline{\smash{\big)}60}} = \text{(gr)} \qquad \frac{3}{4\overline{\smash{\big)}12}} = \text{(gr)} \qquad \frac{0}{1\overline{\smash{\big)}0}} = \text{(b)}$$
$$\begin{array}{r} \underline{48} \\ 12 \end{array} \qquad\qquad \begin{array}{r} \underline{12} \\ 0 \end{array} \qquad\qquad \begin{array}{r} \underline{0} \\ 0 \end{array}$$

$$60 = \text{(gr)} \; \text{(gr)} \; \text{(b)}_4$$

73. a) Each remainder is multiplied by the proper power of 5.

b)
$$\begin{array}{r|r} 5 & 683 \\ \hline 5 & 136 \\ \hline 5 & 27 \\ \hline 5 & 5 \\ \hline 5 & 1 \\ \hline & 0 \end{array} \begin{array}{l} \\ 3 \uparrow \\ 1 \uparrow \\ 2 \uparrow \\ 0 \uparrow \\ 1 \uparrow \end{array}$$

$$683 = 10213_5$$

c)
$$\begin{array}{r|r} 8 & 763 \\ \hline 8 & 95 \\ \hline 8 & 11 \\ \hline 8 & 1 \\ \hline & 0 \end{array} \begin{array}{l} \\ 3 \uparrow \\ 7 \uparrow \\ 3 \uparrow \\ 1 \uparrow \end{array}$$

$$763 = 1373_8$$

75. Answers will vary.

77. $1(b^2) + 1(b) + 1 = 43$

$b^2 + b + 1 = 43$

$b^2 + b - 42 = 0$

$(b+7)(b-6) = 0$

$b + 7 = 0$ or $b - 6 = 0$

$b = -7$ or $b = 6$

   Since the base cannot be negative, $b = 6$.

78. $d(5^2) + d(5) + d(1) = 124$

$25d + 5d + d = 124$

$\dfrac{31d}{31} = \dfrac{124}{31}$

$d = 4$

79. a) $3(4^4) + 1(4^3) + 2(4^2) + 3(4) + 0(1) = 3(256) + 64 + 2(16) + 12 + 0 = 768 + 64 + 32 + 12 + 0 = 876$

b) To convert   ... 256   64   16   4   1

$$\frac{2}{64\overline{\smash{\big)}177}} = \text{(go)} \quad \frac{3}{16\overline{\smash{\big)}49}} = \text{(gr)} \quad \frac{0}{4\overline{\smash{\big)}1}} = \text{(b)} \quad \frac{1}{1\overline{\smash{\big)}1}} = \text{(r)}$$
$$\begin{array}{r} \underline{128} \\ 49 \end{array} \qquad \begin{array}{r} \underline{48} \\ 1 \end{array} \qquad \begin{array}{r} \underline{0} \\ 1 \end{array} \qquad \begin{array}{r} \underline{1} \\ 0 \end{array}$$

$$177 = \text{(go)} \; \text{(gr)} \; \text{(b)} \; \text{(r)}_4$$

Exercise Set 4.4

1. a) $b^0 = 1, b^1 = b, b^2, b^3, b^4$

   b) $6^0 = 1, 6^1 = 6, 6^2, 6^3, 6^4$

3. No; there is no 6 in base 5.

5. Answers will vary.

7.
$$\begin{array}{r} 43_5 \\ \underline{41_5} \\ 134_5 \end{array}$$

9.
$$\begin{array}{r} 2303_4 \\ \underline{232_4} \\ 3201_4 \end{array}$$

11.
$$\begin{array}{r} 799_{12} \\ \underline{218_{12}} \\ 9E5_{12} \end{array}$$

13.
$$\begin{array}{r} 1112_3 \\ \underline{1011_3} \\ 2200_3 \end{array}$$

15.
$$
\begin{array}{r}
14631_7 \\
\underline{6040_7} \\
24001_7
\end{array}
$$

17.
$$
\begin{array}{r}
1110_2 \\
\underline{110_2} \\
10100_2
\end{array}
$$

19.
$$
\begin{array}{r}
322_4 \\
-\ 103_4 \\
\hline
213_4
\end{array}
$$

21.
$$
\begin{array}{r}
2342_5 \\
-\ 1442_5 \\
\hline
400_5
\end{array}
$$

23.
$$
\begin{array}{r}
782_{12} \\
-\ 13T_{12} \\
\hline
644_{12}
\end{array}
$$

25.
$$
\begin{array}{r}
1001_2 \\
-\ 110_2 \\
\hline
11_2
\end{array}
$$

27.
$$
\begin{array}{r}
4223_7 \\
-\ 304_7 \\
\hline
3616_7
\end{array}
$$

29.
$$
\begin{array}{r}
2100_3 \\
-\ 1012_3 \\
\hline
1011_3
\end{array}
$$

31.
$$
\begin{array}{r}
33_5 \\
\times\ 2_5 \\
\hline
121_5
\end{array}
$$

33.
$$
\begin{array}{r}
342_7 \\
\times\ \ 5_7 \\
\hline
2403_7
\end{array}
$$

35.
$$
\begin{array}{r}
512_6 \\
\times\ 23_6 \\
\hline
2340 \\
\underline{1424\ } \\
21020_6
\end{array}
$$

37.
$$
\begin{array}{r}
436_9 \\
\times\ 25_9 \\
\hline
2403 \\
\underline{873\ } \\
12233_9
\end{array}
$$

39.
$$
\begin{array}{r}
111_2 \\
\times 101_2 \\
\hline
111 \\
000 \\
\underline{111\ \ } \\
100011_2
\end{array}
$$

41.
$$
\begin{array}{r}
316_7 \\
\times\ 16_7 \\
\hline
2541 \\
\underline{316\ } \\
6031_7
\end{array}
$$

43. $1_2 \times 1_2 = 1_2$

$$
\begin{array}{r}
110_2 \\
1_2 \overline{\smash{)}110_2} \\
\underline{1}\ \ \ \ \\
01\ \\
\underline{1}\ \\
0 \\
\underline{0} \\
0
\end{array}
$$

45. $3_5 \times 1_5 = 3_5$
$3_5 \times 2_5 = 11_5$
$3_5 \times 3_5 = 14_5$
$3_5 \times 4_5 = 22_5$

$$
\begin{array}{r}
31_5 \\
3_5 \overline{\smash{)}143_5} \\
\underline{14}\ \\
03 \\
\underline{3} \\
0
\end{array}
$$

47. $2_4 \times 1_4 = 2_4$
$2_4 \times 2_4 = 10_4$
$2_4 \times 3_4 = 12_4$

$$
\begin{array}{r}
123_4 \\
2_4 \overline{\smash{)}312_4} \\
\underline{2}\ \ \ \\
11\ \\
\underline{10}\ \\
12 \\
\underline{12} \\
0
\end{array}
$$

49. $\quad 2_4 \times 1_4 = 2_4$

$\qquad 2_4 \times 2_4 = 10_4$

$\qquad 2_4 \times 3_4 = 12_4$

$$\begin{array}{r} 103_4 \quad \text{R1}_4 \\ 2_4 \overline{\smash{)}213_4} \\ \underline{2} \\ 01 \\ \underline{00} \\ 13 \\ \underline{12} \\ 1 \end{array}$$

51. $\quad 3_5 \times 1_5 = 3_5$

$\qquad 3_5 \times 2_5 = 11_5$

$\qquad 3_5 \times 3_5 = 14_5$

$\qquad 3_5 \times 4_5 = 22_5$

$$\begin{array}{r} 41_5 \quad \text{R1}_5 \\ 3_5 \overline{\smash{)}224_5} \\ \underline{22} \\ 04 \\ \underline{3} \\ 1 \end{array}$$

53. $\quad 6_7 \times 1_7 = 6_7$

$\qquad 6_7 \times 2_7 = 15_7$

$\qquad 6_7 \times 3_7 = 24_7$

$\qquad 6_7 \times 4_7 = 33_7$

$\qquad 6_7 \times 5_7 = 42_7$

$\qquad 6_7 \times 6_7 = 51_7$

$$\begin{array}{r} 45_7 \quad \text{R2}_7 \\ 6_7 \overline{\smash{)}404_7} \\ \underline{33} \\ 44 \\ \underline{42} \\ 2 \end{array}$$

55. $\quad\begin{array}{r} 2_5 \\ + 3_5 \\ \hline 10_5 \end{array}$ $= \ominus\bigcirc_5$

57. $\quad\begin{array}{r} 21_5 \\ + 43_5 \\ \hline 114_5 \end{array}$ $= \ominus\ominus\text{\textcircled{\tiny1}}_5$

For #59-65, blue = 0 = b, red = 1 = r, gold = 2 = go, green = 3 = gr

59. $\quad\begin{array}{r} 3_4 \\ + 2_4 \\ \hline 11_4 \end{array}$ $= \text{(r)(r)}_4$

61. $\quad\begin{array}{r} 32_4 \\ + 11_4 \\ \hline 103_4 \end{array}$ $= \text{(r)(b)(gr)}_4$

63. $\quad\begin{array}{r} 33_4 \\ - 12_4 \\ \hline 21_4 \end{array}$ $= \text{(go)(r)}_4$

65. $\quad\begin{array}{r} 231_4 \\ - 103_4 \\ \hline 122_4 \end{array}$ $= \text{(r)(go)(go)}_4$

67. $\quad 2302_5 = 2\left(5^3\right) + 3\left(5^2\right) + 0(5) + 2(1) = 2(125) + 3(25) + 0 + 2 = 250 + 75 + 0 + 2 = 327$

69. $\quad 14_5 \times 1_5 = 14_5$

$\qquad 14_5 \times 2_5 = 33_5$

$\qquad 14_5 \times 3_5 = 102_5$

$\qquad 14_5 \times 4_5 = 121_5$

$$\begin{array}{r} 13_5 \\ 14_5 \overline{\smash{)}242_5} \\ \underline{14} \\ 102 \\ \underline{102} \\ 0 \end{array}$$

71. a) 462_8

    $\times\,35_8$

   2772

   1626

  21252_8

b) $462_8 = 4\left(8^2\right) + 6(8) + 2(1) = 4(64) + 48 + 2 = 256 + 48 + 2 = 306$

    $35_8 = 3(8) + 5(1) = 24 + 5 = 29$

c) $306 \times 29 = 8874$

d) $21252_8 = 2\left(8^4\right) + 1\left(8^3\right) + 2\left(8^2\right) + 5(8) + 2(1)$

    $= 2(4096) + 512 + 2(64) + 40 + 2$

    $= 8192 + 512 + 128 + 40 + 2 = 8874$

e) Yes, in part a), the numbers were multiplied in base 8 and then converted to base 10 in part d).
In part b), the numbers were converted to base 10 first, then multiplied in part c).

72. $b = 5$

73. Orange = 0; purple = 1; turquoise = 2; red = 3

Exercise Set 4.5

1. Duplation and mediation, the galley method and Napier rods

3. a) Answers will vary.

   b)

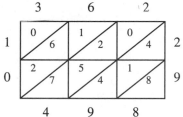

    $362 \times 29 = 10,498$

5.
```
23 –   31
11 –   62
 5 –  124
 2 –  248
 1 –  496
      713
```

7.
```
9 –  162
4 –  324
2 –  648
1 – 1296
     1458
```

9.
```
35 –  236
17 –  472
 8 –  944
 4 – 1888
 2 – 3776
 1 – 7552
      8260
```

11.
```
93 –   93
46 –  186
23 –  372
11 –  744
 5 – 1488
 2 – 2976
 1 – 5952
      8649
```

13.

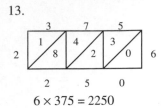

$$6 \times 375 = 2250$$

15.

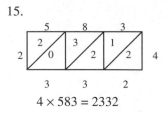

$$4 \times 583 = 2332$$

17.

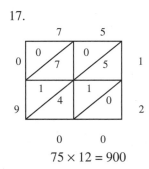

$$75 \times 12 = 900$$

19.

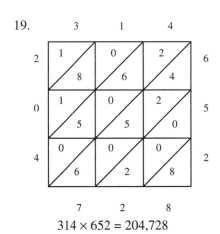

$$314 \times 652 = 204,728$$

21.

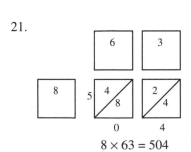

$$8 \times 63 = 504$$

23.

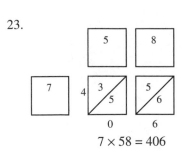

$$7 \times 58 = 406$$

25.

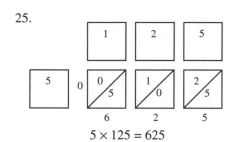

$$5 \times 125 = 625$$

27.

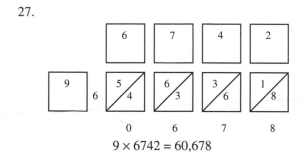

$$9 \times 6742 = 60{,}678$$

29. a) 253×46; Place the factors of 8 until the
       correct factors and placements are found
       so the rest of the rectangle can be completed.

    b)

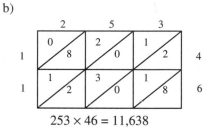

$$253 \times 46 = 11{,}638$$

31. a) 4×382; Place the factors of 12 until the correct
       factors and placements are found so the rest
       can be completed.

    b)

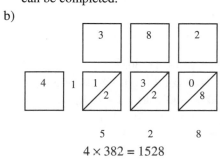

$$4 \times 382 = 1528$$

33. 13 – 22
    ~~6 – 44~~
    3 – 88
    1 – <u>176</u>
    286 =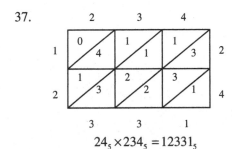

35. Answers will vary.

37.

	2	3	4	
1	0／4	1／1	1／3	2
2	1／3	2／2	3／1	4
	3	3	1	

$24_5 \times 234_5 = 12331_5$

38. a) $1000 + 500 + 100 + 100 + 50 + 10 + 10 + 5 + 1 = 1776$
    b) Answers will vary.

Review Exercises

1. $1000 + 1000 + 1000 + 100 + 1 + 1 + 1 = 3103$
2. $100 + 100 + 10 + 1000 + 1 = 1211$
3. $10 + 100 + 100 + 100 + 1 + 1000 = 1311$
4. $100 + 10 + 1000 + 1 + 1000 + 1 + 1 + 1 = 2114$
5. $1000 + 1000 + 100 + 100 + 100 + 10$
   $+ 1 + 1 + 1 + 1 = 2314$
6. $100 + 100 + 10 + 1 + 1000 + 1000 + 1 + 100 = 2312$
7. *bbbbbaaaaaa*
8. *cbbaaaaa*
9. *ccbbbbbbbbbaaa*
10. *ddaaaaa*
11. *dddddddccccccccbbbbba*
12. *ddcccbaaaa*
13. $4 (10) + 3 = 40 + 3 = 43$
14. $2 (10) + 7 = 20 + 7 = 27$
15. $7 (100) + 4 (10) + 9 = 700 + 40 + 9 = 749$
16. $4 (1000) + 6 (10) + 8 = 4000 + 60 + 8 = 4068$
17. $5 (1000) + 6 (100) + 4 (10) + 8$
    $= 5000 + 600 + 40 + 8 = 5648$
18. $6 (1000) + 9 (100) + 5 = 6000 + 900 + 5 = 6905$
19. *hxb*
20. *byixe*
21. *hyfxb*
22. *czixd*
23. *fzd*
24. *bza*
25. $4 (10) + 5 (1) = 40 + 5 = 45$
26. $3 (100) + 8 (1) = 300 + 8 = 308$
27. $5 (100) + 6 (10) + 8 (1) = 500 + 60 + 8 = 568$
28. $4 (10,000) + 6 (1000) + 8 (100) + 8 (10) + 3(1)$
    $= 40,000 + 6000 + 800 + 80 + 3 = 46,883$
29. $6 (10,000) + 4 (1000) + 4 (100) + 8 (10) + 1$
    $= 60,000 + 4000 + 400 + 80 + 1 = 64,481$
30. $6 (10,000) + 5 (100) + 2 (10) + 9 (1)$
    $= 60,000 + 500 + 20 + 9 = 60,529$
31. *qe*
32. *upb*
33. *vrc*
34. BArg
35. ODvog
36. QFvrf
37. ∫℘9999ᴧᴧᴧᴧᴧᴧ‖‖
38. MCDLXII

39. 一千四百六十二

40. $\alpha' \upsilon \xi \beta$

41.

$$
60 \overline{\smash{\big)}\ 1462} \quad \text{24}
$$

$$
\begin{array}{r} 24 \\ 60\ \overline{\smash{\big)}\ 1462} \\ \underline{1440} \\ 22 \end{array}
$$

⪡IIII　　⪡II

$$1462 = 24\,(60) + 22$$

42.

$$
\begin{array}{r} 4 \\ 360\ \overline{\smash{\big)}\ 1462} \\ \underline{1440} \\ 22 \end{array}
\qquad
\begin{array}{r} 1 \\ 20\ \overline{\smash{\big)}\ 22} \\ \underline{20} \\ 2 \end{array}
\qquad
\begin{array}{l} \cdots \cdot \\ \cdot \\ \cdot\cdot \end{array}
$$

$$1462 = 4\,(18 \times 20) + 1\,(20) + 2\,(1)$$

43. $100{,}000 + 100{,}000 + 10{,}000 + 10{,}000 + 1000 + 1000 + 10 + 10 + 10 + 1 + 1 + 1 + 1 + 1 = 222{,}035$

44. $8\,(1000) + 2\,(100) + 5\,(10) + 4 = 8000 + 200 + 50 + 4 = 8254$

45. $600 + 80 + 5 = 685$

46. $1000 + (1000 - 100) + (100 - 10) + 1 = 1000 + 900 + 90 + 1 = 1991$

47. $21(60) + (20 - 3) = 1260 + 17 = 1277$

48. $7(18 \times 20) + 8(20) + 10(1) = 7(360) + 160 + 10 = 2520 + 160 + 10 = 2690$

49. $47_8 = 4(8) + 7(1) = 32 + 7 = 39$

50. $101_2 = 1(2^2) + 0(2) + 1(1) = 4 + 0 + 1 = 5$

51. $130_4 = 1(4^2) + 3(4) + 0(1) = 16 + 12 + 0 = 28$

52. $3425_7 = 3(7^3) + 4(7^2) + 2(7) + 5(1) = 3(343) + 4(49) + 14 + 5 = 1029 + 196 + 14 + 5 = 1244$

53. $T0E_{12} = 10(12^2) + 0(12) + 11(1) = 10(144) + 0 + 11 = 1440 + 0 + 11 = 1451$

54. $20220_3 = 2(3^4) + 0(3^3) + 2(3^2) + 2(3) + 0(1) = 2(81) + 0 + 2(9) + 6 + 0 = 162 + 0 + 18 + 6 + 0 = 186$

55. To convert 463 to base 4　　　　　… 1024　256　64　16　4　1

$$
\begin{array}{r} 1 \\ 256\ \overline{\smash{\big)}\ 463} \\ \underline{256} \\ 207 \end{array}
\quad
\begin{array}{r} 3 \\ 64\ \overline{\smash{\big)}\ 207} \\ \underline{192} \\ 15 \end{array}
\quad
\begin{array}{r} 0 \\ 16\ \overline{\smash{\big)}\ 15} \\ \underline{0} \\ 15 \end{array}
\quad
\begin{array}{r} 3 \\ 4\ \overline{\smash{\big)}\ 15} \\ \underline{12} \\ 3 \end{array}
\quad
\begin{array}{r} 3 \\ 1\ \overline{\smash{\big)}\ 3} \\ \underline{3} \\ 0 \end{array}
\qquad 463 = 13033_4
$$

56. To convert 463 to base 3　　　　　… 729　243　81　27　9　3　1

$$
\begin{array}{r} 1 \\ 243\ \overline{\smash{\big)}\ 463} \\ \underline{243} \\ 220 \end{array}
\quad
\begin{array}{r} 2 \\ 81\ \overline{\smash{\big)}\ 220} \\ \underline{162} \\ 58 \end{array}
\quad
\begin{array}{r} 2 \\ 27\ \overline{\smash{\big)}\ 58} \\ \underline{54} \\ 4 \end{array}
\quad
\begin{array}{r} 0 \\ 9\ \overline{\smash{\big)}\ 4} \\ \underline{0} \\ 4 \end{array}
\quad
\begin{array}{r} 1 \\ 3\ \overline{\smash{\big)}\ 4} \\ \underline{3} \\ 1 \end{array}
\quad
\begin{array}{r} 1 \\ 1\ \overline{\smash{\big)}\ 1} \\ \underline{1} \\ 0 \end{array}
\qquad 463 = 122011_3
$$

57. To convert 463 to base 2　　　　　… 512　256　128　64　32　16　8　4　2　1

$$
\begin{array}{r} 1 \\ 256\ \overline{\smash{\big)}\ 463} \\ \underline{256} \\ 207 \end{array}
\quad
\begin{array}{r} 1 \\ 128\ \overline{\smash{\big)}\ 207} \\ \underline{128} \\ 79 \end{array}
\quad
\begin{array}{r} 1 \\ 64\ \overline{\smash{\big)}\ 79} \\ \underline{64} \\ 15 \end{array}
\quad
\begin{array}{r} 0 \\ 32\ \overline{\smash{\big)}\ 15} \\ \underline{0} \\ 15 \end{array}
\quad
\begin{array}{r} 0 \\ 16\ \overline{\smash{\big)}\ 15} \\ \underline{0} \\ 15 \end{array}
\quad
\begin{array}{r} 1 \\ 8\ \overline{\smash{\big)}\ 15} \\ \underline{8} \\ 7 \end{array}
\quad
\begin{array}{r} 1 \\ 4\ \overline{\smash{\big)}\ 7} \\ \underline{4} \\ 3 \end{array}
\quad
\begin{array}{r} 1 \\ 2\ \overline{\smash{\big)}\ 3} \\ \underline{2} \\ 1 \end{array}
\quad
\begin{array}{r} 1 \\ 1\ \overline{\smash{\big)}\ 1} \\ \underline{1} \\ 0 \end{array}
$$

$463 = 111001111_2$

58. To convert 463 to base 5   ... 625   125   25   5   1

$$125\overline{)463}\quad 25\overline{)88}\quad 5\overline{)13}\quad 1\overline{)3}$$

$$\begin{array}{c} 3 \\ \underline{375} \\ 88 \end{array}\quad \begin{array}{c} 3 \\ \underline{75} \\ 13 \end{array}\quad \begin{array}{c} 2 \\ \underline{10} \\ 3 \end{array}\quad \begin{array}{c} 3 \\ \underline{3} \\ 0 \end{array}\qquad 463 = 3323_5$$

59. To convert 463 to base 12   ... 1728   144   12   1

$$144\overline{)463}\quad 12\overline{)31}\quad 1\overline{)7}$$

$$\begin{array}{c} 3 \\ \underline{432} \\ 31 \end{array}\quad \begin{array}{c} 2 \\ \underline{24} \\ 7 \end{array}\quad \begin{array}{c} 7 \\ \underline{7} \\ 0 \end{array}\qquad 463 = 327_{12}$$

60. To convert 463 to base 8   ... 512   64   8   1

$$64\overline{)463}\quad 8\overline{)15}\quad 1\overline{)7}$$

$$\begin{array}{c} 7 \\ \underline{448} \\ 15 \end{array}\quad \begin{array}{c} 1 \\ \underline{8} \\ 7 \end{array}\quad \begin{array}{c} 7 \\ \underline{7} \\ 0 \end{array}\qquad 463 = 717_8$$

61.	52_7 $\underline{55_7}$ 140_7	62.	10110_2 $\underline{11001_2}$ 101111_2	63.	TE_{12} $\underline{87_{12}}$ 176_{12}	64.	234_7 $\underline{456_7}$ 1023_7
65.	3024_5 $\underline{4023_5}$ 12102_5	66.	3407_8 $\underline{7014_8}$ 12423_8	67.	4032_7 $\underline{-\ 321_7}$ 3411_7	68.	1001_2 $\underline{-\ 101_2}$ 100_2
69.	$3TT_{12}$ $\underline{-\ E7_{12}}$ $2E3_{12}$	70.	4321_5 $\underline{-\ 442_5}$ 3324_5	71.	1713_8 $\underline{-\ 1243_8}$ 450_8	72.	2021_3 $\underline{-\ 212_3}$ 1102_3

73. 32_6
    $\underline{\times\ 4_6}$
    212_6

74. 34_5
    $\underline{\times\ 21_5}$
    34
    $\underline{123}$
    1314_5

75. 126_{12}
    $\underline{\times\ 47_{12}}$
    856
    $\underline{4T0}$
    5656_{12}

76. 221_3
    $\underline{\times\ 22_3}$
    1212
    $\underline{1212}$
    21102_3

77. 1011_2
    $\underline{\times\ 101_2}$
    1011
    0000
    $\underline{1011}$
    110111_2

78. 476_8
    $\underline{\times\ 23_8}$
    1672
    $\underline{1174}$
    13632_8

79. $1_2 \times 1_2 = 1_2$

$$1_2\overline{)1011_2}$$

$$\begin{array}{r} 1011_2 \\ \hline 1 \\ \hline 00 \\ \underline{00} \\ 01 \\ \underline{1} \\ 01 \\ \underline{1} \\ 0 \end{array}$$

80. $2_4 \times 1_4 = 2_4$
    $2_4 \times 2_4 = 10_4$
    $2_4 \times 3_4 = 12_4$

$$2_4\overline{)320_4}$$

$$\begin{array}{r} 130_4 \\ \hline 2 \\ \hline 12 \\ \underline{12} \\ 0 \\ \underline{0} \\ 0 \end{array}$$

81. $3_5 \times 1_5 = 3_5$

$3_5 \times 2_5 = 11_5$

$3_5 \times 3_5 = 14_5$

$3_5 \times 4_5 = 22_5$

23_5 R1_5

$3_5 \overline{)130_5}$
$\underline{11}$
20
$\underline{14}$
1

82. $4_6 \times 1_6 = 4_6$

$4_6 \times 2_6 = 12_6$

$4_6 \times 3_6 = 20_6$

$4_6 \times 4_6 = 24_6$

$4_6 \times 5_6 = 32_6$

433_6

$4_6 \overline{)3020_6}$
$\underline{24}$
22
$\underline{20}$
20
$\underline{20}$
0

83. $3_6 \times 1_6 = 3_6$

$3_6 \times 2_6 = 10_6$

$3_6 \times 3_6 = 13_6$

$3_6 \times 4_6 = 20_6$

$3_6 \times 5_6 = 23_6$

411_6 R1_6

$3_6 \overline{)2034_6}$
$\underline{20}$
03
$\underline{3}$
04
$\underline{3}$
1

84. $6_8 \times 1_8 = 6_8$

$6_8 \times 2_8 = 14_8$

$6_8 \times 3_8 = 22_8$

$6_8 \times 4_8 = 30_8$

$6_8 \times 5_8 = 36_8$

$6_8 \times 6_8 = 44_8$

$6_8 \times 7_8 = 52_8$

664_8 R2_8

$6_8 \overline{)5072_8}$
$\underline{44}$
47
$\underline{44}$
32
$\underline{30}$
2

85. ~~142~~   ~~24~~

71 - 48

35 - 96

17 - 192

~~8~~   ~~384~~

~~4~~   ~~768~~

~~2~~   ~~1536~~

1 - 3072

3408

86.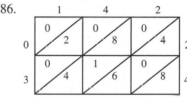

$142 \times 24 = 3408$

87.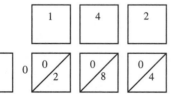

$2 \times 142 = 284$

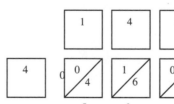

$4 \times 142 = 568$

$2 \times 142 = 284$, therefore $20 \times 142 = 2840$

Therefore, $142 \times 24 = 2840 + 568 = 3408$.

Chapter Test

1. A **number** is a quantity and answers the question "How many?" A **numeral** is a symbol used to represent the number.

2. $1000 + 1000 + 1000 + 500 + 100 + (50 - 10) + 5 + 1 = 3646$

3. $21(60) + 15(1) = 1260 + 15 = 1275$

4. $8 (1000) + 0 + 9(10) = 8000 + 0 + 90 = 8090$

5. $2 (18 \times 20) + 12(20) + 9 (1) = 2 (360) + 240 + 9$
   $= 720 + 240 + 9 = 969$

6. $100,000 + 10,000 + 10,000 + 1000 + 1000 + 100 + 10 + 10 + 10 + 10 + 1 + 1 = 122,142$

7. $9 (1000) + 900 + 90 + 9 = 9000 + 900 + 90 + 9$
   $= 9999$

8. ꝯꝯꝯꝯ𓍢𓍢𓍢||||

9. $\beta' \upsilon o 2$

10.

$$360 \overline{\smash{\big)}\ 1434} \quad \overset{3}{} \qquad 20 \overline{\smash{\big)}\ 354} \quad \overset{17}{}$$

$$\underline{1080} \qquad\qquad \underline{340}$$

$$354 \qquad\qquad\quad 14$$

(Mayan numeral glyphs shown to the right)

$$1434 = 3(18 \times 20) + 17(20) + 14(1)$$

11.

$$60 \overline{\smash{\big)}\ 1596} \quad \overset{26}{}$$

$$\underline{1560}$$

$$36$$

(Babylonian cuneiform numerals shown to the right)

$$1596 = 26(60) + 36(1)$$

12. MMCCCLXXVIII

13. In an additive system, the number represented by a particular set of numerals is the sum of the values of the numerals.

14. In a multiplicative system, there are numerals for each number less than the base and for powers of the base. Each numeral less than the base is multiplied by a numeral for the power of the base, and these products are added to obtain the number.

15. In a ciphered system, the number represented by a particular set of numerals is the sum of the values of the numerals. There are numerals for each number up to and including the base and multiples of the base.

16. In a place-value system, each number is multiplied by a power of the base. The position of the numeral indicates the power of the base by which it is multiplied.

17. $56_7 = 5\,(7) + 6\,(1) = 35 + 6 = 41$

18. $403_5 = 4(5^2) + 0(5) + 3(1) = 4(25) + 0 + 3$
$$= 100 + 0 + 3 = 103$$

19. $101101_2 = 1(2^5) + 0(2^4) + 1(2^3) + 1(2^2) + 0(2) + 1(1) = 32 + 0 + 8 + 4 + 0 + 1 = 45$

20. $368_9 = 3(9^2) + 6(9) + 8(1) = 3(81) + 54 + 8 = 243 + 54 + 8 = 305$

21. To convert 36 to base 2   … 64 32 16 8 4 2 1

$$32 \overline{\smash{\big)}\ 36}^{\,1} \quad 16 \overline{\smash{\big)}\ 4}^{\,0} \quad 8 \overline{\smash{\big)}\ 4}^{\,0} \quad 4 \overline{\smash{\big)}\ 4}^{\,1} \quad 2 \overline{\smash{\big)}\ 0}^{\,0} \quad 1 \overline{\smash{\big)}\ 0}^{\,0}$$

$$\underline{32} \qquad \underline{0} \qquad \underline{0} \qquad \underline{4} \qquad \underline{0} \qquad \underline{0}$$

$$4 \qquad 4 \qquad 4 \qquad 0 \qquad 0 \qquad 0$$

$$36 = 100100_2$$

22. To convert 93 to base 5   … 125 25 5 1

$$25 \overline{\smash{\big)}\ 93}^{\,3} \quad 5 \overline{\smash{\big)}\ 18}^{\,3} \quad 1 \overline{\smash{\big)}\ 3}^{\,3}$$

$$\underline{75} \qquad \underline{15} \qquad \underline{3}$$

$$18 \qquad\ 3 \qquad\ 0 \qquad 93 = 333_5$$

23. To convert 2356 to base 12   … 20,736 1728 144 12 1

$$1728 \overline{\smash{\big)}\ 2356}^{\,1} \quad 144 \overline{\smash{\big)}\ 628}^{\,4} \quad 12 \overline{\smash{\big)}\ 52}^{\,4} \quad 1 \overline{\smash{\big)}\ 4}^{\,4}$$

$$\underline{1728} \qquad\quad \underline{576} \qquad\quad \underline{48} \qquad \underline{4}$$

$$628 \qquad\qquad 52 \qquad\qquad 4 \qquad 0 \qquad 2356 = 1444_{12}$$

24. To convert 2938 to base 7   … 16,807 2401 343 49 7 1

$$2401 \overline{\smash{\big)}\ 2938}^{\,1} \quad 343 \overline{\smash{\big)}\ 537}^{\,1} \quad 49 \overline{\smash{\big)}\ 194}^{\,3} \quad 7 \overline{\smash{\big)}\ 47}^{\,6} \quad 1 \overline{\smash{\big)}\ 5}^{\,5}$$

$$\underline{2401} \qquad\quad \underline{343} \qquad\quad \underline{147} \qquad\quad \underline{42} \qquad \underline{5}$$

$$537 \qquad\qquad 194 \qquad\qquad 47 \qquad\qquad 5 \qquad 0 \qquad 2938 = 11365_7$$

25. 133_5
        $\underline{434_5}$
        1122_5

26. 324_6
        $\underline{-142_6}$
        142_6

27.
$$\begin{array}{r} 45_6 \\ \times\,23_6 \\ \hline 223 \\ 134 \\ \hline 2003_6 \end{array}$$

28.
$3_5 \times 1_5 = 3_5$
$3_5 \times 2_5 = 11_5$
$3_5 \times 3_5 = 14_5$
$3_5 \times 4_5 = 22_5$

$$\begin{array}{r} 220_5 \\ 3_5\,\overline{)\,1210_5} \\ 11 \\ \hline 11 \\ 11 \\ \hline 00 \\ 00 \\ \hline 0 \end{array}$$

29. 35 - 28
     17 - 56
     ~~8~~ ~~112~~
     ~~4~~ ~~224~~
     ~~2~~ ~~448~~
      1 - 896
           980

30.

	1	9	6	
0	0 / 4	3 / 6	2 / 4	4
8	0 / 3	2 / 7	1 / 8	3
	4	2	8	

$43 \times 196 = 8428$

CHAPTER FIVE

NUMBER THEORY AND THE REAL NUMBER SYSTEM

Exercise Set 5.1

1. Number theory is the study of numbers and their properties.

3. a) a divides b means that b divided by a has a remainder of zero.

   b) a is divisible by b means that a divided by b has a remainder of zero.

5. A composite number is a natural number that is divisible by a number other than itself and 1. Any natural number that is not prime is composite.

7. a) The least common multiple (LCM) of a set of natural numbers is the smallest natural number that is divisible (without remainder) by each element of the set.

   b) Determine the prime factorization of each number. Then find the product of the prime factors with the largest exponent that appear in any of the prime factorizations.

   c) The prime factors with the largest exponent that appear in any of the factorizations are 2^4 and 5. The LCM of 16 and 40 is $2^4 \cdot 5$, or 80.

2	16
2	8
2	4
	2

   $16 = 2^4$

9. Mersenne Primes are prime numbers of the form $2^n - 1$ where n is a prime number.

11. Goldbach's conjecture states that every even number greater than or equal to 4 can be represented as the sum of two (not necessarily distinct) prime numbers.

13. The prime numbers between 1 and 100 are: 2, 3, 5, 7, 11, 13, 17, 19, 23, 29, 31, 37, 41, 43, 47, 53, 59, 61, 67, 71, 73, 79, 83, 89, 91, 97.

15. True; since $54 \div 9 = 6$

17. False.

19. False.

21. True; if a number is divisible by 10, then it is also divisible by 5.

23. False; if a number is divisible by 3, then the sum of the number's digits is divisible by 3.

25. True; since $2 \cdot 3 = 6$.

27. Divisible by 2, 3, 4, 6, 8 and 9.

29. Divisible by 3 and 5.

31. Divisible by 2, 3, 4, 5, 6, 8, and 10.

33. $2 \cdot 3 \cdot 4 \cdot 5 \cdot 6 = 720$. (other answers are possible)

35.

5	45
3	9
	3

$45 = 3^2 \cdot 5$

37.

2	196
2	98
7	49
	7

$196 = 2^2 \cdot 7^2$

39.

3	303
	101

$303 = 3 \cdot 101$

41.
$$\begin{array}{c|c} 3 & 513 \\ 3 & 171 \\ 3 & 57 \\ & 19 \end{array}$$
$513 = 3^3 \cdot 19$

43.
$$\begin{array}{c|c} 2 & 1336 \\ 2 & 668 \\ 2 & 334 \\ & 167 \end{array}$$
$1336 = 2^3 \cdot 167$

45.
$$\begin{array}{c|c} 3 & 2001 \\ 23 & 667 \\ & 29 \end{array}$$
$2001 = 3 \cdot 23 \cdot 29$

47. The prime factors of 15 and 18 are: $6 = 3 \cdot 2$, $15 = 3 \cdot 5$
a) The common factor is 3, thus, the GCD = 3.
b) The factors with the greatest exponent that appear in either are 2, 3, 5. Thus, the LCM = $2 \cdot 3 \cdot 5 = 30$.

49. The prime factors of 48 and 54 are: $48 = 2^4 \cdot 3$, $54 = 2 \cdot 3^3$
a) The common factors are: 2,3; thus, the GCD = $2 \cdot 3 = 6$.
b) The factors with the greatest exponent that appear in either are: 2^4, 3^3; thus, the LCM = $2^4 \cdot 3^3 = 432$

51. The prime factors of 40 and 900 are: $40 = 2^3 \cdot 5$, $900 = 2^2 \cdot 3^2 \cdot 5^2$
a) The common factors are: 2^2, 5; thus, the GCD = $2^2 \cdot 5 = 20$.
b) The factors with the greatest exponent that appear in either are: 2^3, 3^2, 5^2; thus, the LCM = $2^2 \cdot 3^2 \cdot 5^2 = 1800$

53. The prime factors of 96 and 212 are: $96 = 2^5 \cdot 3$, $212 = 2^2 \cdot 53$
a) The common factors are: 2^2; thus, the GCD = $2^2 = 4$.
b) The factors with the greatest exponent that appear in either are: 2^5, 3, 53; thus, the LCM = $2^5 \cdot 3 \cdot 53 = 5088$

55. The prime factors of 24, 48, and 128 are: $24 = 2^3 \cdot 3$, $48 = 2^4 \cdot 3$, $128 = 2^7$
a) The common factors are: 2^3; thus, the GCD = $2^3 = 8$.
b) The factors with the greatest exponent that appear in any are: 2^7, 3; thus, LCM = $2^7 \cdot 3 = 384$

57. Use the list of primes generated in exercise 13. The next two sets of twin primes are: 17, 19, 29, 31.

59. (a) 14, 15 Yes; (b) 21, 30 No; (c) 24, 25 Yes; (d) 119, 143 Yes

61. $4 = 2 + 2, 6 = 3 + 3, 8 = 3 + 5, 10 = 3 + 7, 12 = 5 + 7, 14 = 7 + 7, 16 = 3 + 13, 18 = 5 + 13, 20 = 3 + 17$

63. The gcd of 350 and 140 is 70 dolls.

65. The gcd of 432 and 360 is 72 cards.

67. The lcm of 45 and 60 is 180 mins.

69. The least common multiple of 5 and 6 is 30. Thus, it will be 30 days before they both have the same night off again.

71. a) The possible committee sizes are: 4, 5, 10, 20, or 25. b) The number of committees possible are: 25 committees of 4, 20 committees of 5, 10 committees of 10, 5 committees of 20, or 4 committees of 25.

73. A number is divisible by 15 if both 3 and 5 divide the number.

75. $35 \div 15 = 2$ with rem. = 5.
$15 \div 5 = 3$ with rem. = 0.
Thus, gcd of 35 and 15 is 5.

77. $108 \div 36 = 3$ with rem. = 0.
$36 \div 3 = 12$ with rem. = 0.
Thus, gcd of 108 and 36 is 36.

79. $180 \div 150 = 1$ with rem. = 30.
$150 \div 30 = 5$ with rem. = 0.
Thus, the gcd of 150 and 180 is 30.

81. The proper factors of 12 are: .
    1, 2, 3, 4, and 6.
    $1 + 2 + 3 + 4 + 6 = 16 \neq 12$
    Thus, 12 is not a perfect #.

83. The proper factors of 496
    are: 1,2,4,8,16,31,62,124,
    and 248. 1+2+4+8+16
    +31+62+124+248 = 496
    Thus, 496 is a perfect #

85. a) $60 = 2^2 \cdot 3^1 \cdot 5^1$ Adding 1 to each exponent and then multiplying these numbers, we get
    $(2+1)(1+1)(1+1) = 3 \cdot 2 \cdot 2 = 12$ divisors of 60; they are 1, 2, 3, 4, 5, 6, 10, 12, 15, 20, 30, and 60.

87. One of the numbers must be divisible by 3 and at least one must be even, so their product will be
    divisible by 2 and 3 and thus by 6.

89. $36,018 = (36,000 + 18)$; $36,000 \div 18 = 2,000$ and $18 \div 18 = 1$
    Thus, since $18 \mid 36000$ and $18 \mid 18$, $18 \mid 36018$.

91. $8 = 2+3+3$, $9 = 3+3+3$, $10 = 2+3+5$, $11 = 2+2+7$, $12 = 2+5+5$, $13 = 3+3+7$, $14 = 2+5+7$, $15 = 3+5+7$,
    $16 = 2+7+7$, $17 = 5+5+7$, $18 = 2+5+11$, $19 = 3+5+11$, $20 = 2+7+11$.

92. Denmark, Kangaroo, Orange. The sum of the digits minus 5 will always be 4, so the first letter for the
    country will always be D.

Exercise Set 5.2

1. Begin at zero, draw an arrow to the value of the first number. From the tip of that arrow draw another
   arrow by moving a number of spaces equal to the value of the second number. Be sure to move left if
   the number is negative and move right if the number is positive. The sum of the two numbers is at the
   tip of the second arrow.

3. To rewrite a subtraction problem as an addition problem, rewrite the subtraction sign as an addition sign
   and change the second number to its additive inverse.

5. The quotient of two numbers with like signs is a positive number, and the quotient of two numbers with
   unlike signs is a negative number.

7. $-6 + 9 = 3$

9. $(-7) + 9 = 2$

11. $[6+(-11)]+0 = -5+0 = -5$

13. $[(-3)+(-4)]+9 = -7+9 = 2$

15. $[(-23)+(-9)]+11 =$
    $[-32]+11 = -21$

17. $3 - 6 = -3$

19. $-4-6 = -10$

21. $-5 - (-3) = -5+3 = -2$

23. $14 - 20 = 14 + (-20) = -6$

25. $[5+(-3)] -4 = 2-4 = 2+(-4) = -2$

27. $-4 \cdot 5 = -20$

29. $(-12)(-12) = 144$

31. $[(-8)(-2)] \cdot 6 = 16 \cdot 6 = 96$

33. $(5 \cdot 6)(-2) = (30)(-2) = -60$

35. $[(-3)(-6)] \cdot [(-5)(8)] =$
    $(18)(-40) = -720$

37. $-26 \div (-13) = 2$

39. $23 \div (-23) = -1$

41. $\dfrac{56}{-8} = -7$

43. $\dfrac{-210}{14} = -15$

45. $144 \div (-3) = -48$

47. True; every whole
    number is an integer.

49. False; the difference of two negative integers may be positive, negative, or zero.

51. True; the product of two integers with like signs is a positive integer.

53. True; the quotient of two integers with unlike signs is a negative number.

55. False; the sum of a positive integer and a negative integer could be pos., neg., or zero.

57. $(5 + 7) \div 2 = 12 \div 2 = 6$

59. $[6(-2)] - 5 = -12 + (-5) = -17$

61. $(4 - 8)(3) = (-4)(3) = -12$

63. $[2+(-17)] \div 3 = [-15] \div 3 = -5$

65. $[(-22)(-3)] \div (2 - 13) =$
$66 \div (2 + (-13)) = 66 \div (-11)$
$= -6$

67. $-15, -10, -5, 0, 5, 10$

69. $-6, -5, -4, -3, -2, -1$

71. $134 - (-79.8) =$
$134 + 79.8 = 213.8° \text{F.}$

73. $0 + 100 - 40 + 90 - 20 + 80 =$
$60 + 90 - 20 + 80 = 210 \text{ pts.}$

75. $842 - (-927) = 842 + 927 =$
$1,769 \text{ feet}$

77. a) $+1 - (-8) = +1 + 8 = 9.$
There is a 9 hr. time diff.
b) $-5 - (-7) = -5 + 7 = 2.$
There is a 2 hr. time diff.

79.

$$\frac{-1+2-3+4-5+\ldots 99+100}{1-2+3-4+5\ldots+99-100} = \frac{50}{-50} = -1$$

81. $0 + 1 - 2 + 3 + 4 - 5 + 6 - 7 - 8 + 9 = 1$ (other answers are possible)

82. (a) $4+4-4-4=0$ $\dfrac{4+4}{4+4}=1$ $\dfrac{4}{4}+\dfrac{4}{4}=2$ $\dfrac{4+4+4}{4}=3$

$4(4-4)+4=4$ $\dfrac{4 \cdot 4+4}{4}=5$ $4+\dfrac{4+4}{4}=6$ $4+4-\dfrac{4}{4}=7$

$4 \cdot 4 - 4 - 4 = 8$ $4+4+\dfrac{4}{4}=9$

(b) $4\left(4-\dfrac{4}{4}\right)=12$ $4 \bullet 4 - \dfrac{4}{4}=15$ $\dfrac{4 \bullet 4 \bullet 4}{4}=16$ (c) $\dfrac{44-4}{4}=10$

$4 \bullet 4 - \dfrac{4}{4}=17$ $\left(4+\dfrac{4}{4}\right)4=20$

Exercise Set 5.3

1. Rational numbers is the set of all numbers of the form p/q, where p and q are integers, and q ≠ 0.
3. a) Divide both the numerator and the denominator by their greatest common divisor.

   b) $\dfrac{15}{27}=\dfrac{5 \div 3}{9 \div 3}=\dfrac{5}{9}$

5. For positive mixed numbers, multiply the denominator of the fraction by the integer preceding it. Add this product to the numerator. This sum is the numerator of the improper fraction; the denominator is the same as the denominator of the mixed number. For negative mixed numbers, you can temporarily ignore the negative sign, perform the conversion described above, and then reattach the negative sign.

7. a) The reciprocal of a number is 1 divided by the number.

   b) The reciprocal of -2 is $\dfrac{1}{-2} = -\dfrac{1}{2}$

9. a) To add or subtract two fractions with a common denominator, we add or subtract their numerators
       and keep the common denominator.

   b) $\dfrac{11}{36} + \dfrac{13}{36} = \dfrac{24}{36} = \dfrac{24 \div 12}{36 \div 12} = \dfrac{2}{3}$ c) $\dfrac{37}{48} - \dfrac{13}{48} = \dfrac{24}{48} = \dfrac{24 \div 24}{48 \div 24} = \dfrac{1}{2}$

11. We can multiply a fraction by the number one in the form of c/c (where c is a nonzero integer) and
      the number will maintain the same value.

13. GCD of 14 and 21 is 7.

$\dfrac{14}{21} = \dfrac{14 \div 7}{21 \div 7} = \dfrac{2}{3}$

15. GCD of 26 and 91 is 13.

$\dfrac{26}{91} = \dfrac{26 \div 13}{91 \div 13} = \dfrac{2}{7}$

17. GCD of 525 and 800 is 25.

$\dfrac{525}{800} = \dfrac{525 \div 25}{800 \div 25} = \dfrac{21}{32}$

19. GCD of 112 and 176 is 16.

$\dfrac{112}{176} = \dfrac{112 \div 16}{176 \div 16} = \dfrac{7}{11}$

21. GCD of 45 and 495 is 45.

$\dfrac{45}{495} = \dfrac{45 \div 45}{495 \div 45} = \dfrac{1}{11}$

23.

$3\dfrac{4}{7} = \dfrac{(3)(7)+4}{7} = \dfrac{21+4}{7} = \dfrac{25}{7}$

25. $-1\dfrac{15}{16} = -\dfrac{-((1)(16)+15)}{16}$

$= -\dfrac{16+15}{16} = -\dfrac{31}{16}$

27. $-4\dfrac{15}{16} = -\dfrac{(4)(16)+15}{16}$

$= -\dfrac{64+15}{16} = -\dfrac{79}{16}$

29.

$2\dfrac{1}{8} = \dfrac{(2)(8)+1}{8} = \dfrac{16+1}{8} = \dfrac{17}{8}$

31. $1\dfrac{7}{8} = \dfrac{(1)(8)+7}{8} = \dfrac{8+7}{8} = \dfrac{15}{8}$

33. $\dfrac{11}{8} = \dfrac{8+3}{8} = \dfrac{(1)(8)+3}{8} = 1\dfrac{3}{8}$

35. $-\dfrac{73}{6} = \dfrac{-(72+1)}{6}$

$= \dfrac{-(12 \bullet 6 + 1)}{6} = -12\dfrac{1}{6}$

37. $-\dfrac{878}{15} = -\dfrac{870+8}{15} = -\dfrac{(58)(15)+8}{15} = -58\dfrac{8}{15}$

39. $\dfrac{3}{5} = .60$

41. $\dfrac{2}{9} = .\overline{2}$

43. $3 \div 8 = 0.375$

45. $13 \div 3 = 4.\overline{3}$

47. $85 \div 15 = 5.\overline{6}$

49. $0.25 = \dfrac{25}{100} = \dfrac{25 \div 25}{100 \div 25} = \dfrac{1}{4}$

51.

$0.045 = \dfrac{45}{1000} = \dfrac{45 \div 5}{1000 \div 5} = \dfrac{9}{200}$

53. $0.2 = \dfrac{2}{10} = \dfrac{1}{5}$

55. $.452 = \dfrac{452}{1000} = \dfrac{113}{250}$

57. $.0001 = \dfrac{1}{10000}$

59. Let $n = 0.\overline{3}$, $10n = 3.\overline{3}$

$10n = 6.\overline{6}$

$\dfrac{-n = 0.\overline{6}}{9n = 6.0}$ $\dfrac{9n}{9} = \dfrac{6}{9} = \dfrac{2}{3} = n$

61. Let $n = 1.\overline{9}$, $10n = 19.\overline{9}$

$10n = 19.\overline{9}$

$\dfrac{-n = \ 1.\overline{9}}{9n = 18.0}$

$\dfrac{9n}{9} = \dfrac{18}{9} = 2 = n$

63. Let $n = 1.\overline{36}$, $100n = 136.\overline{36}$

$100n = 136.\overline{36}$

$\dfrac{-n = 1.\overline{36}}{99n = 135.0}$

$\dfrac{99n}{99} = \dfrac{135}{99} = \dfrac{15}{11} = n$

65. Let $n = 1.0\overline{2}$, $100n = 102.\overline{2}$

$$100n = 102.\overline{2}$$
$$\underline{-10n = 10.\overline{2}}$$
$$90n = 92.0$$
$$\frac{90n}{90} = \frac{92}{90} = \frac{46}{45} = n$$

67. Let $n = 3.4\overline{78}$,

$$1000n = 3478.\overline{78}$$
$$1000n = 3478.\overline{78}$$
$$\underline{-10n = 34.\overline{78}}$$
$$990n = 3444.0$$
$$\frac{990n}{990} = \frac{3444}{990} = \frac{574}{165} = n$$

69.
$$\frac{4}{11} \cdot \frac{3}{8} = \frac{4 \cdot 3}{11 \cdot 8} = \frac{12}{88} = \frac{12 \div 4}{88 \div 4} = \frac{3}{22}$$

71.
$$\frac{-3}{8} \cdot \frac{-16}{15} = \frac{48}{120} = \frac{2}{5}$$

73.
$$\frac{7}{8} \div \frac{8}{7} = \frac{7}{8} \cdot \frac{7}{8} = \frac{49}{64}$$

75.
$$\left(\frac{3}{5} \cdot \frac{4}{7}\right) \div \frac{1}{3} = \frac{12}{35} \div \frac{1}{3} = \frac{12}{35} \cdot \frac{3}{1} = \frac{36}{35}$$

77.
$$\left[\left(\frac{-3}{4}\right)\left(\frac{-2}{7}\right)\right] \div \frac{3}{5} = \left(\frac{6}{28}\right) \div \frac{3}{5} = \frac{3}{14} \cdot \frac{5}{3} = \frac{15}{42} = \frac{5}{14}$$

79. The lcm of 3 and 5 is 15.
$$\frac{2}{3} + \frac{1}{5} = \left(\frac{2}{3} \cdot \frac{5}{5}\right) + \left(\frac{1}{5} \cdot \frac{3}{3}\right) = \frac{10}{15} + \frac{3}{15} = \frac{13}{15}$$

81. The lcm of 13 and 26 is 26.
$$\frac{5}{13} + \frac{11}{26} = \left(\frac{5}{13} \cdot \frac{2}{2}\right) + \frac{11}{26} = \frac{10}{26} + \frac{11}{26} = \frac{21}{26}$$

83. The lcm of 9 and 54 is 54.
$$\frac{5}{9} - \frac{7}{54} = \left(\frac{5}{9} \cdot \frac{6}{6}\right) - \frac{7}{54} = \frac{30}{54} - \frac{7}{54} = \frac{23}{54}$$

85. The lcm of 12, 48, and 72 is 144.
$$\frac{1}{12} + \frac{1}{48} + \frac{1}{72} = \left(\frac{1}{12} \cdot \frac{12}{12}\right) + \left(\frac{1}{48} \cdot \frac{3}{3}\right) + \left(\frac{1}{72} \cdot \frac{2}{2}\right)$$
$$= \frac{12}{144} + \frac{3}{144} + \frac{2}{144} = \frac{17}{144}$$

87. The lcm of 30, 40, and 50 is 600.
$$\frac{1}{30} - \frac{3}{40} - \frac{7}{50} = \left(\frac{1}{30} \cdot \frac{20}{20}\right)\left(\frac{3}{40} \cdot \frac{15}{15}\right)\left(\frac{7}{50} \cdot \frac{12}{12}\right)$$
$$= \frac{20}{600} - \frac{45}{600} - \frac{84}{600} = -\frac{109}{600}$$

89.
$$\frac{2}{5} + \frac{7}{8} = \frac{2 \cdot 8 + 7 \cdot 5}{8 \cdot 5} = \frac{16 + 35}{40} = \frac{51}{40}$$

91.
$$\frac{5}{6} - \frac{7}{8} = \frac{5 \cdot 4 - 7 \cdot 3}{24} = \frac{20 - 21}{24} = \frac{-1}{24}$$

93.
$$\frac{3}{8} + \frac{5}{12} = \frac{3 \cdot 12 + 8 \cdot 5}{8 \cdot 12} = \frac{36 + 40}{96} = \frac{76}{96} = \frac{19}{24}$$

95.
$$\left(\frac{2}{3} \cdot \frac{9}{10}\right) + \frac{2}{5} = \frac{18}{30} + \frac{2}{5} = \frac{18}{30} + \left(\frac{2}{5} \cdot \frac{6}{6}\right) =$$
$$= \frac{18}{30} + \frac{12}{30} = \frac{30}{30} = 1$$

97.
$$\left(\frac{1}{2} + \frac{3}{10}\right) \div \left(\frac{1}{5} + 2\right) = \left(\frac{1}{2} \cdot \frac{5}{5} + \frac{3}{10}\right) \div \left(\frac{1}{5} + \frac{2}{1} \cdot \frac{5}{5}\right) = \left(\frac{5}{10} + \frac{3}{10}\right) \div \left(\frac{1}{5} + \frac{10}{5}\right) = \frac{8}{10} \div \frac{11}{5} = \frac{4}{5} \cdot \frac{5}{11} = \frac{20}{55} = \frac{4}{11}$$

99.
$$\left(3\frac{4}{9}\right) \div \left(4 + \frac{2}{3}\right) = \left(\frac{3}{1} \cdot \frac{9}{9} - \frac{4}{9}\right) \div \left(\frac{4}{1} \cdot \frac{3}{3} + \frac{2}{3}\right) = \left(\frac{27}{9} - \frac{4}{9}\right) \div \left(\frac{12}{3} + \frac{2}{3}\right) = \frac{23}{9} \div \frac{14}{3} = \frac{23}{9} \cdot \frac{3}{14} = \frac{69}{126} = \frac{23}{42}$$

101. The LCM of 2, 4, 6 is 12. $\frac{1}{2} + \frac{1}{4} + \frac{1}{6} = \left(\frac{1}{2} \cdot \frac{6}{6}\right) + \left(\frac{1}{4} \cdot \frac{3}{3}\right) + \left(\frac{1}{6} \cdot \frac{2}{2}\right) = \frac{6}{12} + \frac{3}{12} + \frac{2}{12} = \frac{11}{12}$;

$1 - \frac{11}{12} = \frac{1}{12}$ musk thistles

103. $14\left(8\dfrac{5}{8}\right) = 14\left(\dfrac{69}{8}\right) = \dfrac{966}{8} = \dfrac{966 \div 2}{8 \div 2} = \dfrac{483}{4} = 120.75''$

105.

$\left(1\dfrac{1}{2}\right)\left(\dfrac{1}{4}\right) = \left(\dfrac{3}{2}\right)\left(\dfrac{1}{4}\right) = \dfrac{3}{8}$ cups of snipped parsley

$\left(1\dfrac{1}{2}\right)\left(\dfrac{1}{8}\right) = \left(\dfrac{3}{2}\right)\left(\dfrac{1}{8}\right) = \dfrac{3}{16}$ tsp of pepper

$\left(1\dfrac{1}{2}\right)\left(\dfrac{1}{2}\right) = \left(\dfrac{3}{2}\right)\left(\dfrac{1}{2}\right) = \dfrac{3}{4}$ cups of sliced carrots

107. The LCM of 4, 5, 3 is 60.

$\dfrac{1}{4} + \dfrac{2}{5} + \dfrac{1}{3} = \left(\dfrac{1}{4}\right)\left(\dfrac{15}{15}\right) + \left(\dfrac{2}{5}\right)\left(\dfrac{12}{12}\right) + \left(\dfrac{1}{3}\right)\left(\dfrac{20}{20}\right)$

$= \dfrac{15}{60} + \dfrac{24}{60} + \dfrac{20}{60} = \dfrac{59}{60}$

109.

$1 - \left(\dfrac{1}{4} + \dfrac{1}{5} + \dfrac{1}{2}\right) = 1 - \left(\dfrac{5}{20} + \dfrac{4}{20} + \dfrac{10}{20}\right)$

$= 1 - \dfrac{19}{20} = \dfrac{20}{20} - \dfrac{19}{20} = \dfrac{1}{20}$

She must proofread .05 of the book or = 27 pages.

111.

$4\dfrac{1}{2} + 30\dfrac{1}{4} + 24\dfrac{1}{8} = 4\dfrac{4}{8} + 30\dfrac{2}{8} + 24\dfrac{1}{8} = 58\dfrac{7}{8}$ inches

113. $\left(24\dfrac{7}{8}\right) \div 2 = \dfrac{199}{8} \cdot \dfrac{1}{2} = \dfrac{199}{16} = 12\dfrac{7}{16}$ in.

115. $8\dfrac{3}{4}$ ft $= \left(\dfrac{35}{4} \cdot \dfrac{12}{1}\right)$ in. $= 105$ in.

$\left[105 - (3)\left(\dfrac{1}{8}\right)\right] \div 4 = \left[\dfrac{840}{8} - \dfrac{3}{8}\right] \div 4 = \dfrac{837}{8} \cdot \dfrac{1}{4} = \dfrac{837}{32} = 26\dfrac{5}{32}$.   The length of each piece is $26\dfrac{5}{32}$ in.

117. width = 8 ft. 3 in. = 96 in. + 3 in. = 99 in.; length = 10 ft. 8 in. = 120 in. + 8 in. = 128 in.

a) perimeter = 2L + 2W = 2(128) + 2(99) = 454 in $\dfrac{454''}{12''/\text{ft.}} = 37\dfrac{10}{12}$ ft. = 37 ft. 10 in.

b) width = 8ft. 3in. = $8\dfrac{3}{12}$ ft. = $8\dfrac{1}{4}$ ft. = $\dfrac{33}{4}$ ft .; length = 10ft. 8in. = $10\dfrac{8}{12}$ ft. = $10\dfrac{2}{3}$ ft. = $\dfrac{32}{3}$ ft

Area = L × w = $\dfrac{32}{3} \times \dfrac{33}{4} = \dfrac{1056}{12} = 88$ sq.ft

117. c) Volume = L · W · H = $\dfrac{32}{3} \times \dfrac{33}{4} \times \dfrac{55}{6} = \dfrac{58080}{72} = 806\dfrac{2}{3}$ cu. ft.

119. $\dfrac{0.10 + 0.11}{2} = \dfrac{0.21}{2} = 0.105$

121. $\dfrac{-2.176 + (-2.175)}{2} = \dfrac{-4.351}{2} = -2.1755$

123. $\dfrac{3.12345 + 3.123451}{2} = \dfrac{6.246901}{2} = 3.1234505$

125. $\dfrac{4.872 + 4.873}{2} = \dfrac{9.745}{2} = 4.8725$

127. $\left(\dfrac{1}{3} + \dfrac{2}{3}\right) \div 2 = \dfrac{3}{3} \cdot \dfrac{1}{2} = \dfrac{3}{6} = \dfrac{1}{2}$

129. $\left(\dfrac{1}{100} + \dfrac{1}{10}\right) \div 2 = \dfrac{11}{100} \cdot \dfrac{1}{2} = \dfrac{11}{200}$

131. $\left(\dfrac{1}{4}+\dfrac{1}{5}\right)\div 2=\left(\dfrac{5}{20}+\dfrac{4}{20}\right)\cdot\dfrac{1}{2}=\dfrac{9}{20}\cdot\dfrac{1}{2}=\dfrac{9}{40}$

133.

$$\left(\dfrac{1}{10}+\dfrac{1}{100}\right)\div 2=\left(\dfrac{10}{100}+\dfrac{1}{100}\right)\cdot\dfrac{1}{2}=\dfrac{11}{100}\cdot\dfrac{1}{2}=\dfrac{11}{200}$$

135. a) Water (or milk): $\left(1+1\dfrac{3}{4}\right)\div 2=\left(\dfrac{4}{4}+\dfrac{7}{4}\right)\cdot\dfrac{1}{2}=\dfrac{11}{4}\cdot\dfrac{1}{2}=\dfrac{11}{8}=1\dfrac{3}{8}$ cup;

   Oats: $\left(\dfrac{1}{2}+1\right)\div 2=\dfrac{3}{2}\cdot\dfrac{1}{2}=\dfrac{3}{4}$ cup

   b) Water (or milk): $1+\dfrac{1}{2}=1\dfrac{1}{2}$ cup;

   Oats: $\dfrac{1}{2}+\dfrac{1}{2}\cdot\dfrac{1}{2}=\dfrac{3}{4}$ cup

137. a) $\dfrac{1}{8}$ b) $\dfrac{1}{16}$ c) 5 times   d) 5 times

Exercise Set 5.4

1. A rational number can be written as a ratio of two integers, p/q, with q not equal to zero. Numbers that cannot be written as the ratio of two integers are called irrational numbers.
3. A perfect square number is any number that is the square of a natural number.
5. a) To add or subtract two or more square roots with the same radicand, add or subtract their coefficients and then multiply by the common radical.
   b) $3\sqrt{6}+5\sqrt{6}-9\sqrt{6}=8\sqrt{6}-9\sqrt{6}=-1\sqrt{6}=-\sqrt{6}$

7. a) Multiply both the numerator and denominator by the same number that will result in the radicand in the denominator becoming a perfect square.

   b) $\dfrac{7}{\sqrt{3}}=\dfrac{7}{\sqrt{3}}\cdot\dfrac{\sqrt{3}}{\sqrt{3}}=\dfrac{7\sqrt{3}}{\sqrt{9}}=\dfrac{7\sqrt{3}}{3}$

9. $\sqrt{36}=6$ rational

11. $\dfrac{2}{3}$ rational

13. Irrational; non-terminating, non-repeating decimal

15. Rational; quotient of two integers

17. Irrational; non-terminating, non-repeating decimal

19. $\sqrt{64}=8$

21. $\sqrt{100}=10$

23. $-\sqrt{169}=-13$

25. $-\sqrt{225}=-15$

27. $-\sqrt{100}=-10$

29. 1, rational, integer, natural

31. $\sqrt{49}=7$, rat'l, integer., nat'l

33. irrational

35. rational

37. rational

39. $\sqrt{18}=\sqrt{2}\sqrt{9}=3\sqrt{2}$

41. $\sqrt{48}=\sqrt{3}\sqrt{16}=4\sqrt{3}$

43. $\sqrt{63}=\sqrt{9}\sqrt{7}=3\sqrt{7}$

45. $\sqrt{80}=\sqrt{16}\sqrt{5}=4\sqrt{5}$

47. $\sqrt{162}=\sqrt{81}\sqrt{2}=9\sqrt{2}$

49. $2\sqrt{6}+5\sqrt{6}=(2+5)\sqrt{6}=7\sqrt{6}$

51.

$5\sqrt{12}-\sqrt{75}=5\sqrt{4}\sqrt{3}-\sqrt{25}\sqrt{3}$

$=5\cdot2\sqrt{3}-5\sqrt{3}=10\sqrt{3}-5\sqrt{3}=5\sqrt{3}$

53.

$4\sqrt{12}-7\sqrt{27}=4\sqrt{4}\sqrt{3}-7\sqrt{9}\sqrt{3}$

$=4\cdot2\sqrt{3}-7\cdot3\sqrt{3}=8\sqrt{3}-21\sqrt{3}$

$=-13\sqrt{3}$

55.

$5\sqrt{3}+7\sqrt{12}-3\sqrt{75}$

$=5\sqrt{3}+7\cdot2\sqrt{3}-3\cdot5\sqrt{3}$

$=5\sqrt{3}+14\sqrt{3}-15\sqrt{3}$

$=(5+14-15)\sqrt{3}=4\sqrt{3}$

57.

$\sqrt{8}-3\sqrt{50}+9\sqrt{32}$

$=2\sqrt{2}-3\cdot5\sqrt{2}+9\cdot4\sqrt{2}$

$=2\sqrt{2}-15\sqrt{2}+36\sqrt{2}$

$=(2-15+36)\sqrt{2}=23\sqrt{2}$

59.

$\sqrt{2}\cdot\sqrt{8}=\sqrt{2}\sqrt{4}\sqrt{2}$

$=2\sqrt{2}\sqrt{2}=2\sqrt{4}$

$=2\cdot2=4$

61. $\sqrt{6}\cdot\sqrt{10}=\sqrt{2}\sqrt{3}\sqrt{2}\sqrt{5}$

$=\sqrt{4}\sqrt{15}=2\sqrt{15}$

63. $\sqrt{10}\cdot\sqrt{20}=\sqrt{200}$

$=\sqrt{100}\cdot\sqrt{2}=10\sqrt{2}$

65. $\dfrac{\sqrt{8}}{\sqrt{4}}=\sqrt{2}$

67. $\dfrac{\sqrt{72}}{\sqrt{8}}=\sqrt{9}=3$

69. $\dfrac{1}{\sqrt{2}}=\dfrac{1}{\sqrt{2}}\dfrac{\sqrt{2}}{\sqrt{2}}=\dfrac{1\sqrt{2}}{\sqrt{4}}=\dfrac{\sqrt{2}}{2}$

71. $\dfrac{\sqrt{3}}{\sqrt{7}}\cdot\dfrac{\sqrt{7}}{\sqrt{7}}\cdot\dfrac{\sqrt{21}}{7}$

73. $\dfrac{\sqrt{20}}{\sqrt{3}}=\dfrac{\sqrt{20}}{\sqrt{3}}\dfrac{\sqrt{3}}{\sqrt{3}}=\dfrac{\sqrt{60}}{\sqrt{9}}$

$=\dfrac{\sqrt{4}\sqrt{15}}{3}=\dfrac{2\sqrt{15}}{3}$

75. $\dfrac{\sqrt{9}}{\sqrt{2}}\cdot\dfrac{\sqrt{2}}{\sqrt{2}}=\dfrac{3\sqrt{2}}{2}$

77. $\dfrac{\sqrt{10}}{\sqrt{6}}\cdot\dfrac{\sqrt{6}}{\sqrt{6}}=\dfrac{\sqrt{60}}{6}$

$=\dfrac{2\sqrt{15}}{6}=\dfrac{\sqrt{15}}{3}$

79. $\sqrt{7}$ is between 2 and 3 since $\sqrt{7}$ is between $\sqrt{4}=2$ and $\sqrt{9}=3$. $\sqrt{7}$ is between 2.5 and 3 since 7 is closer to 9 than to 4. Using a calculator $\sqrt{7}\approx2.6$.

81. $\sqrt{107}$ is between 10 and 11 since $\sqrt{107}$ is between $\sqrt{100}=10$ and $\sqrt{121}=11$. $\sqrt{107}$ is between 10 and 10.5 since 107 is closer to 100 than to 121. Using a calculator $\sqrt{107}\approx10.3$.

83. $\sqrt{170}$ is between 13 and 14 since $\sqrt{170}$ is between $\sqrt{169}=13$ and $\sqrt{196}=14$. $\sqrt{170}$ is between 13 and 13.5 since 170 is closer to 169 than to 196. Using a calculator $\sqrt{170}\approx13.04$.

85. False. $\sqrt{p}$ is an irrational number for any prime number p.

87. True

89. False. The result may be a rational number or an irrational number.

91. $\sqrt{2}+(-\sqrt{2})=0$

93. $\sqrt{2}\cdot\sqrt{3}=\sqrt{6}$

95. No. $\sqrt{5}\neq2.236$ since $\sqrt{5}$ is an irrational number and 2.236 is a rational number.

97. No. 3.14 and $\dfrac{22}{7}$ are rational numbers, π is an irrational number.

99. $\sqrt{4\cdot16}=\sqrt{4}\sqrt{16}$

$\sqrt{64}\ =\ 2\cdot4$

$8\ =\ \ 8$

101. a) $s = \sqrt{\dfrac{4}{0.04}} = \sqrt{100} = 10$ mph

 b) $s = \sqrt{\dfrac{16}{0.04}} = \sqrt{400} = 20$ mph

 c) $s = \sqrt{\dfrac{64}{0.04}} = \sqrt{1600} = 40$ mph

 d) $s = \sqrt{\dfrac{256}{0.04}} = \sqrt{6400} = 80$ mph

103. a) You cannot. $\sqrt{\left(\dfrac{100}{97}\right)^2}$ is rational, but you won't see a repeat unless your calculator shows 97 places.

 b) Using a calculator, $\sqrt{0.04} = 0.2$ a terminating decimal and thus it is rational.

 c) Using a calculator, $\sqrt{0.07} = 0.264575131\ldots$, thus it is irrational.

105. a) $\left(44 \div \sqrt{4}\right) \div \sqrt{4} = \left(44 \div 2\right) \div 2 = 22 \div 2 = 11$

 b) $\left(44 \div 4\right) + \sqrt{4} = 11 + 2 = 13$

 c) $4 + 4 + 4 + \sqrt{4} = 12 + 2 = 14$

 d) $\sqrt{4}\left(4 + 4\right) + \sqrt{4} = 2(8) + 2 = 16 + 2 = 18$

Exercise Set 5.5

1. The set of real numbers is the union of the rational numbers and the irrational numbers.

3. If the given operation is preformed on any two elements of the set and the result is an element of the set, then the set is <u>closed</u> under the given operation.

5. The order in which two numbers are added does not make a difference in the result.   Ex. a+b = b+a

7. The associative property of multiplication states that when multiplying three real numbers, parentheses may be placed around any two adjacent numbers.   Ex. $(2 \bullet 3) \bullet 4 = 2 \bullet (3 \bullet 4)$.

9. Closed. The sum of two natural numbers is a natural number.

11. Not closed. (i.e. $3 \div 5 = \dfrac{3}{5} = 0.6$ is not a natural number).

13. Closed. The difference of two integers is an integer.

15. Not closed. (i.e. $2 \div 5 = \dfrac{2}{5} = 0.4 = 0.4$ is not an integer).

17. Closed

19. Closed

21. Not closed

23. Not closed

25. Closed

27. Not closed

29. Commutative property. The order is changed from $(x) + (3+4) = (3+4) + x$.

31. $(-4) \bullet (-5) = 20 = (-5) \bullet (-4)$

33. No. $6 \div 3 = 2$, but $3 \div 6 = \dfrac{1}{2}$

35. $[(-3) \bullet (-5]) \bullet (-7) = (15) \bullet (-7) = -105$

 $(-3) \bullet [(-5]) \bullet (-7)] = (-3) \bullet (35) = -105$

37. No.

 $(8 \div 4) \div 2 = 2 \div 2 = 1$, but $8 \div (4 \div 2) = 8 \div 2 = 4$

39. No. $(8 \div 4) \div 2 = 2 \div 2 = 1$,
    but $8 \div (4 \div 2) = 8 \div 2 = 4$

41. $24 + 7 = 7 + 24$
    Commutative property of addition

43. $(7 \cdot 4) \cdot 5 = 7 \cdot (4 \cdot 5)$
    Associative property of multiplication

45. $(24 + 7) + 3 = 24 + (7 + 3)$
    Associative property of addition

47. $\sqrt{3} \cdot 7 = 7 \cdot \sqrt{3}$
    Commutative property of multiplication

49. $8(7 + \sqrt{2}) = 8 \cdot 7 + 8 \cdot \sqrt{2}$
    Distributive property

51. Commutative property of addition

53. Distributive property

55. Commutative property of addition

57. $2(c + 7) = 2c + 14$

59. $\frac{2}{3}(x - 6) = \frac{2}{3}x - \frac{12}{3} = \frac{2}{3}x - 4$

61. $6\left(\frac{x}{2} + \frac{2}{3}\right) = \frac{6x}{2} + \frac{12}{3} = 3x + 4$

63. $32\left(\frac{1}{16}x - \frac{1}{32}\right) = \frac{32x}{16} - \frac{32}{32} = 2x - 1$

65. $3(5 - \sqrt{5}) = 15 - 3\sqrt{5}$

67. $\sqrt{2}(\sqrt{2} + \sqrt{3}) = \sqrt{4} + \sqrt{6} = 2 + \sqrt{6}$

69. a) Distributive property
    b) Associative property of addition

71. a) Distributive property
    b) Associative property of addition;
    c) Commutative property of addition
    d) Associative property of addition

73. a) Distributive property
    b) Commutative property of addition;
    c) Associative property of addition
    d) Commutative property of addition

75. Yes. You can either lock your door first or put on your seat belt first.

77. No. The clothes must be washed first before being dried.

79. Yes. Can be done in either order; either fill the car with gas or wash the windshield

81. Yes. The order of events does not matter.

83. Yes. The order does not matter.

85. Yes. The order does not matter

87. Yes. The final result will be the same regardless of the order of the events.

89. Baking pizzelles: mixing eggs into the batter, or mixing sugar into the batter.; Yard work: mowing the lawn, or trimming the bushes

91. No. $0 \div a = 0$ but $a \div 0$ is undefined.

92. a) No. (Man eating) tiger is a tiger that eats men, and man (eating tiger) is a man that is eating a tiger.
    b) No. (Horse riding) monkey is a monkey that rides a horse, and horse (riding monkey) is a horse that rides a monkey.
    c) Answers will vary.

Exercise Set 5.6

1. 2 is the base and 3 is the exponent.

3. a) If m and n are natural numbers and a is any real number, then $a^m a^n = a^{m+n}$

    b) $2^3 \cdot 2^4 = 2^{3+4} = 2^7 = 128$

5. a) If a is any real number except 0, then $a^0 = 1$.

    b) $7^0 = 1$

7. a) If m and n are natural numbers and a is any real number, then $\left(a^m\right)^n = a^{m \cdot n}$

    b) $\left(3^2\right)^4 = 3^{2 \cdot 4} = 3^8 = 6561$

9. a) Since 1 raised to any exponent equals $+1$, then $-1^{500} = (-1)\left(1^{500}\right) = (-1)(1) = -1$

    b) Since -1 raised to an even exponent equals 1, then number $(-1)^{500} = \left((-1)^2\right)^{250} = (1)^{250} = 1$

    c) In -1^{501}, -1 is not raised to the 501st power, but $+1$ is; so $-1^{501} = (-1)\left(1^{501}\right) = (-1)(1) = -1$

    d) Since -1 is raised to an odd exponent is -1, then $(-1)^{501} = -1$

11. a) If the exponent is positive, move the decimal point in the number to the right the same number of places as the exponent adding zeros where necessary. If the exponent is negative, move the decimal point in the number to the left the same number of places as the exponent adding zeros where necessary.

    b) $5.76 \times 10^{-4} = 0.000576$

13. $5^2 = 5 \cdot 5 = 25$

15. $(-2)^4 = (-2) \cdot (-2) \cdot (-2) \cdot (-2) = 16$

17. $-3^2 = -(3) \cdot (3) = -9$

19. $\left(\dfrac{2}{3}\right)^2 = \left(\dfrac{2}{3}\right)\left(\dfrac{2}{3}\right) = \dfrac{4}{9}$

21. $(-5)^2 = (-5) \cdot (-5) = 25$

23. $2^3 \cdot 3^2 = (2) \cdot (2) \cdot (2) \cdot (3) \cdot (3) = 72$

25. $\dfrac{5^7}{5^5} = 5^{7-5} = 5^2 = 5 \cdot 5 = 25$

27. $\dfrac{7}{7^3} = 7^{1-3} = 7^{-2} = \dfrac{1}{7^2} = \dfrac{1}{7 \cdot 7} = \dfrac{1}{49}$

29. $(-13)^0 = 1$

31. $3^4 = (3)(3)(3)(3) = 81$

33. $3^{-2} = \dfrac{1}{3^2} = \dfrac{1}{9}$

35. $\left(2^3\right)^4 = 2^{3 \cdot 4} = 2^{12} = 4096$

37. $\dfrac{11^{25}}{11^{23}} = 11^{25-23} = 11^2 = 121$

39. $(-4)^2 = (-4) \cdot (-4) = 16$

41. $-(4)^2 = -(4) \cdot (4) = -16$

43. $\left(2^2\right)^{-3} = 2^{2(-3)} = 2^{-6} = \dfrac{1}{2^6} = \dfrac{1}{64}$

45. $231000 = 2.31 \times 10^5$

47. $15 = 1.5 \times 10^1$

49. $0.56 = 5.6 \times 10^{-1}$

51. $19000 = 1.9 \times 10^4$

53. $0.000186 = 1.86 \times 10^{-4}$

55. $0.00000423 = 4.23 \times 10^{-6}$

57. $711 = 7.11 \times 10^2$

59. $0.153 = 1.53 \times 10^{-1}$

61. $2.3 \times 10^3 = 2300$

63. $3.901 \times 10^{-3} = 0.003901$

65. $8.62 \times 10^{-5} = 0.0000862$

67. $3.12 \times 10^{-1} = 0.312$

69. $9.0 \times 10^6 = 9000000$

71. $2.31 \times 10^2 = 231$

73. $3.5 \times 10^4 = 35000$

75. $1.0 \times 10^4 = 10000$

77. $\left(2.0 \times 10^3\right)\left(4.0 \times 10^2\right) = 8.0 \times 10^5 = 800000$

79. $\left(5.1 \times 10^1\right)\left(3.0 \times 10^{-4}\right) = 15.3 \times 10^{-3} = 0.0153$

81. $\dfrac{6.4 \times 10^5}{2.0 \times 10^3} = 3.2 \times 10^2 = 320$

83. $\dfrac{8.4 \times 10^{-6}}{4.0 \times 10^{-3}} = 2.1 \times 10^{-3} = 0.0021$

85. $\dfrac{4.0 \times 10^5}{2.0 \times 10^4} = 2.0 \times 10^1 = 20$

87. $\begin{aligned}(300000)(2000000) &= \left(3.0 \times 10^5\right)\left(2.0 \times 10^6\right) \\ &= 6.0 \times 10^{11}\end{aligned}$

89. $\left(3.0 \times 10^{-3}\right)\left(1.5 \times 10^{-4}\right) = 4.5 \times 10^{-7}$

91. $\dfrac{1.4 \times 10^6}{7.0 \times 10^2} = 0.2 \times 10^4 = 2.0 \times 10^3$

93. $\dfrac{4.0 \times 10^{-5}}{2.0 \times 10^2} = 2.0 \times 10^{-7}$

95. $\dfrac{1.5 \times 10^5}{5.0 \times 10^{-4}} = 0.3 \times 10^9 = 3.0 \times 10^8$

97. $8.3 \times 10^{-4},\ 3.2 \times 10^{-1},\ 4.6,\ 5.8 \times 10^5$

99. 8.3×10^{-5}; 0.00079; 4.1×10^3; $40,000$; Note: $0.00079 = 7.9 \times 10^{-4}$, $40,000 = 4 \times 10^4$

101. $\dfrac{\$10.1432 \times 10^{12}}{285.0 \times 10^6} = 0.03559017548 \times 10^6$ a) $\$35,590.18$ b) 3.559018×10^4 GDP/person

103. $\dfrac{7.69 \times 10^{33}}{36.6 \times 10^{12}} = 0.2101092896 \times 10^{21}$

    a) $210,109,000,000,000,000,000$ seconds   b) 2.1011×10^{20} seconds

105. $t = \dfrac{d}{r} = \dfrac{4.5 \times 10^8}{2.5 \times 10^4} = 1.8 \times 10^4$ a) $18,000$ hrs.   b) 1.8×10^4 hrs

107. $(500,000)(40,000,000,000) = (5 \times 10^5)(4 \times 10^{10}) = 20 \times 10^{15} = 2 \times 10^{16}$
 a) 20,000,000,000,000,000 drops   b) 2.0 x 10^{16} drops

109. $\dfrac{4.5 \times 10^9}{2.5 \times 10^5} = 1.8 \times 10^4$ a) 18,000 times   b) 1.8 x 10^4 times

111. a) $\dfrac{\$4.41 \times 10^{12}}{258.0 \times 10^6} = 0.01709302 \times 10^6$; \$17,093.02   b) \$23,082.19 − 17,093.02 = \$5989.17

113. a) (0.60) (1,200,000,000) = \$720,000,000   b) (0.25) (1,200,000,000) = \$300,000,000
 c) (0.10) (1,200,000,000) = \$120,000,000   d) (0.05) (1,200,000,000) = \$60,000,000

115. 1,000 times, since 1 meter = 10^3 millimeters = 1,000 millimeters

117. $\dfrac{2 \times 10^{30}}{6 \times 10^{24}} = 0.\overline{3} \times 10^6 = 300,000$ times

119. $\dfrac{897,000,000,000,000,000}{3,900,000,000,000} = \dfrac{8.97 \times 10^{17}}{3.9 \times 10^{12}} = 2.3 \times 10^5 = 230,000$ seconds or about 2.66 days

121. a) $(1.86 \times 10^5$ mi/sec) (60 sec/min) (60 min/hr) (24 hr/day) (365 days/yr) (1 yr)
 $= (1.86 \times 10^5)(6 \times 10^1)(6 \times 10^1)(2.4 \times 10^1)(3.65 \times 10^2) = 586.5696 \times 10^{10} \approx 5.87 \times 10^{12}$ miles

 b) $t = \dfrac{d}{r} = \dfrac{9.3 \times 10^7}{1.86 \times 10^5} = 5.0 \times 10^2 = 500$ seconds or 8 min. 20 sec.

Exercise Set 5.7

1. A sequence is a list of numbers that are related to each other by a given rule. One example is 2, 4, 6, 8....

3. a) An arithmetic sequence is a sequence in which each term differs from the preceding term by a constant amount. One example is 1, 4, 7, 10,....
 b) A geometric sequence is one in which the ratio of any term to the term that directly precedes it is a constant. One example is 1, 3, 9, 27,....

5. a) $a_n = n^{th}$ term of the sequence   b) $a_1 = $ first term of a sequence   c) d = common difference in a sequence
 d) $s_n = $ the sum of the first n terms of the arithmetic sequence

7. $a_1 = 3, d = 2$ 3, 5, 7, 9, 11

9. $a_1 = -5, d = 3$ -5, -2, 1, 4, 7

11. 5, 3, 1, − 1, − 3

13. 1/2, 1, 3/2, 2, 5/2

15. $a_6 = 2 + (6-1)3 = 2 + 15 = 17$

17. $a_{10} = -5 + (10-1)(2) = -5 + 18 = 13$

19. $a_{20} = \dfrac{4}{5} + (19)(-1) = \dfrac{4}{5} - 19 = \dfrac{4}{5} - \dfrac{95}{5} = -\dfrac{91}{5}$

21. $a_{11} = 4 + (10)\left(\dfrac{1}{2}\right) = 4 + 5 = 9$

23. $a_n = n$ $a_n = 1 + (n - 1)1 = 1 + n - 1 = n$

25. $a_n = 2n$ $a_n = 2 + (n-1)2 = 2 + 2n - 2 = 2n$

27. $a_n = \dfrac{-5}{3} + (n-1)\left(\dfrac{1}{3}\right) = \dfrac{-5}{3} + \dfrac{1}{3}n - \dfrac{1}{3} = \dfrac{1}{3}n - 2$

29. $a_n = -3 + (n-1)\left(\dfrac{3}{2}\right) = -3 + \dfrac{3}{2}n - \dfrac{3}{2} = \dfrac{3}{2}n - \dfrac{9}{2}$

31. $s_n = \dfrac{n(a_1 + a_n)}{2} = \dfrac{50(1+50)}{2} = \dfrac{50(51)}{2}$
$= (25)(51) = 1275$

33. $s_n = \dfrac{50(1+99)}{2} = \dfrac{50(100)}{2} = (25)(100) = 2500$

35. $s_8 = \dfrac{8(11 + (-24))}{2} = \dfrac{8 \cdot (-13)}{2} = -52$

37. $s_8 = \dfrac{8\left(\dfrac{1}{2} + \dfrac{29}{2}\right)}{2} = \dfrac{8 \cdot \left(\dfrac{30}{2}\right)}{2} = \dfrac{8 \cdot 15}{2} = 60$

39. $a_1 = 3$, $r = 2$, $a_n = a_1 r^{n-1} = 3(2)^{n-1}$
3, 6, 12, 24, 48

41. $a_1 = 2$, $r = -2$,
2, -4, 8, -16, 32

43. –3, 3, –3, 3, –3

45. $-16, 8, -4, 2, -1$

47. $a_6 = 3(4)^5 = (3)(1024) = 3072$

49. $a_3 = 3\left(\dfrac{1}{2}\right)^2 = 3\left(\dfrac{1}{4}\right) = \dfrac{3}{4}$

51. $a_5 = \left(\dfrac{1}{2}\right) \cdot 2^4 = \left(\dfrac{1}{2}\right)(16) = 8$

53. $a_{10} = (-2)(3)^9 = -39,366$

55. 1, 2, 4, 8 $a_n = 1(2)^{n-1} = 2^{n-1}$

57. 3, -3, 3, -3 $a_n = 3(-1)^{n-1}$

59. $a_n = a_1 r^{n-1} = \left(\dfrac{1}{4}\right)(2)^{n-1}$

61. $a_n = a_1 r^{n-1} = (9)\left(\dfrac{1}{3}\right)^{n-1}$

63. $s_4 = \dfrac{a_1(1-r^4)}{1-r} = \dfrac{3(1-2^4)}{1-2} = \dfrac{3(-15)}{-1} = 45$

65. $s_7 = \dfrac{a_1(1-r^7)}{1-r} = \dfrac{5(1-4^7)}{1-4} = \dfrac{5(-16383)}{-3}$
$= 27,305$

67. $s_{11} = \dfrac{a_1(1-r^{11})}{1-r} = \dfrac{-7(1-3^{11})}{1-3} = \dfrac{-7(-177146)}{-2}$
$= -620,011$

69. n = 15, a_1 = -1, r = -2
$s_{13} = \dfrac{(-1)(1-(-2)^{15})}{1-(-2)} = \dfrac{(-1)(1+32768)}{3}$
$= \dfrac{(-1)(32769)}{3} = -10923$

71.
$s_{100} = \dfrac{(100)(1+100)}{2} = \dfrac{(100)(101)}{2} = 50(101) = 5050$

73.
$s_{100} = \dfrac{(100)(1+199)}{2} = \dfrac{(100)(200)}{2} = 50(200) = 10000$

75. a) Using the formula $a_n = a_1 + (n-1)d$, we get
$a_8 = 20,200 + (8-1)(1200) = \$28,600$

b) $\dfrac{8(20200 + 28600)}{2} = \dfrac{8(48800)}{2} = \$195,200$

77. $a_{11} = 72 + (10)(-6) = 72 - 60 = 12$ in.

79. 1, 2, 3,.... n=31

$$s_{31} = \frac{31(1+31)}{2} = \frac{31(32)}{2} = 31(16) = 496 \text{ PCs}$$

81. $a_6 = 200(0.8)^6 = 200(0.262144)^1 = 52.4288$ g

83. $a_{15} = 20,000(1.06)^{14} = \$45,218$

85. This is a geometric sequence where $a_1 = 2000$ and $r = 3$. In ten years the stock will triple its value 5 times.
$a_6 = a_1r^{6-1} = 2000(3)^5 = \$486,000$

87. $\dfrac{82[1-(1/2)^6]}{1-(1/2)} = \dfrac{82[1-(1/64)]}{1/2} = \dfrac{82}{1} \bullet \dfrac{63}{64} \bullet \dfrac{2}{1} = 161.4375$

89. 12, 18, 24, ... ,1608 is an arithmetic sequence with $a_1 = 12$ and $d = 6$. Using the expression for the nth term
of an arithmetic sequence $a_n = a_1 + (n-1)d$ or $1608 = 12 + (n-1)6$ and dividing both sides by 6 gives
$268 = 2 + n - 1$ or $n = 267$

91. The total distance is 30 plus twice the sum of the terms of the geometric sequence having $a_1 = (30)(0.8) = 24$
and $r = 0.8$. Thus $s_5 = \dfrac{24[1-(0.8)^5]}{(1-0.8)} = \dfrac{24[1-0.32768]}{0.2} = \dfrac{24(0.67232)}{0.2} = 80.6784$.

So the total distance is $30 + 2(80.6784) = 191.3568$ ft.

92. The sequence of bets during a losing streak is geometric.

a) $a_6 = a_1r^{n-1} = 1(2)^{6-1} = 1(32) = \$32$ $\quad s_5 = \dfrac{a_1(1-r^n)}{1-r} = \dfrac{1(1-2^5)}{1-2} = \dfrac{-31}{-1} = \$31$

b) $a_6 = a_1r^{n-1} = 10(2)^{6-1} = 10(32) = \$320$ $\quad s_5 = \dfrac{a_1(1-r^n)}{1-r} = \dfrac{10(1-2^5)}{1-2} = \dfrac{10(-31)}{-1} = \$310$

c) $a_{11} = a_1r^{n-1} = 1(2)^{11-1} = 1(1024) = \$1,024$ $\quad s_{10} = \dfrac{a_1(1-r^n)}{1-r} = \dfrac{1(1-2^{10})}{1-2} = \dfrac{1(-1023)}{-1} = \$1,023$

d) $a_{11} = a_1r^{n-1} = 10(2)^{11-1} = 10(1024) = \$10,240$ $\quad s_{10} = \dfrac{a_1(1-r^n)}{1-r} = \dfrac{10(1-2^{10})}{1-2} = \dfrac{10(-1023)}{-1} = \$10,230$

e) If you lose too many times in a row, then you will run out of money.

Exercise Set 5.8

1. Begin with the numbers 1, 1, then add 1 and 1 to get 2 and continue to add the previous two numbers in the
sequence to get the next number in the sequence.

3. a) Golden number $= \dfrac{\sqrt{5}+1}{2}$

b) 1.618 = golden ratio   When a line segment AB is divided at a point C, such that the ratio of the whole,
AB, to the larger part, AC, is equal to the ratio of the larger part, AC, to the smaller part, CB, then each
of the ratios $\dfrac{AB}{AC}$ and $\dfrac{AC}{CB}$ is known as the golden ratio.

c) The golden proportion is: $\dfrac{AB}{AC} = \dfrac{AC}{CB}$

d) The golden rectangle: $\dfrac{L}{W} = \dfrac{a+b}{a} = \dfrac{a}{b} = \dfrac{\sqrt{5}+1}{2} = $ golden number

5. a) Flowering head of a sunflower b) Great Pyramid

7. a) $\dfrac{\sqrt{5}+1}{2} = 1.618033989$ b) $\dfrac{\sqrt{5}-1}{2} = .618033989$

c) Differ by 1

9. $1/1 = 1, 2/1 = 2, 3/2 = 1.5, 5/3 = 1.6, 8/5 = 1.6, 13/8 = 1.625, 21/13 = 1.6154, 34/21 = 1.619, 55/34 = 1.6176$
   $89/55 = 1.61818$. The consecutive ratios alternate increasing then decreasing about the golden ratio.

11.

Fib. No.	prime factors	Fib. No.	prime factors
1	-------	34	$2 \bullet 17$
1	-------	55	$5 \bullet 11$
2	prime	89	prime
3	prime	144	$2^4 \bullet 3^2$
5	prime	233	prime
8	2^3	377	$13 \bullet 29$
13	prime	610	$2 \bullet 5 \bullet 61$

13. If 5 is selected the result is $2(5) - 8 = 10 - 8 = 2$ which is the second number preceding 5.

15. Answers will vary.   17. Answers will vary   19. Answers will vary.

21. Answers will vary.

23. Fibonacci type; $11 + 18 = 29$ $18 + 29 = 47$

25. Not Fibonacci. Each term is not the sum of the two preceding terms.

27. Fibonacci type; $40 + 65 = 105$; $65 + 105 = 170$ 29. Fibonacci type; $-1 + 0 = -1$; $0 + (-1) = -1$

31. a) If 6 and 10 are selected the sequence is 6, 10, 16, 26, 42, 68, 110, …
    b) $10/6 = 1.666, 16/10 = 1.600, 26/16 = 1.625, 42/26 = 1.615, 68/42 = 1.619, 110/68 = 1.618,$ …

33. a) If 5, 8, and 13 are selected the result is $8^2 - (5)(13) = 64 - 65 = -1$.

    b) If 21, 34, and 55 are selected the result is $34^2 - (21)(55) = 1156 - 1155 = 1$.

    c) The square of the middle term of three consecutive terms in a Fibonacci sequence differs from the
       product of the 1st and 2nd term by 1.

35. a) Lucas sequence: 1, 3, 4, 7, 11, 18, 29, 47, …   b) $8 + 21 = 29$; $13 + 34 = 47$
    c) The first column is a Fibonacci-type sequence.

37. $\dfrac{(a+b)}{a} = \dfrac{a}{b}$ Let $x = \dfrac{a}{b}$ $\dfrac{b}{a} = \dfrac{1}{x}$ $1 + \dfrac{b}{a} = \dfrac{a}{b}$ $1 + \dfrac{1}{x} = x$ multiply by x $x\left(1 + \dfrac{1}{x}\right) = x(x)$

$x + 1 = x^2$ $x^2 - x - 1 = 0$

Solve for x using the quadratic formula, $x = \dfrac{-b \pm \sqrt{b^2 - 4ac}}{2a} = \dfrac{1 \pm \sqrt{1 - 4(1)(-1)}}{2(1)} = \dfrac{1 \pm \sqrt{5}}{2}$

39. Answers will vary.   {5, 12, 13}   {16, 30, 34}   {105, 208, 233}   {272, 546, 610}

Review Exercises

1. Use the divisibility rules in section 5.1.
   670,920 is divisible by 2, 3, 4, 5, 6, and 9.

2. Use the divisibility rules in section 5.1.
   400,644 is divisible by 2, 3, 4, 6, and 9

3.
```
2 | 252
2 | 126
3 | 63
3 | 21
    7
```
$252 = 2^2 \cdot 3^2 \cdot 7$

4.
```
5 | 385
7 | 77
    11
```
$385 = 5 \cdot 7 \cdot 11$

5.
```
2 | 840
2 | 420
2 | 210
5 | 105
3 | 21
    7
```
$840 = 2^3 \cdot 3 \cdot 5 \cdot 7$

6.
```
2 | 882
3 | 441
3 | 147
7 | 49
    7
```
$882 = 2 \cdot 3^2 \cdot 7^2$

7.
```
2 | 1452
2 | 726
3 | 363
11 | 121
     11
```
$1452 = 2^2 \cdot 3 \cdot 11^2$

8. $15 = 3 \cdot 5, \quad 60 = 2^2 \cdot 3 \cdot 5$
   gcd $= 15$ lcm $= 60$

9. $63 = 3 \cdot 3 \cdot 5, \quad 108 = 3 \cdot 4 \cdot 9$
   gcd $= 9$;   lcm $= 756$

10. $45 = 3^2 \cdot 5, \quad 250 = 2 \cdot 5^3; \quad$ gcd $= 5$; lcm $= 2 \cdot 3^2 \cdot 5^3 = 2250$

11. $840 = 2^3 \cdot 3 \cdot 5 \cdot 7, \quad 320 = 2^6 \cdot 5; \quad$ gcd $= 2^3 \cdot 5 = 40$; lcm $= 2^6 \cdot 3 \cdot 5 \cdot 7 = 6720$

12. $60 = 2^2 \cdot 3 \cdot 5, \quad 40 = 2^3 \cdot 5, \quad 96 = 2^5 \cdot 3; \quad$ gcd $= 2^2 = 4$; lcm $= 2^5 \cdot 3 \cdot 5 = 480$

13. $36 = 2^2 \cdot 3^2, \quad 108 = 2^2 \cdot 3^3, \quad 144 = 2^4 \cdot 3^2; \quad$ gcd $= 2^2 \cdot 3^2 = 36$; lcm $= 2^4 \cdot 3^3 = 432$

14. $15 = 3 \cdot 5, \quad 9 = 3^2; \quad$ lcm $= 3^2 \cdot 5 = 45$. In 45 days the train stopped in both cities.

15. $-2 + 5 = 3$

16. $4 + (-7) = -3$

17. $4 - 8 = 4 + (-8) = -4$

18. $(-2) + (-4) = -6$

19. $-5 - 4 = -5 + (-4) = -9$

20. $-3 - (-6) = -3 + 6 = 3$

21. $(-3 + 7) - 4 = 4 + (-4) = 0$

22. $-1 + (9 - 4) = -1 + 5 = 4$

23. $(-3)(-11) = 33$

24. $(-4)(9) = -36$

25. $14(-4) = -56$

26. $-35/-7 = 5$

27. $12/-6 = -2$

28. $[8 \div (-4)](-3) = (-2)(-3) = 6$

29. $[(-4)(-3)] \div 2 = 12 \div 2 = 6$

30. $[-30 \div (10)] \div (-1) = -3 \div (-1) = 3$

31. $3/10 = 0.3$

32. $3/5 = 0.6$

33. $15/40 = 3/8 = 0.375$

34. $13/4 = 3.25$

35. $3/7 = 0.\overline{428571}$

36. $7/12 = 0.58\overline{3}$

37. $3/8 = 0.375$

38. $7/8 = 0.875$

39. $5/7 = 0.\overline{714285}$

40. $0.225 = \dfrac{225}{1000} = \dfrac{45}{200} = \dfrac{9}{40}$

41. $4.5 = 4\dfrac{5}{10} = \dfrac{45}{10} = \dfrac{9}{2}$

42. $0.6666\ldots$ $10n = 6.6666\ldots$

   $10n = 6.\overline{6}$ $\dfrac{9n}{9} = \dfrac{6}{9}$

   $\underline{-n = 0.\overline{6}}$

   $9n = 6.0$ $n = \dfrac{2}{3}$

43. $2.373737\ldots$ $100n = 237.373737\ldots$
$100n = 237.\overline{37}$
$\underline{-\ n = \ \ \ 2.\overline{37}}$ $\dfrac{99n}{99} = \dfrac{235}{99} = n$
$99n = 235.00$

44. $0.083 = \dfrac{83}{1000}$

45. $0.0042 = \dfrac{42}{10000} = \dfrac{21}{5000}$

46. $2.344444\ldots$ $100n = 234.444444\ldots$
$100n = 234.\overline{4}$
$\underline{-10n = \ \ 23.\overline{4}}$ $\dfrac{90n}{90} = \dfrac{211}{90} = n$
$90n = 211.00$

47. $2\dfrac{5}{7} = \dfrac{19}{7}$

48. $4\dfrac{1}{6} = \dfrac{25}{6}$

49. $-3\,{}^{1}\!/_{4} = \dfrac{((-3)(4))-1}{4} = \dfrac{-13}{4}$

50. $-35\,{}^{3}\!/_{8} = \dfrac{((-35)(8))-3}{8} = \dfrac{-283}{8}$

51. $\dfrac{11}{5} = \dfrac{2\bullet 5+1}{5} = 2\dfrac{1}{5}$

52. $\dfrac{27}{15} = \dfrac{1\bullet 15+12}{15} = 1\dfrac{12}{15} = 1\dfrac{4}{5}$

53. $\dfrac{-12}{7} = \dfrac{(-1)(7)-5}{7} = -1\,{}^{5}\!/_{7}$

54. $\dfrac{-136}{5} = \dfrac{(-27)(5)-1}{5} = -27\,{}^{1}\!/_{5}$

55. $\dfrac{1}{2} + \dfrac{4}{5} = \dfrac{1}{2}\bullet\dfrac{5}{5} + \dfrac{4}{5}\bullet\dfrac{2}{2} = \dfrac{5}{10} + \dfrac{8}{10} = \dfrac{13}{10}$

56. $\dfrac{7}{8} - \dfrac{3}{4} = \dfrac{7}{8} - \dfrac{3}{4}\bullet\dfrac{2}{2} = \dfrac{7}{8} - \dfrac{6}{8} = \dfrac{1}{8}$

57. $\dfrac{1}{6} + \dfrac{5}{4} = \dfrac{1}{6}\bullet\dfrac{2}{2} + \dfrac{5}{4}\bullet\dfrac{3}{3} = \dfrac{2}{12} + \dfrac{15}{12} = \dfrac{17}{12}$

58. $\dfrac{4}{5}\bullet\dfrac{15}{16} = \dfrac{60}{80} = \dfrac{6}{8} = \dfrac{3}{4}$

59. $\dfrac{5}{9} \div \dfrac{6}{7} = \dfrac{5}{9} \div \dfrac{7}{6} = \dfrac{35}{54}$

60. $\left(\dfrac{4}{5} + \dfrac{5}{7}\right) \div \dfrac{4}{5} = \dfrac{28+25}{35}\bullet\dfrac{5}{4} = \dfrac{53}{35}\bullet\dfrac{5}{4} = \dfrac{53}{28}$

61. $\left(\dfrac{2}{3}\bullet\dfrac{1}{7}\right) \div \dfrac{4}{7} = \dfrac{2}{21}\bullet\dfrac{7}{4} = \dfrac{1}{6}$

62. $\left(\dfrac{1}{5} + \dfrac{2}{3}\right)\bullet\dfrac{3}{8} = \dfrac{3+10}{15}\bullet\dfrac{3}{8} = \dfrac{13}{15}\bullet\dfrac{3}{8} = \dfrac{13}{40}$

63. $\left(\dfrac{1}{5}\right)\left(\dfrac{2}{3}\right) + \left(\dfrac{1}{5} \div \dfrac{1}{2}\right) = \dfrac{2}{15} + \left(\dfrac{1}{5}\right)\left(\dfrac{2}{1}\right) = \dfrac{2}{15} + \dfrac{2}{5}$
$= \dfrac{2}{15} + \dfrac{6}{15} = \dfrac{8}{15}$

64. $\left(\dfrac{1}{8}\right)(17\,{}^{3}\!/_{4}) = \left(\dfrac{1}{8}\right)\left(\dfrac{71}{4}\right) = \dfrac{71}{32} = 2\,{}^{7}\!/_{32}$ teaspoons

65. $\sqrt{50} = \sqrt{25\cdot 2} = \sqrt{25}\cdot\sqrt{2} = 5\sqrt{2}$

66. $\sqrt{200} = \sqrt{100\cdot 2} = \sqrt{100}\cdot\sqrt{2} = 10\sqrt{2}$

67. $\sqrt{5} + 7\sqrt{5} = 8\sqrt{5}$

68. $\sqrt{3} - 4\sqrt{3} = -3\sqrt{3}$

69. $\sqrt{8} + 6\sqrt{2} = 2\sqrt{2} + 6\sqrt{2} = 8\sqrt{2}$

70. $\sqrt{3} - 7\sqrt{27} = \sqrt{3} - 21\sqrt{3} = -20\sqrt{3}$

71. $\sqrt{75} + \sqrt{27} = 5\sqrt{3} + 3\sqrt{3} = 8\sqrt{3}$

72. $\sqrt{3}\cdot\sqrt{6} = \sqrt{18} = \sqrt{9\cdot 2} = \sqrt{9}\cdot\sqrt{2} = 3\sqrt{2}$

73. $\sqrt{8}\cdot\sqrt{6} = \sqrt{48} = \sqrt{16\cdot 3} = \sqrt{16}\cdot\sqrt{3} = 4\sqrt{3}$

74. $\dfrac{\sqrt{18}}{\sqrt{2}} = \sqrt{\dfrac{18}{2}} = \sqrt{9} = 3$

75. $\dfrac{\sqrt{56}}{\sqrt{2}} = \sqrt{\dfrac{56}{2}} = \sqrt{28} = 2\sqrt{7}$

76. $\dfrac{4}{\sqrt{3}} \cdot \dfrac{\sqrt{3}}{\sqrt{3}} = \dfrac{4\sqrt{3}}{3}$

77. $\dfrac{\sqrt{3}}{\sqrt{5}} \cdot \dfrac{\sqrt{5}}{\sqrt{5}} = \dfrac{\sqrt{15}}{5}$

78. $3(2 + \sqrt{7}) = 6 + 3\sqrt{7}$

79. $\sqrt{3}(4 + \sqrt{6}) = 4\sqrt{3} + \sqrt{18} = 4\sqrt{3} + 3\sqrt{2}$

80. $\sqrt{3}(\sqrt{6} + \sqrt{15}) = \sqrt{18} + \sqrt{45} = 3\sqrt{2} + 3\sqrt{5}$

81. $x + 2 = 2 + x$ Commutative property of addition

82. $5 \cdot m = m \cdot 5$ Commutative property of multiplication

83. Associative property of addition

84. Distributive property

85. Commutative property of addition

86. Commutative property of addition

87. Associative property of multiplication

88. Commutative property of multiplication

89. Distributive property

90. Commutative property of multiplication

91. Natural numbers – closed for addition

92. Whole numbers – not closed for subtraction
$2 - 3 = -1$ and -1 is not a whole number.

93. Not closed; $1 \div 2$ is not an integer

94. Closed

95. Not closed; $\sqrt{2} \bullet \sqrt{2} = 2$ is not irrational

96. Not closed; $1 \div 0$ is undefined

97. $3^2 = 3 \bullet 3 = 9$

98. $3\,3^{-2} = \dfrac{1}{3^2} = \dfrac{1}{3 \bullet 3} = \dfrac{1}{9}$

99. $\dfrac{9^5}{9^3} = 9^{5-3} = 9^2 = 81$

100. $5^2 \bullet 5^1 = 5^3 = 125$

101. $7^0 = 1$

102. $4^{-3} = \dfrac{1}{4^3} = \dfrac{1}{64}$

103. $(2^3)^2 = 2^{3 \bullet 2} = 2^6 = 64$

104. $(3^2)^2 = 3^{2 \bullet 2} = 3^4 = 81$

105. $230{,}000 = 2.3 \times 10^5$

106. $0.0000158 = 1.58 \times 10^{-5}$

107. $0.00275 = 2.75 \times 10^{-3}$

108. $4{,}950{,}000 = 4.95 \times 10^6$

109. $4.3 \times 10^7 = 43{,}000{,}000$

110. $1.39 \times 10^{-4} = 0.000139$

111. $1.75 \times 10^{-4} = 0.000175$

112. $1 \times 10^5 = 100{,}000$

113. a) $(7 \times 10^3)(2 \times 10^{-5})$
$(14) \times 10^{-2} = 1.4 \times 10^{-1}$

114. a) $(4 \times 10^2)(2.5 \times 10^2)$
$(4)(2.5) \times (10^2 \bullet 10^2)$
$10.0 \times 10^4 = 1.0 \times 10^5$

115. $\dfrac{8.4 \times 10^3}{4 \times 10^2} = \dfrac{8.4}{4} \times \dfrac{10^3}{10^2} = 2.1 \times 10^1$

116. $\dfrac{1.5 \times 10^{-3}}{5 \times 10^{-4}} = \dfrac{1.5}{5} \times \dfrac{10^{-3}}{10^{-4}} = 0.3 \times 10^1 = 3.0 \times 10^0$

117. a) $(4{,}000{,}000)(2{,}000) = (4.0 \times 10^6)(2.0 \times 10^3)$
$= (4)(2) \times 10^6 \bullet 10^3 = 8{,}000{,}000{,}000$
 b) 8.0 E 09

118. a) $(35{,}000)(0.00002) = (3.5 \times 10^4)(2.0 \times 10^{-5})$
$= (3.5)(2) \times 10^4 \bullet 10^{-5} = 7.0 \times 10^{-1} = 0.7$
 b) 7.0 E -01

119. $\dfrac{9600000}{3000} = \dfrac{9.6 \times 10^6}{3 \times 10^3} = 3.2 \times 10^3 = 3{,}200$

120. $\dfrac{0.000002}{0.0000004} = \dfrac{2 \times 10^{-6}}{4 \times 10^{-7}} = 0.5 \times 10^1 = 5.0$

121. $\dfrac{1.49\times10^{11}}{3.84\times10^{8}} = .3880208333\times10^{3} \approx 388.02$

     388 times

122. $\dfrac{20,000,000}{3,600} = \dfrac{2.0\times10^{7}}{3.6\times10^{3}}$

     $\approx 0.555556\times10^{4} = \$5,555.56$

123. Arithmetic   14, 17

124. Geometric   8, 16

125. Arithmetic $-15, -18$

126. Geometric   1/32, 1/64

127. Arithmetic   16, 19

128. Geometric $-2, 2$

     $a_8 = -6 + 7(-4) = -6 - 28 = -34$

129. 3, 7, 11, 15 $a_4 = 15$

130. $-4, -10, -14, -18, -22, -26, -30, -34$ $a_8 = -34$

131. $a_{10} = -20 + 9(5) = -20 + 45 = 25$

132. 3, 6, 12, 24, 48 $a_4 = 48$

133. $a_5 = 4(1/2)^{5-1} = 4(1/2)^4 = 4(1/16) = 1/4$

134. $a_4 = -6(2)^{4-1} = -6(2)^3 = -6(8) = -48$

135. $s_{30} = \dfrac{30(2+89)}{2} = (15)(91) = 1365$

136. $s_8 = \dfrac{8\left(-4+(-2\frac{1}{4})\right)}{2} = \dfrac{(8)(-6\frac{1}{4})}{2} = -25$

137. $s_8 = \dfrac{8(100+58)}{2} = \dfrac{(8)(158)}{2} = 632$

138. $s_{20} = \dfrac{20(0.5+5.25)}{2} = \dfrac{(20)(5.75)}{2} = 57.5$

139. $s_3 = \dfrac{5\left(1-3^4\right)}{1-3} = \dfrac{(5)(1-81)}{-2} = \dfrac{(5)(-80)}{-2} = 200$

140. $s_4 = \dfrac{2\left(1-3^4\right)}{1-3} = \dfrac{(2)(1-81)}{-2} = \dfrac{(2)(-80)}{-2} = 80$

141. $s_5 = \dfrac{3\left(1-(-2)^5\right)}{1-(-2)} = \dfrac{(3)(1+32)}{3} = \dfrac{(3)(33)}{3} = 33$

142. $s_6 = \dfrac{1\left(1-(-2)^6\right)}{1-(-2)} = \dfrac{(1)(1-64)}{3} = \dfrac{(1)(-63)}{3} = -21$

143. Arithmetic: $a_n = -3n + 10$

144. Arithmetic: $a_n = 3 + (n-1)3 = 3 + 3n - 3 = 3n$

145. Arithmetic: $a_n = -(3/2)n + (11/2)$

146. Geometric: $a_n = 3(2)^{n-1}$

147. Geometric: $a_n = 2(-1)^{n-1}$

148. Geometric: $a_n = 5(1/3)^{n-1}$

149. Yes;   13, 21   150. Yes;   17, 28   151. No   152. No

Chapter Test

1. 38,610 is divisible by: 2, 3, 5, 6, 9, 10

2.

$$
\begin{array}{r|r}
2 & 840 \\
2 & 420 \\
2 & 210 \\
5 & 105 \\
3 & 35 \\
& 7
\end{array}
$$

$840 = 2^3 \bullet 3 \bullet 5 \bullet 7$

3. $[(-6) + (-9)] + 8 = -15 + 8 = -7$

4. $-7 - 13 = -20$

5. $[(-70)(-5)] \div (8-10) = 350 \div [8 + (-10)]$
   $= 350 \div (-2) = -175$

6. $4\tfrac{5}{8} = \dfrac{(8)(4)+5}{8} = \dfrac{32+5}{8} = \dfrac{37}{8}$

7. $\dfrac{176}{9} = \dfrac{(19)(9)+5}{9} = 19\tfrac{5}{9}$

8. $\dfrac{5}{8} = 0.625$

9. $6.45 = \dfrac{645}{100} = \dfrac{129}{20}$

10. $\left(\dfrac{5}{16} \div 3\right) + \left(\dfrac{4}{5} \bullet \dfrac{1}{2}\right) = \left(\dfrac{5}{16} \bullet \dfrac{1}{3}\right) + \dfrac{4}{10}$

$= \dfrac{5}{48} + \dfrac{4}{10} = \dfrac{50}{480} + \dfrac{192}{480} = \dfrac{242}{480} = \dfrac{121}{240}$

11. $\dfrac{11}{12} - \dfrac{3}{8} = \left(\dfrac{11}{12}\right)\left(\dfrac{2}{2}\right) - \left(\dfrac{3}{8}\right)\left(\dfrac{3}{3}\right) = \dfrac{22}{24} - \dfrac{9}{24} = \dfrac{13}{24}$

12. $\sqrt{75} + \sqrt{48} = \sqrt{25}\sqrt{3} + \sqrt{16}\sqrt{3} = 5\sqrt{3} + 4\sqrt{3} = 9\sqrt{3}$

13. $\dfrac{\sqrt{2}}{\sqrt{7}} = \dfrac{\sqrt{2}}{\sqrt{7}} \bullet \dfrac{\sqrt{7}}{\sqrt{7}} = \dfrac{\sqrt{14}}{\sqrt{49}} = \dfrac{\sqrt{14}}{7}$

14. The integers are closed under multiplication since the product of two integers is always an integer.

15. Associative property of addition

16. Distributive property

17. $\dfrac{4^5}{4^2} = 4^{5-2} = 4^3 = 64$

18. $4^3 \bullet 4^2 = 4^5 = 4 \cdot 4 \cdot 4 \cdot 4 \cdot 4 = 1024$

19. $3^{-4} = \dfrac{1}{3^4} = \dfrac{1}{81}$

20. $\dfrac{7.2 \times 10^6}{9.0 \times 10^{-6}} = 0.8 \times 10^{12} = 8.0 \times 10^{11}$

21. $a_n = -4n + 2$

22. $\dfrac{11[-2 + (-32)]}{2} = \dfrac{11(-34)}{2} = -187$

23. $a_5 = 3(3)^4 = 3^5 = 243$

24. $\dfrac{3(1 - 4^5)}{1-4} = \dfrac{3(1 - 1024)}{-3} = 1023$

25. $a_n = 3 \bullet (2)^{n-1}$

26. 1, 1, 2, 3, 5, 8, 13, 21, 34, 55

CHAPTER SIX

ALGEBRA, GRAPHS, AND FUNCTIONS

Exercise Set 6.1

1. **Variables** are letters of the alphabet used to represent numbers.

3. An **algebraic expression** is a collection of variables, numbers, parentheses, and operation symbols.

   An example is $5x^2y - 11$.

5. a) Base: 4, exponent: 5
   b) Multiply 4 by itself 5 times.

7. $8 + 16 \div 4 = 8 + 4 = 12$

9. $x = 7, \ x^2 = (7)^2 = 49$

11. $x = -3, -x^2 = -(-3)^2 = -9$

13. $x = -7, -2x^3 = -2(-7)^3 = -2(-343) = 686$

15. $x = 4, x - 7 = 4 - 7 = -3$

17. $x = -2, \ -7x + 4 = -7(-2) + 4 = 14 + 4 = 18$

19. $x = -2, \ -x^2 + 5x - 13 = -(-2)^2 + 5(-2) - 13$
$$= -4 - 10 - 13 = -27$$

21. $x = \dfrac{2}{3}, \dfrac{1}{2}x^2 - 5x + 2 = \dfrac{1}{2}\left(\dfrac{2}{3}\right)^2 - 5\left(\dfrac{2}{3}\right) + 2$
$$= \dfrac{1}{2}\left(\dfrac{4}{9}\right) - \dfrac{10}{3} + 2$$
$$= \dfrac{4}{18} - \dfrac{10}{3} + 2$$
$$= \dfrac{4}{18} - \dfrac{60}{18} + \dfrac{36}{18} = -\dfrac{20}{18} = -\dfrac{10}{9}$$

23. $x = \dfrac{1}{2}, 8x^3 - 4x^2 + 7 = 8\left(\dfrac{1}{2}\right)^3 - 4\left(\dfrac{1}{2}\right)^2 + 7$
$$= 8\left(\dfrac{1}{8}\right) - 4\left(\dfrac{1}{4}\right) + 7$$
$$= 1 - 1 + 7 = 7$$

25. $x = -2, \ y = 1, \ 2x^2 + xy + 3y^2$
$$= 2(-2)^2 + (-2)(1) + 3(1)^2 = 8 - 2 + 3 = 9$$

27. $x = 3, \ y = 2, \ 4x^2 - 12xy + 9y^2$
$$= 4(3)^2 - 12(3)(2) + 9(2)^2 = 36 - 72 + 36 = 0$$

29. $7x + 3 = 23, \ x = 3$

$7(3) + 3 = 21 + 3 = 24$

$24 \neq 23, x = 3$ is not a solution.

31. $x - 3y = 0$, $x = 6$, $y = 3$

   $6 - 3(3) = 6 - 9 = -3$

   $-3 \neq 0, x = 6, y = 3$ is not a solution.

33. $x^2 + 3x - 4 = 5, x = 2$

   $(2)^2 + 3(2) - 4 = 4 + 6 - 4 = 6$

   $6 \neq 5, x = 2$ is not a solution.

35. $2x^2 + x = 28, x = -4$

   $2(-4)^2 + (-4) = 2(16) - 4 = 32 - 4 = 28$

   $28 = 28, x = -4$ is a solution.

37. $y = -x^2 + 3x - 1$, $x = 3$, $y = -1$

   $-(3)^2 + 3(3) - 1 = -9 + 9 - 1 = -1$

   $-1 = -1, x = 3, y = -1$ is a solution.

39. $d = \$175, 0.07d = 0.07(\$175) = \$12.25$

41. $x = 75, 220 + 2.75x = 220 + 2.75(75)$

   $= 220 + 206.25 = \$426.25$

43. $n = 8{,}000{,}000{,}000{,}000$

   $0.000002n = 0.000002(8{,}000{,}000{,}000{,}000)$

   $= 16{,}000{,}000 \text{ sec}$

45. $R = 2, T = 70, 0.2R^2 + 0.003RT + 0.0001T^2 = 0.2(2)^2 + 0.003(2)(70) + 0.0001(70)^2 = 0.8 + 0.42 + 0.49 = 1.71 \text{ in.}$

47.

x	y	$(x+y)^2$	$x^2 + y^2$
2	3	$5^2 = 25$	$4 + 9 = 13$
-2	-3	$(-5)^2 = 25$	$4 + 9 = 13$
-2	3	$1^2 = 1$	$4 + 9 = 13$
2	-3	$(-1)^2 = 1$	$4 + 9 = 13$

   The two expressions are not equal.

Exercise Set 6.2

1. The parts that are added or subtracted in an algebraic expression are called **terms**.

   In $3x - 2y$, the $3x$ and $-2y$ are terms.

3. The numerical part of a term is called its **numerical coefficient.**

   For the term $3x$, 3 is the numerical coefficient.

5. To **simplify** an expression means to combine like terms by using the commutative, associative, and distributive properties. Example: $12 + x + 7 - 3x = x - 3x + 12 + 7 = -2x + 19$

7. If $a = b$, then $a - c = b - c$ for all real numbers a, b, and c. Example: If $2x + 3 = 5$, then $2x + 3 - 3 = 5 - 3$.

9. If $a = b$, then $\dfrac{a}{c} = \dfrac{b}{c}$ for all real numbers a, b, and c, where $c \neq 0$. Example: If $4x = 8$ then $\dfrac{4x}{4} = \dfrac{8}{4}$.

11. A **ratio** is a quotient of two quantities. Example: $\dfrac{7}{9}$

13. Yes. They have the same variable and the same exponent on the variable.

15. $2x + 9x = 11x$

17. $5x - 3x + 12 = 2x + 12$

19. $7x + 3y - 4x + 8y = 3x + 11y$

21. $-3x + 2 - 5x = -8x + 2$

23. $2 - 3x - 2x + 1 = -5x + 3$

25. $6.2x - 8.3 + 7.1x = 13.3x - 8.3$

27. $\dfrac{1}{5}x - \dfrac{1}{3}x - 4 = \dfrac{3}{15}x - \dfrac{5}{15}x - 4 = -\dfrac{2}{15}x - 4$

29. $5x - 4y - 3y + 8x + 3 = 13x - 7y + 3$

31. $2(s+3) + 6(s-4) + 1 = 2s + 6 + 6s - 24 + 1 = 8s - 17$

33. $0.3(x+2) + 1.2(x-4) = 0.3x + 0.6 + 1.2x - 4.8$
$\qquad = 1.5x - 4.2$

35. $\dfrac{2}{3}x + \dfrac{3}{7} - \dfrac{1}{4}x = \dfrac{8}{12}x - \dfrac{3}{12}x + \dfrac{3}{7} = \dfrac{5}{12}x + \dfrac{3}{7}$

37. $0.5(2.6x - 4) + 2.3(1.4x - 5) = 1.3x - 2 + 3.22 - 11.5$
$\qquad = 4.52x - 13.5$

39.
$$y + 8 = 13$$
$$y + 8 - 8 = 13 - 8 \qquad \text{Subtract 8 from both sides of the equation.}$$
$$y = 5$$

41.
$$9 = 12 - 3x$$
$$9 - 12 = 12 - 12 - 3x \qquad \text{Subtract 12 from both sides of the equation.}$$
$$-3 = -3x$$
$$\dfrac{-3}{-3} = \dfrac{-3x}{-3} \qquad \text{Divide both sides of the equation by -3.}$$
$$1 = x$$

43.
$$\dfrac{3}{x} = \dfrac{7}{8}$$
$$3(8) = 7x \qquad \text{Cross multiplication}$$
$$24 = 7x$$
$$\dfrac{24}{7} = \dfrac{7x}{7} \qquad \text{Divide both sides of the equation by 7.}$$
$$\dfrac{24}{7} = x$$

45.
$$\dfrac{1}{2}x + \dfrac{1}{3} = \dfrac{2}{3}$$
$$6\left(\dfrac{1}{2}x + \dfrac{1}{3}\right) = 6\left(\dfrac{2}{3}\right) \qquad \text{Multiply both sides of the equation by the LCD.}$$
$$3x + 2 = 4 \qquad \text{Distributive Property}$$
$$3x + 2 - 2 = 4 - 2 \qquad \text{Subtract 2 from both sides of the equation.}$$
$$3x = 2$$
$$\dfrac{3x}{3} = \dfrac{2}{3} \qquad \text{Divide both sides of the equation by 3.}$$
$$x = \dfrac{2}{3}$$

47.
$$0.7x - 0.3 = 1.8$$
$$0.7x - 0.3 + 0.3 = 1.8 + 0.3 \qquad \text{Add 0.3 to both sides of the equation.}$$
$$0.7x = 2.1$$
$$\frac{0.7x}{0.7} = \frac{2.1}{0.7} \qquad \text{Divide both sides of the equation by 0.7.}$$
$$x = 3$$

49.
$$6t - 8 = 4t - 2$$
$$6t - 4t - 8 = 4t - 4t - 2 \qquad \text{Subtract } 4t \text{ from both sides of the equation.}$$
$$2t - 8 = -2$$
$$2t - 8 + 8 = -2 + 8 \qquad \text{Add 8 to both sides of the equation.}$$
$$2t = 6$$
$$\frac{2t}{2} = \frac{6}{2} \qquad \text{Divide both sides of the equation by 2.}$$
$$t = 3$$

51.
$$\frac{x-3}{2} = \frac{x+4}{3}$$
$$3(x-3) = 2(x+4) \qquad \text{Cross multiplication}$$
$$3x - 9 = 2x + 8 \qquad \text{Distributive Property}$$
$$3x - 2x - 9 = 2x - 2x + 8 \qquad \text{Subtract } 2x \text{ from both sides of the equation.}$$
$$x - 9 = 8$$
$$x - 9 + 9 = 8 + 9 \qquad \text{Add 9 to both sides of the equation.}$$
$$x = 17$$

53.
$$6t - 7 = 8t + 9$$
$$6t - 6t - 7 = 8t - 6t + 9 \qquad \text{Subtract } 6t \text{ from both sides of the equation.}$$
$$-7 = 2t + 9$$
$$-7 - 9 = 2t + 9 - 9 \qquad \text{Subtract 9 from both sides of the equation.}$$
$$-16 = 2t$$
$$\frac{-16}{2} = \frac{2t}{2} \qquad \text{Divide both sides of the equation by 2.}$$
$$-8 = t$$

55.
$$2(x+3) - 4 = 2(x-4)$$
$$2x + 6 - 4 = 2x - 8 \qquad \text{Distributive Property}$$
$$2x + 2 = 2x - 8$$
$$2x - 2x + 2 = 2x - 2x - 8 \qquad \text{Subtract } 2x \text{ from both sides of the equation.}$$
$$2 = -8 \qquad \text{False}$$
$$\text{No solution}$$

57. $$4(x-4)+12=4(x-1)$$
$$4x-16+12=4x-4 \qquad \text{Distributive Property}$$
$$4x-4=4x-4$$

This equation is an identity. Therefore, the solution is all real numbers.

59. $$\frac{1}{4}(x+4)=\frac{2}{5}(x+2)$$
$$20\left(\frac{1}{4}\right)(x+4)=20\left(\frac{2}{5}\right)(x+2) \qquad \text{Multiply both sides of the equation by the LCD.}$$
$$5(x+4)=8(x+2)$$
$$5x+20=8x+16 \qquad \text{Distributive Property}$$
$$5x-8x+20=8x-8x+16 \qquad \text{Subtract } 8x \text{ from both sides of the equation.}$$
$$-3x+20=16$$
$$-3x+20-20=16-20 \qquad \text{Subtract 20 from both sides of the equation.}$$
$$-3x=-4$$
$$\frac{-3x}{-3}=\frac{-4}{-3} \qquad \text{Divide both sides of the equation by -3.}$$
$$x=\frac{4}{3}$$

61. $$3x+2-6x=-x-15+8-5x$$
$$-3x+2=-6x-7$$
$$-3x+6x+2=-6x+6x-7 \qquad \text{Add } 6x \text{ to both sides of the equation.}$$
$$3x+2=-7$$
$$3x+2-2=-7-2 \qquad \text{Subtract 2 from both sides of the equation.}$$
$$3x=-9$$
$$\frac{3x}{3}=\frac{-9}{3} \qquad \text{Divide both sides of the equation by 3.}$$
$$x=-3$$

63. $$2(t-3)+2=2(2t-6)$$
$$2t-6+2=4t-12 \qquad \text{Distributive Property}$$
$$2t-4=4t-12$$
$$2t-4t-4=4t-4t-12 \qquad \text{Subtract } 4t \text{ from both sides of the equation.}$$
$$-2t-4=-12$$
$$-2t-4+4=-12+4 \qquad \text{Add 4 to both sides of the equation.}$$
$$-2t=-8$$
$$\frac{-2t}{-2}=\frac{-8}{-2} \qquad \text{Divide both sides of the equation by -2.}$$
$$t=4$$

65.

$$\frac{2.05}{1000} = \frac{x}{35,300}$$

$$2.05(35,300) = 1000x$$

$$72,365 = 1000x$$

$$\frac{72,365}{1000} = \frac{1000x}{1000}$$

$$x = 72.365 \approx \$72.37$$

67.

$$\frac{x}{354} = \frac{1}{6}$$

$$6x = 354$$

$$\frac{6x}{6} = \frac{354}{6}$$

$$x = 59 \text{ times}$$

69.

$$\frac{1}{1,022,000} = \frac{20.3}{x}$$

$$x = 1,022,000(20.3)$$

$$x = 20,746,600 \text{ households}$$

71. a)

$$\frac{50}{80} = \frac{1}{x}$$

$$50x = 80$$

$$\frac{50x}{50} = \frac{80}{50}$$

$$x = 1.6 \text{ kph}$$

b)

$$\frac{50}{80} = \frac{x}{90}$$

$$80x = 50(90)$$

$$80x = 4500$$

$$\frac{80x}{80} = \frac{4500}{80}$$

$$x = 56.25 \text{ mph}$$

73.

$$\frac{40}{1} = \frac{12}{x}$$

$$40x = 12$$

$$\frac{40x}{40} = \frac{12}{40}$$

$$x = 0.3 \text{ cc}$$

75. a) Answers will vary.

b)

$$2(x+3) = 4x+3-5x$$

$$2x+6 = -x+3 \qquad \text{Distributive Property}$$

$$2x+x+6 = -x+x+3 \qquad \text{Add } x \text{ to both sides of the equation.}$$

$$3x+6 = 3$$

$$3x+6-6 = 3-6 \qquad \text{Subtract 6 from both sides of the equation.}$$

$$3x = -3$$

$$\frac{3x}{3} = \frac{-3}{3} \qquad \text{Divide both sides of the equation by 3.}$$

$$x = -1$$

77. a) An **inconsistent equation** is an equation that has no solution.

b) When solving an equation, if you obtain a false statement, then the equation is inconsistent.

79. a) 2:5; There are $2x$ males and a total of $2x + 3x = 5x$ students.

b) $m : m+n$

Exercise Set 6.3

1. A **formula** is an equation that typically has a real-life application.

3. **Subscripts** are numbers (or letters) placed below and to the right of variables. They are used to help clarify a formula.

5. An **exponential equation** is of the form $y = a^x, a > 0, a \neq 1$.

7. $P = 4s = 4(5) = 20$

9. $P = 2l + 2w$
   $P = 2(12) + 2(16) = 24 + 32 = 56$

11. $E = mc^2$
    $400 = m(4)^2$
    $400 = 16m$
    $\dfrac{400}{16} = \dfrac{16m}{16}$
    $25 = m$

13. $A = \pi(R^2 - r^2)$
    $A = 3.14\left((6)^2 - (4)^2\right)$
    $A = 3.14(36 - 16)$
    $A = 3.14(20)$
    $A = 62.8$

15. $z = \dfrac{x - \mu}{\sigma}$
    $\dfrac{2.5}{1} = \dfrac{42.1 - \mu}{2}$
    $2.5(2) = 42.1 - \mu$
    $5 = 42.1 - \mu$
    $5 - 42.1 = 42.1 - 42.1 - \mu$
    $-37.1 = -\mu$
    $\dfrac{-37.1}{-1} = \dfrac{-\mu}{-1}$
    $37.1 = \mu$

17. $T = \dfrac{PV}{k}$
    $\dfrac{80}{1} = \dfrac{P(20)}{0.5}$
    $80(0.5) = 20P$
    $40 = 20P$
    $\dfrac{40}{20} = \dfrac{20P}{20}$
    $2 = P$

19. $A = P(1 + rt)$
    $3600 = P(1 + 0.04(5))$
    $3600 = P(1 + 0.2)$
    $3600 = 1.2P$
    $\dfrac{3600}{1.2} = \dfrac{1.2P}{1.2}$
    $3000 = P$

21. $V = \dfrac{1}{2}at^2$
    $576 = \dfrac{1}{2}a(12)^2$
    $\dfrac{576}{1} = \dfrac{144a}{2}$
    $576(2) = 144a$
    $1152 = 144a$
    $\dfrac{1152}{144} = \dfrac{144a}{144}$
    $8 = a$

23. $C = \dfrac{5}{9}(F - 32)$
    $C = \dfrac{5}{9}(77 - 32)$
    $C = \dfrac{5}{9}(45) = 25$

25. $m = \dfrac{y_2 - y_1}{x_2 - x_1}$
    $m = \dfrac{8 - (-4)}{-3 - (-5)}$
    $m = \dfrac{8 + 4}{-3 + 5} = \dfrac{12}{2} = 6$

27.
$$S = R - rR$$
$$186 = 1R - 0.07R$$
$$186 = 0.93R$$
$$\frac{186}{0.93} = \frac{0.93R}{0.93}$$
$$200 = R$$

29.
$$E = a_1 p_1 + a_2 p_2 + a_3 p_3$$
$$E = 5(0.2) + 7(0.6) + 10(0.2)$$
$$E = 1 + 4.2 + 2 = 7.2$$

31.
$$s = -16t^2 + v_0 t + s_0$$
$$s = -16(4)^2 + 30(4) + 150$$
$$s = -16(16) + 120 + 150$$
$$s = -256 + 120 + 150 = 14$$

33.
$$P = \frac{f}{1+i}$$
$$3000 = \frac{f}{1+0.08}$$
$$\frac{3000}{1} = \frac{f}{1.08}$$
$$3000(1.08) = f$$
$$3240 = f$$

35.
$$F = \frac{Gm_1 m_2}{r^2}$$
$$625 = \frac{G(100)(200)}{(4)^2}$$
$$625 = 1250G$$
$$\frac{625}{1250} = \frac{1250G}{1250}$$
$$0.5 = G$$

37.
$$S_n = \frac{a_1\left(1 - r^n\right)}{1 - r}$$
$$S_n = \frac{8\left(1 - \left(\frac{1}{2}\right)^3\right)}{1 - \frac{1}{2}}$$
$$S_n = \frac{8\left(1 - \frac{1}{8}\right)}{1 - \frac{1}{2}}$$
$$S_n = \frac{8\left(\frac{7}{8}\right)}{\frac{1}{2}} = \frac{7}{\frac{1}{2}} = 7(2) = 14$$

39.
$$10x - 4y = 13$$
$$10x - 10x - 4y = -10x + 13 \qquad \text{Subtract } 10x \text{ from both sides of the equation.}$$
$$-4y = -10x + 13$$
$$\frac{-4y}{-4} = \frac{-10x + 13}{-4} \qquad \text{Divide both sides of the equation by -4.}$$
$$y = \frac{-10x + 13}{-4} = \frac{-(-10x + 13)}{4}$$
$$= \frac{10x - 13}{4} = \frac{10x}{4} - \frac{13}{4} = \frac{5}{2}x - \frac{13}{4}$$

41.
$$4x + 7y = 14$$
$$-4x + 4x + 7y = -4x + 14 \qquad \text{Subtract } 4x \text{ from both sides of the equation.}$$
$$7y = -4x + 14$$
$$\frac{7y}{7} = \frac{-4x + 14}{7} \qquad \text{Divide both sides of the equation by 7.}$$
$$y = \frac{-4x + 14}{7} = \frac{-4x}{7} + \frac{14}{7} = -\frac{4}{7}x + 2$$

43.
$$2x - 3y + 6 = 0$$
$$2x - 3y + 6 - 6 = 0 - 6 \qquad \text{Subtract 6 from both sides of the equation.}$$
$$2x - 3y = -6$$
$$-2x + 2x - 3y = -2x - 6 \qquad \text{Subtract } 2x \text{ from both sides of the equation.}$$
$$-3y = -2x - 6$$
$$\frac{-3y}{-3} = \frac{-2x - 6}{-3} \qquad \text{Divide both sides of the equation by -3.}$$
$$y = \frac{-2x - 6}{-3} = \frac{-(-2x - 6)}{3} = \frac{2x + 6}{3} = \frac{2x}{3} + \frac{6}{3} = \frac{2}{3}x + 2$$

45.
$$-2x + 3y + z = 15$$
$$-2x + 2x + 3y + z = 2x + 15 \qquad \text{Add } 2x \text{ to both sides of the equation.}$$
$$3y + z = 2x + 15$$
$$3y + z - z = 2x - z + 15 \qquad \text{Subtract } z \text{ from both sides of the equation.}$$
$$3y = 2x - z + 15$$
$$\frac{3y}{3} = \frac{2x - z + 15}{3} \qquad \text{Divide both sides of the equation by 3.}$$
$$y = \frac{2x - z + 15}{3} = \frac{2}{3}x - \frac{1}{3}z + 5$$

47.
$$9x + 4z = 7 + 8y$$
$$9x + 4z - 7 = 7 - 7 + 8y \qquad \text{Subtract 7 from both sides of the equation.}$$
$$9x + 4z - 7 = 8y$$
$$\frac{9x + 4z - 7}{8} = \frac{8y}{8} \qquad \text{Divide both sides of the equation by 8.}$$
$$y = \frac{9x + 4z - 7}{8} = \frac{9}{8}x + \frac{1}{2}z - \frac{7}{8}$$

49.
$$E = IR$$
$$\frac{E}{I} = \frac{IR}{I} \qquad \text{Divide both sides of the equation by } I.$$
$$R = \frac{E}{I}$$

51.
$$p = a + b + c$$
$$p - b = a + b - b + c \qquad \text{Subtract } b \text{ from both sides of the equation.}$$
$$p - b = a + c$$
$$p - b - c = a + c - c \qquad \text{Subtract } c \text{ from both sides of the equation.}$$
$$a = p - b - c$$

53. $$V = \frac{1}{3}Bh$$

$$3V = 3\left(\frac{1}{3}Bh\right)$$ Multiply both sides of the equation by 3.

$$3V = Bh$$

$$\frac{3V}{h} = \frac{Bh}{h}$$ Divide both sides of the equation by h.

$$B = \frac{3V}{h}$$

55. $$C = 2\pi r$$

$$\frac{C}{2} = \frac{2\pi r}{2}$$ Divide both sides of the equation by 2.

$$\frac{C}{2} = \pi r$$

$$\frac{C}{2\pi} = \frac{\pi r}{\pi}$$ Divide both sides of the equation by π.

$$r = \frac{C}{2\pi}$$

57. $$y = mx + b$$

$$y - mx = mx - mx + b$$ Subtract mx from both sides of the equation.

$$b = y - mx$$

59. $$P = 2l + 2w$$

$$P - 2l = 2l - 2l + 2w$$ Subtract $2l$ from both sides of the equation.

$$P - 2l = 2w$$

$$\frac{P - 2l}{2} = \frac{2w}{2}$$ Divide both sides of the equation by 2.

$$w = \frac{P - 2l}{2}$$

61. $$A = \frac{a + b + c}{3}$$

$$3A = 3\left(\frac{a + b + c}{3}\right)$$ Multiply both sides of the equation by 3.

$$3A = a + b + c$$

$$3A - a = a - a + b + c$$ Subtract a from both sides of the equation.

$$3A - a = b + c$$

$$3A - a - b = b - b + c$$ Subtract b from both sides of the equation.

$$c = 3A - a - b$$

63.
$$P = \frac{KT}{V}$$

$$PV = \left(\frac{KT}{V}\right)V$$ Multiply both sides of the equation by V.

$$PV = KT$$

$$\frac{PV}{K} = \frac{KT}{K}$$ Divide both sides of the equation by K.

$$T = \frac{PV}{K}$$

65.
$$F = \frac{9}{5}C + 32$$

$$F - 32 = \frac{9}{5}C + 32 - 32$$ Subtract 32 from both sides of the equation.

$$F - 32 = \frac{9}{5}C$$

$$\frac{5}{9}(F - 32) = \frac{5}{9}\left(\frac{9}{5}C\right)$$ Multiply both sides of the equation by $\frac{5}{9}$.

$$C = \frac{5}{9}(F - 32)$$

67.
$$S = \pi r^2 + \pi rs$$

$$S - \pi r^2 = \pi r^2 - \pi r^2 + \pi rs$$ Subtract πr^2 from both sides of the equation.

$$S - \pi r^2 = \pi rs$$

$$\frac{S - \pi r^2}{\pi} = \frac{\pi rs}{\pi}$$ Divide both sides of the equation by π.

$$\frac{S - \pi r^2}{\pi} = rs$$

$$\frac{S - \pi r^2}{\pi r} = \frac{rs}{r}$$ Divide both sides of the equation by r.

$$s = \frac{S - \pi r^2}{\pi r}$$

69. a) $i = prt$

$i = 600(0.02)(1) = \$12$

b) $\$600 + \$12 = \$612$

71. Radius $= \dfrac{2.5}{2} = 1.25$ in.

$V = \pi r^2 h$

$V = \pi(1.25)^2 (3.75)$

$V = \pi(1.5625)(3.75)$

$V = 18.40776945$ in.$^3 \approx 18.4$ in.^{3}

73. $y = 2000(3)^x$

$y = 2000(3)^5$

$y = 2000(243)$

$y = 486,000$ bacteria

75. $V = 24e^{0.08t}$

$V = 24e^{0.08(377)}$

$V = 24e^{30.16}$

$V = \$300,976,658,300,000$

77. $V = lwh - \pi r^2 h$

   $V = 12(8)(12) - \pi(2)^2(8)$

   $V = 1152 - 100.5309649$

   $V = 1051.469035 \text{ in.}^3 \approx 1051.47 \text{ in.}^3$

78.

Exercise Set 6.4

1. A **mathematical expression** is a collection of variables, numbers, parentheses, and operation symbols.
   An **equation** is two algebraic expressions joined by an equal sign.

3. $4 + 3x$

5. $6r + 5$

7. $15 - 2r$

9. $2m + 9$

11. $\dfrac{18 - s}{4}$

13. $(5y - 6) + 3$

15.   Let $x = $ the number

   $x - 6 = $ the number decreased by 6

   $x - 6 = 5$

   $x - 6 + 6 = 5 + 6$

   $x = 11$

17.   Let $x = $ the number

   $x - 4 = $ the difference between the number and 4

   $x - 4 = 20$

   $x - 4 + 4 = 20 + 4$

   $x = 24$

19.   Let $x = $ the number

   $12 + 5x = 12$ increased by 5 times the number

   $12 + 5x = 47$

   $12 - 12 + 5x = 47 - 12$

   $5x = 35$

   $\dfrac{5x}{5} = \dfrac{35}{5}$

   $x = 7$

21.   Let $x = $ the number

   $8x + 16 = 16$ more than 8 times the number

   $8x + 16 = 88$

   $8x + 16 - 16 = 88 - 16$

   $8x = 72$

   $\dfrac{8x}{8} = \dfrac{72}{8}$

   $x = 9$

23.   Let $x = $ the number

   $x + 11 = $ the number increased by 11

   $3x + 1 = 1$ more than 3 times the number

   $x + 11 = 3x + 1$

   $x - x + 11 = 3x - x + 1$

   $11 = 2x + 1$

   $11 - 1 = 2x + 1 - 1$

   $10 = 2x$

   $\dfrac{10}{2} = \dfrac{2x}{2}$

   $5 = x$

25.   Let $x = $ the number

   $x + 10 = $ the number increased by 10

   $2(x + 3) = 2$ times the sum of the number and 3

   $x + 10 = 2(x + 3)$

   $x + 10 = 2x + 6$

   $x - x + 10 = 2x - x + 6$

   $10 = x + 6$

   $10 - 6 = x + 6 - 6$

   $4 = x$

27.   Let x = the number of tickets sold
         to nonstudents
       $3x$ = the number of tickets sold
         to students
       $x + 3x = 600$
       $4x = 600$
       $\dfrac{4x}{4} = \dfrac{600}{4}$
       $x = 150$ tickets to nonstudents
       $3x = 3(150) = 450$ tickets to students

29.   Let x = the number filing electronically
         in 1999
       $0.116x$ = the amount of the increase
       $x + 0.116x = 34.20$
       $1.116x = 34.20$
       $\dfrac{1.116x}{1.116} = \dfrac{34.20}{1.116}$
       $x = 30.64516129$
       ≈ 30.65 million taxpayers

31.   Let x = the original price before tax
       $0.10x$ = the amount saved on
         spending x dollars
       $x - 0.10x = 15.72$
       $0.9x = 15.72$
       $\dfrac{0.9x}{0.9} = \dfrac{15.72}{0.9}$
       $x = 17.4\overline{6} \approx \$17.47$

33.   Let x = the number of compact discs
         for Samantha
       $3x$ = the number of compact discs
         for Josie
       $x + 3x = 12$
       $4x = 12$
       $\dfrac{4x}{4} = \dfrac{12}{4}$
       $x = 3$ compact discs for Samantha
       $3x = 3(3) = 9$ compact discs for Josie

35.   Let x = the amount charged
         to each homeowner
       $50x$ = the total amount charged
         to homeowners
       $2000 + 50x$ = the total cost for
         the repairs
       $2000 + 50x = 13,350$
       $2000 - 2000 + 50x = 13,350 - 2000$
       $50x = 11,350$
       $\dfrac{50x}{50} = \dfrac{11,350}{50}$
       $x = \$227$

37. a)   Let x = area of smaller ones
       $3x$ = area of largest one
       $x + x + 3x = 45,000$
       $5x = 45,000$
       $\dfrac{5x}{5} = \dfrac{45,000}{5}$
       $x = 9000$ ft^2 for the two smaller barns
       $3x = 3(9000)$
       $= 27,000$ ft^2 for the largest barn

b) Yes

39.   Let x = the number of vacation days
in the U.S.

$3x+3$ = the number of vacation days
in Italy

$$x+3x+3 = 55$$
$$4x+3 = 55$$
$$4x+3-3 = 55-3$$
$$4x = 52$$
$$\frac{4x}{4} = \frac{52}{4}$$
$$x = 13 \text{ in the U.S.}$$
$$3x+3 = 3(13)+3 = 39+3 = 42 \text{ in Italy}$$

41.   Let w = width

$2w$ = length of entire enclosed region

$3w+2(2w)$ = total amount of fencing

$$3w+2(2w) = 140$$
$$3w+4w = 140$$
$$7w = 140$$
$$\frac{7w}{7} = \frac{140}{7}$$
$$\text{width} = 20 \text{ ft}$$
$$\text{length} = 2w = 2(20) = 40 \text{ ft}$$

43.   Let x = the number of months

$70x$ = cost of laundry for x months

$$70x = 760$$
$$\frac{70x}{70} = \frac{760}{70}$$
$$x = 10.85714286 \text{ months} \approx 11 \text{ months}$$

45.   Let r = regular fare

$$\frac{r}{2} = \text{half off regular fare}$$
$$0.07r = \text{tax on regular fare}$$
$$\frac{r}{2}+0.07r = 257$$
$$2\left(\frac{r}{2}+0.07r\right) = 2(257)$$
$$r+0.14r = 514$$
$$1.14r = 514$$
$$\frac{1.14r}{1.14} = \frac{514}{1.14}$$
$$r = \$450.877193$$
$$\approx \$450.88$$

47.   Let x = amount of tax reduction to be
deducted from Mr. McAdam's income

$3640-x$ = amount of tax reduction to be
deducted from Mrs. McAdam's income

$$24,200-x = 26,400-(3640-x)$$
$$24,200-x = 26,400-3640+x$$
$$24,200-x = 22,760+x$$
$$24,200-x+x = 22,760+x+x$$
$$24,200 = 22,760+2x$$
$$24,200-22,760 = 22,760-22,760+2x$$
$$1440 = 2x$$
$$\frac{1440}{2} = \frac{2x}{2}$$
$$x = \$720 \text{ deducted from}$$
Mr. McAdam's income
$$3640-x = 3640-720 = \$2920 \text{ deducted}$$
from Mrs. McAdam's income

49.   Let x = the first integer

$x+1$ = the second integer

$x+2$ = the third integer (the largest)

$$x+(x+1)+(x+2) = 3(x+2)-3$$
$$3x+3 = 3x+6-3$$
$$3x+3 = 3x+3$$

51. $F = \dfrac{9}{5}C + 32$

The thermometers will read the same when $F = C$.

Substitute C for F in the above equation.

$$C = \dfrac{9}{5}C + 32$$

$$5C = 5\left(\dfrac{9}{5}C + 32\right)$$

$$5C = 9C + 160$$

$$5C - 9C = 9C - 9C + 160$$

$$-4C = 160$$

$$\dfrac{-4C}{-4} = \dfrac{160}{-4}$$

$$C = -40°$$

Exercise Set 6.5

1. **Inverse variation** — y varies inversely with x if $y = \frac{k}{x}$.

3. **Joint variation** — One quantity varies directly as the product of two or more other quantities.

5.	Direct	7.	Inverse
9.	Direct	11.	Inverse
13.	Inverse	15.	Inverse
17.	Direct	19.	Direct

21. Answers will vary.

23. a) $y = kx$

    b) $y = 3(5) = 15$

25. a) $m = \dfrac{k}{n^2}$

    b) $m = \dfrac{16}{(8)^2} = \dfrac{16}{64} = 0.25$

27. a) $R = \dfrac{k}{W}$

    b) $R = \dfrac{8}{160} = 0.05$

29. a) $F = kDE$
    b) $F = 7(3)(10) = 210$

31. a) $t = \dfrac{kd^2}{f}$

    b) $192 = \dfrac{k(8)^2}{4}$

    $192 = \dfrac{64k}{4}$

    $768 = 64k$

    $\dfrac{768}{64} = \dfrac{64k}{64}$

    $k = 12$

    $t = \dfrac{12d^2}{f}$

    $t = \dfrac{12(10)^2}{6} = \dfrac{12(100)}{6} = \dfrac{1200}{6} = 200$

33. a) $Z = kWY$
    b) $12 = k(9)(4)$

    $12 = 36k$

    $\dfrac{12}{36} = \dfrac{36k}{36}$

    $k = \dfrac{1}{3}$

    $Z = \dfrac{1}{3}WY$

    $Z = \dfrac{1}{3}(50)(6) = \dfrac{300}{3} = 100$

35. a) $H = kL$
    b) $15 = k(50)$

    $\dfrac{15}{50} = \dfrac{50k}{50}$

    $k = 0.3$

    $H = 0.3L$

    $H = 0.3(10) = 3$

37. a) $A = kB^2$
    b) $245 = k(7)^2$

    $245 = 49k$

    $\dfrac{245}{49} = \dfrac{49k}{49}$

    $k = 5$

    $A = 5B^2$

    $A = 5(12)^2 = 5(144) = 720$

39. a) $F = \dfrac{kq_1q_2}{d^2}$

    b) $8 = \dfrac{k(2)(8)}{(4)^2}$

    $8 = \dfrac{16k}{16}$

    $k = 8$

    $F = \dfrac{8q_1q_2}{d^2}$

    $F = \dfrac{8(28)(12)}{(2)^2} = \dfrac{2688}{4} = 672$

41. a) $R = kL$
    b) $0.24 = k(30)$

    $\dfrac{0.24}{30} = \dfrac{30k}{30}$

    $k = 0.008$

    $R = 0.008L$

    $R = 0.008(40) = 0.32$ ohm

43. a) $l = \dfrac{k}{d^2}$

    b) $20 = \dfrac{k}{(6)^2}$

    $k = 20(36) = 720$

    $l = \dfrac{720}{d^2} = \dfrac{720}{(3)^2} = \dfrac{720}{9} = 80$ dB

45. a) $R = \dfrac{kA}{P}$

    b) $4800 = \dfrac{k(600)}{3}$

    $600k = 14,400$

    $k = \dfrac{14,400}{600} = 24$

    $R = \dfrac{24A}{P}$

    $R = \dfrac{24(700)}{3.50} = \dfrac{16,800}{3.50} = 4800$ tapes

47. a) $s = kwd^2$

    b) $2250 = k(2)(10)^2$

    $2250 = 200k$

    $\dfrac{2250}{200} = \dfrac{200k}{200}$

    $k = \dfrac{2250}{200} = 11.25$

    $s = 11.25wd^2$

    $s = 11.25(4)(12)^2 = 11.25(4)(144)$

    $= 6480$ pounds per square inch

49. a) $N = \dfrac{kp_1 p_2}{d}$

    b) $100,000 = \dfrac{k(60,000)(200,000)}{300}$

    $12,000,000,000k = 30,000,000$

    $k = \dfrac{30,000,000}{12,000,000,000} = 0.0025$

    $N = \dfrac{0.0025p_1 p_2}{d}$

    $N = \dfrac{0.0025(125,000)(175,000)}{450}$

    $N = \dfrac{54,687,500}{450}$

    $= 121,527.7778 \approx 121,528$ calls

51. a) $y = \dfrac{k}{x}$

    $y = \dfrac{0.3}{x}$

    $xy = 0.3$

    $\dfrac{xy}{y} = \dfrac{0.3}{y}$

    $x = \dfrac{0.3}{y}$

    Inversely

    b) k stays 0.3

53. $W = \dfrac{kTA\sqrt{F}}{R}$

    $72 = \dfrac{k(78)(1000)\sqrt{4}}{5.6}$

    $156,000k = 403.2$

    $k = \dfrac{403.2}{156,000} = 0.0025846154$

    $W = \dfrac{0.0025846154TA\sqrt{F}}{R}$

    $W = \dfrac{0.0025846154(78)(1500)\sqrt{6}}{5.6}$

    $W = \dfrac{740.7256982}{5.6} = 132.2724461 \approx \$132.27$

Exercise Set 6.6

1. $a < b$ means that a is less than b, $a \leq b$ means that a is less than or equal to b, $a > b$ means that a is greater than b, $a \geq b$ means that a is greater than or equal to b.

   b) $2 < 7$, $3 > -1$, $5x + 2 \geq 9$

3. When both sides of an inequality are multiplied or divided by a negative number, the direction of the inequality symbol must be reversed.

5. Yes, the inequality symbol points to the -3 in both cases.

7. $x > 6$

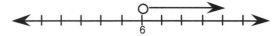

9. $x + 4 \geq 7$

   $x + 4 - 4 \geq 7 - 4$

   $x \geq 3$

11. $-3x \leq 18$

    $\dfrac{-3x}{-3} \geq \dfrac{18}{-3}$

    $x \geq -6$

13. $\dfrac{x}{6} < -2$

    $6\left(\dfrac{x}{6}\right) < 6(-2)$

    $x < -12$

15. $\dfrac{-x}{3} \geq 3$

    $-3\left(\dfrac{-x}{3}\right) \leq -3(3)$

    $x \leq -9$

17.
$$2x + 6 \geq 14$$
$$2x + 6 - 6 \geq 14 - 6$$
$$2x \geq 8$$
$$\frac{2x}{2} \geq \frac{8}{2}$$
$$x \geq 4$$

19.
$$4(x-1) < 6$$
$$4x - 4 < 6$$
$$4x - 4 + 4 < 6 + 4$$
$$4x < 10$$
$$\frac{4x}{4} < \frac{10}{4}$$
$$x < \frac{5}{2}$$

21.
$$3(x+4) - 2 < 3x + 10$$
$$3x + 12 - 2 < 3x + 10$$
$$3x + 10 < 3x + 10$$
False, no solution

23.
$$3 < x - 7 \leq 6$$
$$3 + 7 < x - 7 + 7 \leq 6 + 7$$
$$10 < x \leq 13$$

25. $x \geq 2$

27.
$$-3x \leq 27$$
$$\frac{-3x}{-3} \geq \frac{27}{-3}$$
$$x \geq -9$$

29.
$$x - 2 < 4$$
$$x - 2 + 2 < 4 + 2$$
$$x < 6$$

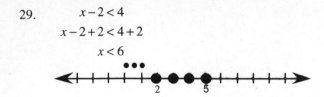

31.
$$\frac{x}{3} \le -2$$
$$3\left(\frac{x}{3}\right) \le 3(-2)$$
$$x \le -6$$

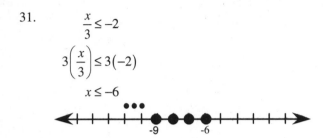

33.
$$\frac{-x}{6} \ge 3$$
$$-6\left(\frac{-x}{6}\right) \le -6(3)$$
$$x \le -18$$

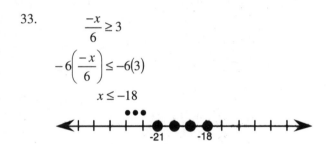

35.
$$-11 < -5x + 4$$
$$-11 - 4 < -5x + 4 - 4$$
$$-15 < -5x$$
$$\frac{-15}{-5} > \frac{-5x}{-5}$$
$$3 > x$$
$$x < 3$$

37.
$$3(x + 4) \ge 4x + 13$$
$$3x + 12 \ge 4x + 13$$
$$3x - 4x + 12 \ge 4x - 4x + 13$$
$$-x + 12 \ge 13$$
$$-x + 12 - 12 \ge 13 - 12$$
$$-x \ge 1$$
$$\frac{-x}{-1} \le \frac{1}{-1}$$
$$x \le -1$$

39. $5(x+4)-6 \leq 2x+8$

      $5x+20-6 \leq 2x+8$

        $5x+14 \leq 2x+8$

    $5x-2x+14 \leq 2x-2x+8$

         $3x+14 \leq 8$

    $3x+14-14 \leq 8-14$

          $3x \leq -6$

         $\dfrac{3x}{3} \leq \dfrac{-6}{3}$

           $x \leq -2$

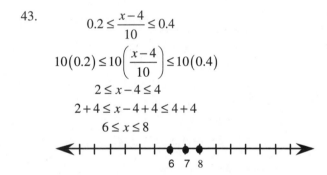

41. $1 > -x > -5$

     $\dfrac{1}{-1} < \dfrac{-x}{-1} < \dfrac{-5}{-1}$

      $-1 < x < 5$

43. $0.2 \leq \dfrac{x-4}{10} \leq 0.4$

   $10(0.2) \leq 10\left(\dfrac{x-4}{10}\right) \leq 10(0.4)$

          $2 \leq x-4 \leq 4$

       $2+4 \leq x-4+4 \leq 4+4$

          $6 \leq x \leq 8$

45.   a) 2000, 2001
      b) 1997, 1998
      c) 1997, 1998, 1999, 2000
      d) 1998, 1999, 2000, 2001

47. Let x = the number of videos

No Fee Plan cost: $2.99x$

Annual Fee Plan: $30 + 1.49x$

$$2.99x < 30 + 1.49x$$
$$2.99x - 1.49x < 30 + 1.49x - 1.49x$$
$$1.50x < 30$$
$$\frac{1.50x}{1.50} < \frac{30}{1.50}$$
$$x < 20$$

The maximum number of videos that can be rented for the No Fee Plan to cost less than the Annual Fee Plan is 19.

49. Let x = the number of miles

$110 + 0.25x$ = cost of renting from Fred's

$$110 + 0.25x < 200$$
$$110 - 110 + 0.25x < 200 - 110$$
$$0.25x < 90$$
$$\frac{0.25x}{0.25} < \frac{90}{0.25}$$
$$x < 360 \text{ mi}$$

51. Let x = the cost of the meal

$0.07x$ = the tax on the meal

$0.15x$ = the tip on the meal

$$x + 0.07x + 0.15x \le 19$$
$$1.22x \le 19$$
$$\frac{1.22x}{1.22} \le \frac{19}{1.22}$$
$$x \le 15.57377049$$

Mrs. Franklin can select a meal for $x \le \$15.57$.

53.
$$36 < 84 - 32t < 68$$
$$36 - 84 < 84 - 84 - 32t < 68 - 84$$
$$-48 < -32t < -16$$
$$\frac{-48}{-32} > \frac{-32t}{-32} > \frac{-16}{-32}$$
$$1.5 > t > 0.5$$
$$0.5 < t < 1.5$$

The velocity will be between $36\dfrac{\text{ft}}{\text{sec}}$ and $68\dfrac{\text{ft}}{\text{sec}}$ when t is between $0.5\,\text{sec}$ and $1.5\,\text{sec}$.

55. Let x = Devon's grade on the fifth test

$$80 \le \frac{78 + 64 + 88 + 76 + x}{5} < 90$$
$$80 \le \frac{306 + x}{5} < 90$$
$$5(80) \le 5\left(\frac{306 + x}{5}\right) < 5(90)$$
$$400 \le 306 + x < 450$$
$$400 - 306 \le 306 - 306 + x < 450 - 306$$
$$94 \le x < 144$$

Devon must have a score of $94 \le x \le 100$, assuming 100 is the highest grade possible.

57. Let $x =$ the number of gallons

$250x = 2750$ and $400x = 2750$

$x = \dfrac{2750}{250}$, $x = \dfrac{2750}{400}$

$x = 11$, $x = 6.875$

$6.875 \le x \le 11$

59. Student's answer: $-\dfrac{1}{3}x \le 4$

$-3\left(-\dfrac{1}{3}x\right) \le -3(4)$

$x \le -12$

Correct answer: $-\dfrac{1}{3}x \le 4$

$-3\left(-\dfrac{1}{3}x\right) \ge -3(4)$

$x \ge -12$

Yes, -12 is in both solution sets.

Exercise Set 6.7

1. A **graph** is an illustration of all the points whose coordinates satisfy an equation.

3. To find the **y-intercept**, set $x = 0$ and solve the equation for y.

5. a) Divide the difference between the y-coordinates by the difference between the x-coordinates.

   b) $m = \dfrac{5-2}{-3-6} = \dfrac{3}{-9} = -\dfrac{1}{3}$

7. a) First

   b) Second

9. - 15.   17. - 23.

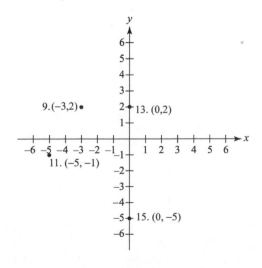

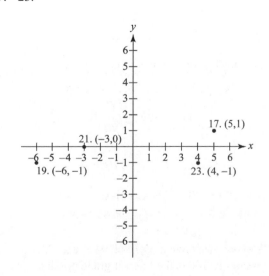

25. (0, 2)

27. (-2, 0)

29. (-5, -3)

31. (2, -3)

33. (2, 2)

35. Substituting (1, 3) into $3x + y = 7$, we have

$$3(1) + 3 = 7$$
$$3 + 3 = 7$$
$$6 \neq 7$$

Therefore, (1, 3) does not satisfy $3x + y = 7$.

Substituting (1, 4) into $3x + y = 7$, we have

$$3(1) + 4 = 7$$
$$3 + 4 = 7$$
$$7 = 7$$

Therefore, (1, 4) satisfies $3x + y = 7$.

Substituting (-1, 10) into $3x + y = 7$, we have

$$3(-1) + 10 = 7$$
$$-3 + 10 = 7$$
$$7 = 7$$

Therefore, (-1, 10) satisfies $3x + y = 7$.

37. Substituting (5, 0) into $2x - 3y = 10$, we have

$$2(5) - 3(0) = 10$$
$$10 - 0 = 10$$
$$10 = 10$$

Therefore, (5, 0) satisfies $2x - 3y = 10$.

Substituting (0, 3) into $2x - 3y = 10$, we have

$$2(0) - 3(3) = 10$$
$$0 - 9 = 10$$
$$-9 \neq 10$$

Therefore, (0, 3) does not satisfy $2x - 3y = 10$.

Substituting $\left(0, -\dfrac{10}{3}\right)$ into $2x - 3y = 10$, we have

$$2(0) - 3\left(-\dfrac{10}{3}\right) = 10$$
$$0 + 10 = 10$$
$$10 = 10$$

Therefore, $\left(0, -\dfrac{10}{3}\right)$ satisfies $2x - 3y = 10$.

39. Substituting (1, -1) into $7y = 3x - 5$, we have

$$7(-1) = 3(1) - 5$$
$$-7 = 3 - 5$$
$$-7 \neq -2$$

Therefore, (1, -1) does not satisfy $7y = 3x - 5$.

Substituting (-3, -2) into $7y = 3x - 5$, we have

$$7(-2) = 3(-3) - 5$$
$$-14 = -9 - 5$$
$$-14 = -14$$

Therefore, (-3, -2) satisfies $7y = 3x - 5$.

Substituting (2, 5) into $7y = 3x - 5$, we have

$$7(5) = 3(2) - 5$$
$$35 = 6 - 5$$
$$35 \neq 1$$

Therefore, (2, 5) does not satisfy $7y = 3x - 5$.

41. Substituting $\left(0, \frac{8}{3}\right)$ into $\frac{x}{2} + \frac{3y}{4} = 2$, we have

$$\frac{0}{2} + \frac{8}{4} = 2$$
$$0 + 2 = 2$$
$$2 = 2$$

Therefore, $\left(0, \frac{8}{3}\right)$ satisfies $\frac{x}{2} + \frac{3y}{4} = 2$.

Substituting $\left(1, \frac{11}{4}\right)$ into $\frac{x}{2} + \frac{3y}{4} = 2$, we have

$$\frac{1}{2} + \frac{33}{16} = 2$$
$$\frac{8}{16} + \frac{33}{16} = 2$$
$$\frac{41}{16} \neq 2$$

Therefore, $\left(1, \frac{11}{4}\right)$ does not satisfy $\frac{x}{2} + \frac{3y}{4} = 2$.

Substituting (4, 0) into $\frac{x}{2} + \frac{3y}{4} = 2$, we have

$$\frac{4}{2} + \frac{0}{4} = 2$$
$$2 + 0 = 2$$
$$2 = 2$$

Therefore, (4, 0) satisfies $\frac{x}{2} + \frac{3y}{4} = 2$.

43. Since the line is vertical, its slope is undefined.

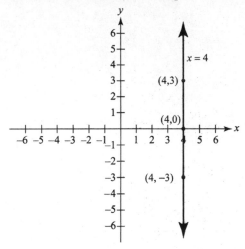

45. Since the line is horizontal, its slope is 0.

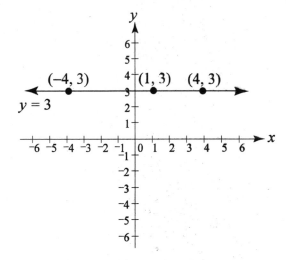

47.

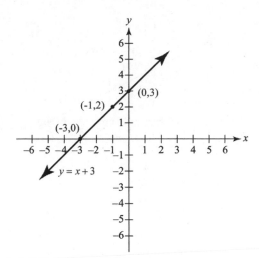

49.

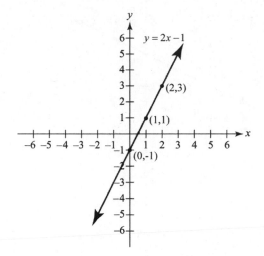

51.

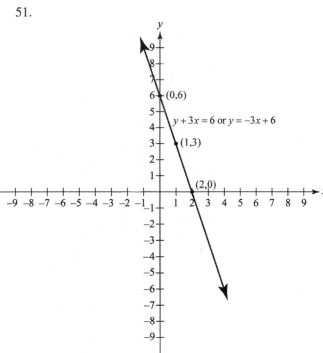

53.

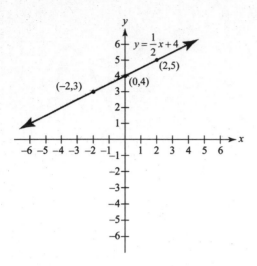

55.

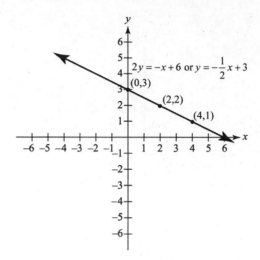

57.

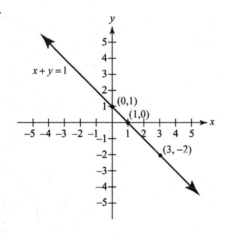

59.

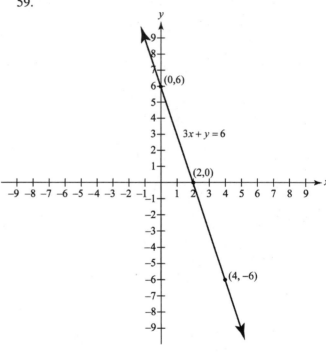

61.

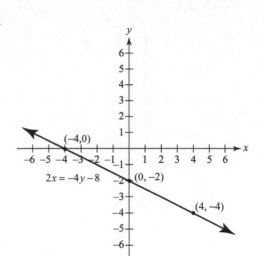

63.

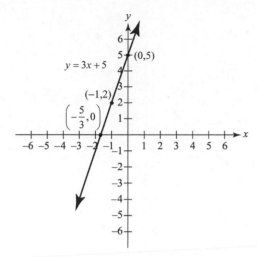

65.

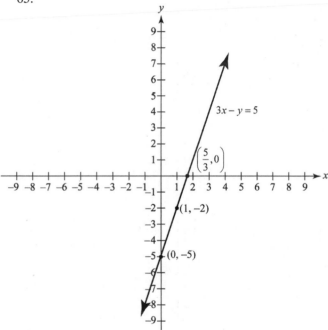

67. $(3,7),(10,21)$ $m = \dfrac{21-7}{10-3} = \dfrac{14}{7} = 2$

69. $(2,6),(-5,-9)$ $m = \dfrac{-9-6}{-5-2} = \dfrac{-15}{-7} = \dfrac{15}{7}$

71. $(5,2),(-3,2)$ $m = \dfrac{2-2}{-3-5} = \dfrac{0}{-8} = 0$

73. $(8,-3),(8,3)$ $m = \dfrac{3-(-3)}{8-8} = \dfrac{6}{0}$ Undefined

75. $(-2,3),(1,-1)$ $m = \dfrac{-1-3}{1-(-2)} = \dfrac{-4}{3} = -\dfrac{4}{3}$

77.

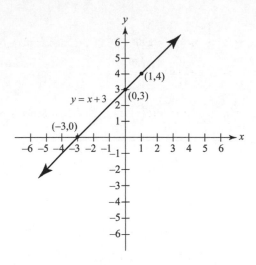

79.

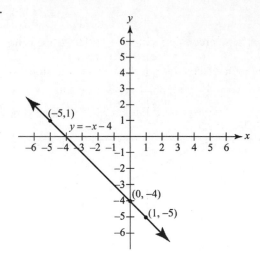

81.

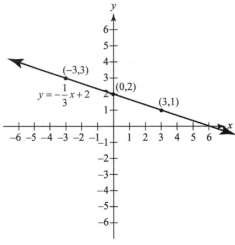

83.

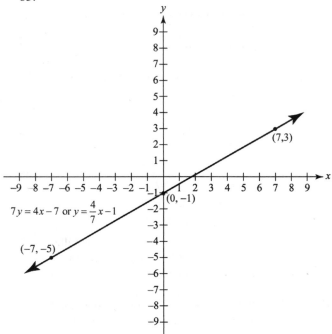

85.

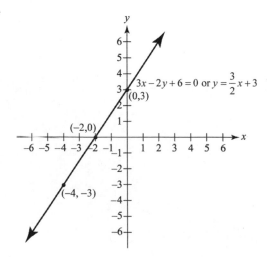

87. The y-intercept is 3; thus $b = 3$. The slope is negative since the graph falls from left to right. The change in y is 3, while the change in x is 4. Thus m, the slope, is $-\dfrac{3}{4}$. The equation is $y = -\dfrac{3}{4}x + 3$.

89. The y-intercept is 2; thus $b = 2$. The slope is positive since the graph rises from left to right. The change in y is 3, while the change in x is 1. Thus m, the slope, is $\dfrac{3}{1} = 3$. The equation is $y = 3x + 2$.

91. a)

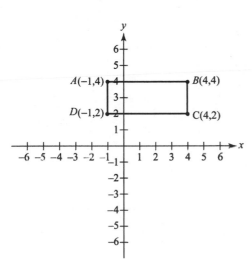

93.

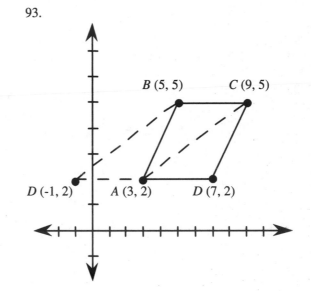

b) $A = lw = 5(2) = 10$ square units

95. For the line joining points P and Q to be parallel to the x-axis, both ordered pairs must have the same y-value. Thus, $b = 3$.

97. For the line joining points P and Q to be parallel to the x-axis, both ordered pairs must have the same y-value.

$$2b + 1 = 7$$
$$2b + 1 - 1 = 7 - 1$$
$$2b = 6$$
$$b = 3$$

99. a)

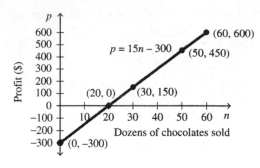

b) $300

c) To break even, profit must equal zero.

$$15n - 300 = 0$$

$$15n - 300 + 300 = 0 + 300$$

$$15n = 300$$

$$n = 20 \text{ dozens of chocolates}$$

101. a)

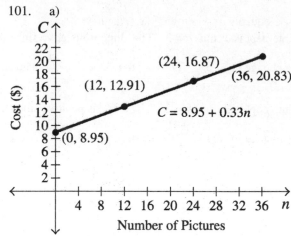

b) $8.95 + 0.33(20) = \$15.55$

c) $8.95 + 0.33n = 20.83$

$$0.33n = 11.88$$

$$n = 36 \text{ pictures}$$

103. a) $m = \dfrac{96 - 53}{4 - 0} = \dfrac{43}{4} = 10.75$

b) $y = 10.75x + 53$

c) $y = 10.75(3) + 53 = 32.25 + 53 = 85.25$

d) $80 = 10.75x + 53$

$$27 = 10.75x$$

$$x = 2.511627907 \approx 2.5 \text{ hours}$$

105. a) $m = \dfrac{24 - 40}{30 - 0} = \dfrac{-16}{30} = -\dfrac{8}{15}$

b) $y = -\dfrac{8}{15}x + 40$

c) $y = -\dfrac{8}{15}(15) + 40 = -8 + 40$

$$= 32\%$$

d) $30 = -\dfrac{8}{15}x + 40$

$$30 - 40 = -\dfrac{8}{15}x + 40 - 40$$

$$-10 = -\dfrac{8}{15}x$$

$$-10\left(-\dfrac{15}{8}\right) = -\dfrac{8}{15}x\left(-\dfrac{15}{8}\right)$$

$$x = \dfrac{150}{8}$$

$$= 18.75 \text{ years after 1970, or in 1988}$$

107. a) Solve the equations for y to put them in slope-intercept form. Then compare the slopes and y-intercepts. If the slopes are equal but the y-intercepts are different, then the lines are parallel.

   b)
   $$2x - 3y = 6$$
   $$2x - 2x - 3y = -2x + 6$$
   $$-3y = -2x + 6$$
   $$\frac{-3y}{-3} = \frac{-2x}{-3} + \frac{6}{-3}$$
   $$y = \frac{2}{3}x - 2$$

   $$4x = 6y + 6$$
   $$4x - 6 = 6y + 6 - 6$$
   $$4x - 6 = 6y$$
   $$\frac{4x}{6} - \frac{6}{6} = \frac{6y}{6}$$
   $$\frac{2}{3}x - 1 = y$$

   Since the two equations have the same slope, $m = \dfrac{2}{3}$, the graphs of the equations are parallel lines.

Exercise Set 6.8

1. (1) Mentally substitute the equal sign for the inequality sign and plot points as if you were graphing the equation. (2) If the inequality is $<$ or $>$, draw a dashed line through the points. If the inequality is $\leq$ or $\geq$, draw a solid line through the points. (3) Select a test point not on the line and substitute the x- and y- coordinates into the inequality. If the substitution results in a true statement, shade in the area on the same side of the line as the test point. If the substitution results in a false statement, shade in the area on the opposite side of the line as the test point.

3. Graph $x = 1$. Since the original statement is less than or equal to, a solid line is drawn. Since the point (0, 0) satisfies the inequality $x \leq 1$, all points on the line and in the half-plane to the left of the line $x = 1$ are in the solution set.

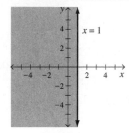

5. Graph $y = x + 3$. Since the original statement is strictly greater than, a dashed line is drawn. Since the point (0, 0) does not satisfy the inequality $y > x + 3$, all points in the half-plane above the line $y = x + 3$ are in the solution set.

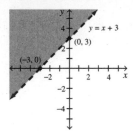

7. Graph $y = 2x - 6$. Since the original statement is greater than or equal to, a solid line is drawn. Since the point (0, 0) satisfies the inequality $y \geq 2x - 6$, all points on the line and in the half-plane above the line $y = 2x - 6$ are in the solution set.

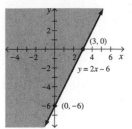

9. Graph $3x - 4y = 12$. Since the original statement is strictly greater than, a dashed line is drawn. Since the point (0, 0) does not satisfy the inequality $3x - 4y > 12$, all points in the half-plane below the line $3x - 4y = 12$ are in the solution set.

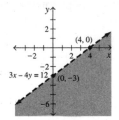

11. Graph $3x - 4y = 9$. Since the original statement is less than or equal to, a solid line is drawn. Since the point (0, 0) satisfies the inequality $3x - 4y \leq 9$, all points on the line and in the half-plane above the line $3x - 4y = 9$ are in the solution set.

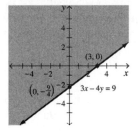

13. Graph $3x + 2y = 6$. Since the original statement is strictly less than, a dashed line is drawn. Since the point (0, 0) satisfies the inequality $3x + 2y < 6$, all points in the half-plane to the left of the line $3x + 2y = 6$ are in the solution set.

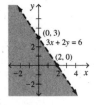

15. Graph $x + y = 0$. Since the original statement is strictly greater than, a dashed line is drawn. Since the point (1, 1) satisfies the inequality $x + y > 0$, all points in the half-plane above the line $x + y = 0$ are in the solution set.

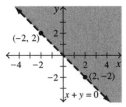

17. Graph $5x - 2y = 10$. Since the original statement is less than or equal to, a solid line is drawn. Since the point (0, 0) satisfies the inequality $5x - 2y \leq 10$, all points on the line and in the half-plane above the line $5x - 2y = 10$ are in the solution set.

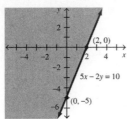

19. Graph $3x + 2y = 12$. Since the original statement is strictly greater than, a dashed line is drawn. Since the point (0, 0) does not satisfy the inequality $3x + 2y > 12$, all points in the half-plane above the line $3x + 2y = 12$ are in the solution set.

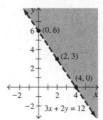

21. Graph $\dfrac{2}{5}x - \dfrac{1}{2}y = 1$. Since the original statement is less than or equal to, a solid line is drawn. Since the point (0, 0) satisfies the inequality $\dfrac{2}{5}x - \dfrac{1}{2}y \leq 1$, all points on the line and in the half-plane above the line $\dfrac{2}{5}x - \dfrac{1}{2}y = 1$ are in the solution set.

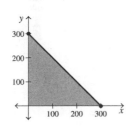

23. Graph $0.2x + 0.5y = 0.3$. Since the original statement is less than or equal to, a solid line is drawn. Since the point (0, 0) satisfies the inequality $0.2x + 0.5y \leq 0.3$, all points on the line and in the half-plane below the line $0.2x + 0.5y = 0.3$ are in the solution set.

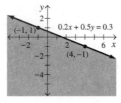

25. a) $x + y \leq 300$

    b)

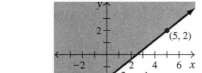

27. a) $x =$ the number of acres of land, $y =$ the number of square feet in the house

  b)

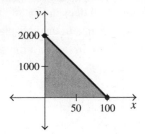

  c) $1500x + 75(1950) = 150,000$

  $$1500x + 146,250 = 150,000$$

  $$1500x = 3750$$

  $$x = 2.5 \text{ acres or less}$$

  d) $1500(5) + 75y = 150,000$

  $$7500 + 75y = 150,000$$

  $$75y = 142,500$$

  $$y = 1900 \text{ ft}^2 \text{ or less}$$

29. a) $3x - y < 6$

  $$3x - 3x - y < -3x + 6$$

  $$-y < -3x + 6$$

  $$\frac{-y}{-1} > \frac{-3x}{-1} + \frac{6}{-1}$$

  $$y > 3x - 6$$

  b) $-3x + y > -6$

  $$-3x + 3x + y > 3x - 6$$

  $$y > 3x - 6$$

  c) $3x - 2y < 12$

  $$3x - 3x - 2y < -3x + 12$$

  $$-2y < -3x + 12$$

  $$\frac{-2y}{-2} > \frac{-3x}{-2} + \frac{12}{-2}$$

  $$y > \frac{3}{2}x - 6$$

  d) $y > 3x - 6$

  a, b, and d

Exercise Set 6.9

1. A **binomial** is an expression that contains two terms in which each exponent that appears on the variable is a whole number. Examples: $2x+3$, $x-7$, x^2-9

3. The **FOIL method** is a method that obtains the products of the **F**irst, **O**uter, **I**nner, and **L**ast terms of the binomials.

5. $ax^2+bx+c=0$, $a\neq 0$

7. $x^2+9x+18=(x+6)(x+3)$

9. $x^2-x-6=(x-3)(x+2)$

11. $x^2+2x-24=(x+6)(x-4)$

13. $x^2-2x-3=(x+1)(x-3)$

15. $x^2-10x+21=(x-7)(x-3)$

17. $x^2-25=(x-5)(x+5)$

19. $x^2+3x-28=(x+7)(x-4)$

21. $x^2+2x-63=(x+9)(x-7)$

23. $2x^2-x-10=(2x-5)(x+2)$

25. $4x^2+13x+3=(4x+1)(x+3)$

27. $5x^2+12x+4=(5x+2)(x+2)$

29. $4x^2+11x+6=(4x+3)(x+2)$

31. $4x^2-11x+6=(4x-3)(x-2)$

33. $3x^2-14x-24=(3x+4)(x-6)$

35. $(x-1)(x+2)=0$
$x-1=0$ or $x+2=0$
$x=1$ $\qquad x=-2$

37. $(3x+4)(2x-1)=0$
$3x+4=0$ or $2x-1=0$
$3x=-4$ $\qquad 2x=1$
$x=-\dfrac{4}{3}$ $\qquad x=\dfrac{1}{2}$

39. $x^2+10x+21=0$
$(x+7)(x+3)=0$
$x+7=0$ or $x+3=0$
$x=-7$ $\qquad x=-3$

41. $x^2-4x+3=0$
$(x-3)(x-1)=0$
$x-3=0$ or $x-1=0$
$x=3$ $\qquad x=1$

43. $x^2-15=2x$
$x^2-2x-15=0$
$(x-5)(x+3)=0$
$x-5=0$ or $x+3=0$
$x=5$ $\qquad x=-3$

45. $x^2=4x-3$
$x^2-4x+3=0$
$(x-3)(x-1)=0$
$x-3=0$ or $x-1=0$
$x=3$ $\qquad x=1$

47. $x^2-81=0$
$(x-9)(x+9)=0$
$x-9=0$ or $x+9=0$
$x=9$ $\qquad x=-9$

49. $x^2+5x-36=0$
$(x+9)(x-4)=0$
$x+9=0$ or $x-4=0$
$x=-9$ $\qquad x=4$

51. $3x^2 + 10x = 8$

$3x^2 + 10x - 8 = 0$

$(3x - 2)(x + 4) = 0$

$3x - 2 = 0 \quad \text{or} \quad x + 4 = 0$

$3x = 2 \qquad\qquad x = -4$

$x = \dfrac{2}{3}$

53. $5x^2 + 11x = -2$

$5x^2 + 11x + 2 = 0$

$(5x + 1)(x + 2) = 0$

$5x + 1 = 0 \text{ or } x + 2 = 0$

$5x = -1 \qquad x = -2$

$x = -\dfrac{1}{5}$

55. $3x^2 - 4x = -1$

$3x^2 - 4x + 1 = 0$

$(3x - 1)(x - 1) = 0$

$3x - 1 = 0 \text{ or } x - 1 = 0$

$3x = 1 \qquad x = 1$

$x = \dfrac{1}{3}$

57. $4x^2 - 9x + 2 = 0$

$(4x - 1)(x - 2) = 0$

$4x - 1 = 0 \text{ or } x - 2 = 0$

$4x = 1 \qquad x = 2$

$x = \dfrac{1}{4}$

59. $x^2 + 2x - 15 = 0$

$a = 1, \ b = 2, \ c = -15$

$x = \dfrac{-2 \pm \sqrt{(2)^2 - 4(1)(-15)}}{2(1)}$

$x = \dfrac{-2 \pm \sqrt{4 + 60}}{2} = \dfrac{-2 \pm \sqrt{64}}{2} = \dfrac{-2 \pm 8}{2}$

$x = \dfrac{6}{2} = 3 \ \text{ or } \ x = \dfrac{-10}{2} = -5$

61. $x^2 - 3x - 18 = 0$

$a = 1, \ b = -3, \ c = -18$

$x = \dfrac{-(-3) \pm \sqrt{(-3)^2 - 4(1)(-18)}}{2(1)}$

$x = \dfrac{3 \pm \sqrt{9 + 72}}{2} = \dfrac{3 \pm \sqrt{81}}{2} = \dfrac{3 \pm 9}{2}$

$x = \dfrac{12}{2} = 6 \ \text{ or } \ x = \dfrac{-6}{2} = -3$

63. $x^2 - 8x = 9$

$x^2 - 8x - 9 = 0$

$a = 1, \ b = -8, \ c = -9$

$x = \dfrac{-(-8) \pm \sqrt{(-8)^2 - 4(1)(-9)}}{2(1)}$

$x = \dfrac{8 \pm \sqrt{64 + 36}}{2} = \dfrac{8 \pm \sqrt{100}}{2} = \dfrac{8 \pm 10}{2}$

$x = \dfrac{18}{2} = 9 \ \text{ or } \ x = \dfrac{-2}{2} = -1$

65. $x^2 - 2x + 3 = 0$

$a = 1, \ b = -2, \ c = 3$

$x = \dfrac{-(-2) \pm \sqrt{(-2)^2 - 4(1)(3)}}{2(1)}$

$x = \dfrac{2 \pm \sqrt{4 - 12}}{2} = \dfrac{2 \pm \sqrt{-8}}{2}$

No real solution

67. $x^2 - 4x + 2 = 0$

$a = 1, \ b = -4, \ c = 2$

$x = \dfrac{-(-4) \pm \sqrt{(-4)^2 - 4(1)(2)}}{2(1)}$

$x = \dfrac{4 \pm \sqrt{16 - 8}}{2} = \dfrac{4 \pm \sqrt{8}}{2} = \dfrac{4 \pm 2\sqrt{2}}{2}$

$x = 2 \pm \sqrt{2}$

69. $3x^2 - 8x + 1 = 0$

$a = 3, \ b = -8, \ c = 1$

$x = \dfrac{-(-8) \pm \sqrt{(-8)^2 - 4(3)(1)}}{2(3)}$

$x = \dfrac{8 \pm \sqrt{64 - 12}}{6} = \dfrac{8 \pm \sqrt{52}}{6} = \dfrac{8 \pm 2\sqrt{13}}{6}$

$x = \dfrac{4 \pm \sqrt{13}}{3}$

71. $4x^2 - x - 1 = 0$

$a = 4,\ b = -1,\ c = -1$

$$x = \frac{-(-1) \pm \sqrt{(-1)^2 - 4(4)(-1)}}{2(4)}$$

$$x = \frac{1 \pm \sqrt{1+16}}{8} = \frac{1 \pm \sqrt{17}}{8}$$

73. $2x^2 + 7x + 5 = 0$

$a = 2,\ b = 7,\ c = 5$

$$x = \frac{-7 \pm \sqrt{(7)^2 - 4(2)(5)}}{2(2)}$$

$$x = \frac{-7 \pm \sqrt{49 - 40}}{4} = \frac{-7 \pm \sqrt{9}}{4} = \frac{-7 \pm 3}{4}$$

$$x = \frac{-4}{4} = -1 \ \text{ or } \ x = \frac{-10}{4} = -\frac{5}{2}$$

75. $3x^2 - 10x + 7 = 0$

$a = 3,\ b = -10,\ c = 7$

$$x = \frac{-(-10) \pm \sqrt{(-10)^2 - 4(3)(7)}}{2(3)}$$

$$x = \frac{10 \pm \sqrt{100 - 84}}{6} = \frac{10 \pm \sqrt{16}}{6} = \frac{10 \pm 4}{6}$$

$$x = \frac{14}{6} = \frac{7}{3} \ \text{ or } \ x = \frac{6}{6} = 1$$

77. $4x^2 - 11x + 13 = 0$

$a = 4,\ b = -11,\ c = 13$

$$x = \frac{-(-11) \pm \sqrt{(-11)^2 - 4(4)(13)}}{2(4)}$$

$$x = \frac{11 \pm \sqrt{121 - 208}}{8} = \frac{11 \pm \sqrt{-87}}{8}$$

No real solution

79. Area of backyard $= lw = 30(20) = 600$ m^2

Let $x =$ width of grass around all sides of the flower garden

Width of flower garden $= 20 - 2x$

Length of flower garden $= 30 - 2x$

Area of flower garden $= lw = (30 - 2x)(20 - 2x)$

Area of grass $= 600 - (30 - 2x)(20 - 2x)$

$600 - (30 - 2x)(20 - 2x) = 336$

$600 - \left(600 - 100x + 4x^2\right) = 336$

$600 - 600 + 100x - 4x^2 = 336$

$-4x^2 + 100x = 336$

$4x^2 - 100x + 336 = 0$

$x^2 - 25x + 84 = 0$

$(x - 21)(x - 4) = 0$

$x - 21 = 0 \quad \text{or} \quad x - 4 = 0$

$x = 21 \qquad\qquad x = 4$

$x \neq 21$ since the width of the backyard is 20 m.

Width of grass $= 4$ m

Width of flower garden

$= 20 - 2x = 20 - 2(4) = 20 - 8 = 12$ m

Length of flower garden

$= 30 - 2x = 30 - 2(4) = 30 - 8 = 22$ m

81. a) Since the equation is equal to 6 and not 0, the zero-factor property cannot be used.

b) $(x - 4)(x - 7) = 6$

$x^2 - 11x + 28 = 6$

$x^2 - 11x + 22 = 0$

$a = 1,\ b = -11,\ c = 22$

$$x = \frac{-(-11) \pm \sqrt{(-11)^2 - 4(1)(22)}}{2(1)}$$

$$x = \frac{11 \pm \sqrt{121 - 88}}{2} = \frac{11 \pm \sqrt{33}}{2}$$

$x \approx 8.37 \ \text{ or } \ x \approx 2.63$

83. $(x + 1)(x - 3) = 0$

$x^2 - 2x - 3 = 0$

84.

```
R I A A S N E Y
A V T R R I U P
U I L O N R A B
R R L U A L G D
U S O T N A E C
A L U I O R B U
X S M S E I I N
F O R M M O N S
C G E M I A L P
```

Exercise Set 6.10

1. A **function** is a special type of relation where each value of the independent variable corresponds to a unique value of the dependent variable.

3. The **domain** of a function is the set of values that can be used for the independent variable.

5. The vertical line test can be used to determine if a graph represents a function. If a vertical line can be drawn so that it intersects the graph at more than one point, then each value of x does not have a unique value of y and the graph does not represent a function. If a vertical line cannot be made to intersect the graph in at least two different places, then the graph represents a function.

7. Not a function since $x = 2$ is not paired with a unique value of y.

9. Function since each vertical line intersects the graph at only one point.
D: all real numbers   R: all real numbers

11. Function since each vertical line intersects the graph at only one point.
D: all real numbers   R: $y = 2$

13. Function since each vertical line intersects the graph at only one point.
D: all real numbers   R: $y \geq -4$

15. Not a function since it is possible to draw a vertical line that intersects the graph at more than one point.

17. Function since each vertical line intersects the graph at only one point.
D: $0 \leq x < 12$ R: $y = 1, 2, 3$

19. Not a function since it is possible to draw a vertical line that intersects the graph at more than one point.

21. Function since each vertical line intersects the graph at only one point.
D: all real numbers   R: $y > 0$

23. Not a function since it is possible to draw a vertical line that intersects the graph at more than one point.

25. Function since each value of x is paired with a unique value of y.

27. Not a function since $x = 4$ is paired with two different values of y.

29. Function since each value of x is paired with a unique value of y.

31. $f(x) = x + 3, \quad x = 2$
$f(2) = 2 + 3 = 5$

33. $f(x) = -2x - 7, \quad x = -4$
$f(-4) = -2(-4) - 7 = 8 - 7 = 1$

35. $f(x) = 10x - 6, \quad x = 0$
$f(0) = 10(0) - 6 = 0 - 6 = -6$

37. $f(x) = x^2 - 3x + 1, \quad x = 4$
$f(4) = (4)^2 - 3(4) + 1 = 16 - 12 + 1 = 5$

39. $f(x) = 2x^2 - 2x - 8, \quad x = -2$
$f(-2) = 2(-2)^2 - 2(-2) - 8 = 8 + 4 - 8 = 4$

41. $f(x) = -3x^2 + 5x + 4, \quad x = -3$
$f(-3) = -3(-3)^2 + 5(-3) + 4 = -27 - 15 + 4 = -38$

43. $f(x) = -5x^2 + 3x - 9, \quad x = -1$
$f(-1) = -5(-1)^2 + 3(-1) - 9 = -5 - 3 - 9 = -17$

45.

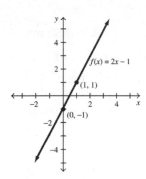

47.

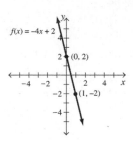

49.

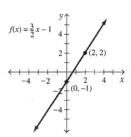

51. $y = x^2 - 16$

a) $a = 1 > 0$, opens upward

b) $x = 0$ c) $(0, -16)$ d) $(0, -16)$

e) $(4, 0), (-4, 0)$

f)

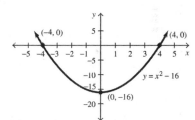

g) D: all real numbers   R: $y \geq -16$

53. $y = -x^2 + 4$

a) $a = -1 < 0$, opens downward

b) $x = 0$ c) $(0, 4)$ d) $(0, 4)$

e) $(-2, 0), (2, 0)$

f)

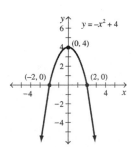

g) D: all real numbers   R: $y \leq 4$

55. $f(x) = -x^2 - 4$

a) $a = -1 < 0$, opens downward

b) $x = 0$ c) $(0, -4)$ d) $(0, -4)$

e) no x-intercepts

f)

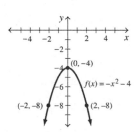

g) D: all real numbers   R: $y \leq -4$

57. $y = 2x^2 - 3$

a) $a = 2 > 0$, opens upward

b) $x = 0$ c) $(0, -3)$ d) $(0, -3)$

e) $(-1.22, 0), (1.22, 0)$

f)

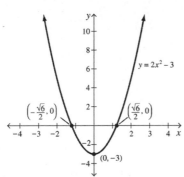

g) D: all real numbers   R: $y \geq -3$

59. $f(x) = x^2 + 2x + 6$

a) $a = 1 > 0$, opens upward

b) $x = -1$ c) $(-1, 5)$ d) $(0, 6)$

e) no x-intercepts

f)

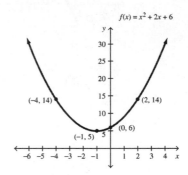

g) D: all real numbers   R: $y \geq 5$

61. $y = x^2 + 5x + 6$

a) $a = 1 > 0$, opens upward

b) $x = -\dfrac{5}{2}$ c) $(-2.5, -0.25)$ d) $(0, 6)$

e) $(-3, 0), (-2, 0)$

f)

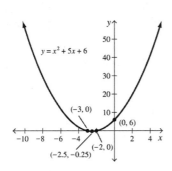

g) D: all real numbers   R: $y \geq -0.25$

63. $y = -x^2 + 4x - 6$

a) $a = -1 < 0$, opens downward

b) $x = 2$ c) $(2, -2)$ d) $(0, -6)$

e) no x-intercepts

f)

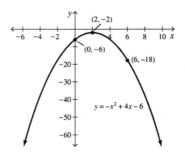

g) D: all real numbers   R: $y \leq -2$

65. $y = -3x^2 + 14x - 8$

a) $a = -3 < 0$, opens downward

b) $x = \dfrac{7}{3}$ c) $\left(\dfrac{7}{3}, \dfrac{25}{3}\right)$ d) $(0, -8)$

e) $\left(\dfrac{2}{3}, 0\right), (4, 0)$

f)

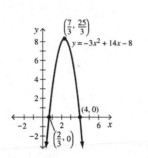

g) D: all real numbers   R: $y \le \dfrac{25}{3}$

67.

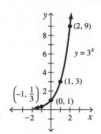

D: all real numbers   R: $y > 0$

69.

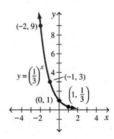

D: all real numbers   R: $y > 0$

71.

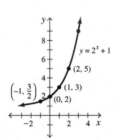

D: all real numbers   R: $y > 1$

73.

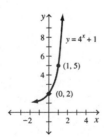

D: all real numbers   R: $y > 1$

75.

D: all real numbers   R: $y > 0$

77.

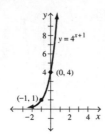

D: all real numbers   R: $y > 0$

79. $m(s) = 300 + 0.10s$

$$m(20,000) = 300 + 0.10(20,000)$$
$$= 300 + 2000 = \$2300$$

81. a) $2000 \rightarrow x = 8$

$$f(8) = 0.56(8)^2 - 5.43(8) + 59.83$$
$$= 35.84 - 43.44 + 59.83 = 52.23\%$$

b) 1997

c) $x = \dfrac{-b}{2a} = \dfrac{-(-5.43)}{2(0.56)} = \dfrac{5.43}{1.12} = 4.848214286$

≈ 4.85

$$f(4.85) = 0.56(4.85)^2 - 5.43(4.85) + 59.83$$
$$= 13.1726 - 26.3355 + 59.83$$
$$= 46.6671 \approx 46.67\%$$

85. a) Yes

b) ≈ 6500 scooter injuries

83. $P(x) = 4000(1.3)^{0.1x}$

a) $x = 10$, $P(10) = 4000(1.3)^{0.1(10)}$

$$= 4000(1.3) = 5200 \text{ people}$$

b) $x = 50$, $P(50) = 4000(1.3)^{0.1(50)}$

$$= 4000(3.71293)$$
$$= 14,851.72 \approx 14,852 \text{ people}$$

87. $d = (21.9)(2)^{(20-x)/12}$

a) $x = 19$, $d = (21.9)(2)^{(20-19)/12}$

$$= (21.9)(1.059463094) \approx 23.2 \text{ cm}$$

b) $x = 4$, $d = (21.9)(2)^{(20-4)/12}$

$$= (21.9)(2.519842099) \approx 55.2 \text{ cm}$$

c) $x = 0$, $d = (21.9)(2)^{(20-0)/12}$

$$= (21.9)(3.174802105) \approx 69.5 \text{ cm}$$

89. $f(x) = -0.85x + 187$

a) $f(20) = -0.85(20) + 187 = 170$ beats per minute

b) $f(30) = -0.85(30) + 187 = 161.5 \approx 162$ beats per minute

c) $f(50) = -0.85(50) + 187 = 144.5 \approx 145$ beats per minute

d) $f(60) = -0.85(60) + 187 = 136$ beats per minute

e) $-0.85x + 187 = 85$

$-0.85x = -102$

$x = 120$ years of age

Review Exercises

1. $x = 3, \quad x^2 + 12 = (3)^2 + 12 = 9 + 12 = 21$

2. $x = -1, \quad -x^2 - 9 = -(-1)^2 - 9 = -1 - 9 = -10$

3. $x = 2, 4x^2 - 2x + 5 = 4(2)^2 - 2(2) + 5$
   $= 16 - 4 + 5 = 17$

4. $x = \dfrac{1}{2}, \quad -x^2 + 7x - 3 = -\left(\dfrac{1}{2}\right)^2 + 7\left(\dfrac{1}{2}\right) - 3$
   $$= -\frac{1}{4} + \frac{14}{4} - \frac{12}{4} = \frac{1}{4}$$

5. $x = -2, \ 4x^3 - 7x^2 + 3x + 1$
   $= 4(-2)^3 - 7(-2)^2 + 3(-2) + 1$
   $= -32 - 28 - 6 + 1 = -65$

6. $x = 1, y = -2, \quad 3x^2 - xy + 2y^2$
   $= 3(1)^2 - 1(-2) + 2(-2)^2$
   $= 3 + 2 + 8 = 13$

7. $3x - 4 + x + 5 = 4x + 1$

8. $3x + 4(x - 2) + 6x = 3x + 4x - 8 + 6x = 13x - 8$

9. $4(x - 1) + \dfrac{1}{3}(9x + 3) = 4x - 4 + 3x + 1 = 7x - 3$

10.
$$4s + 10 = -30$$
$$4s + 10 - 10 = -30 - 10$$
$$4s = -40$$
$$\frac{4s}{4} = \frac{-40}{4}$$
$$s = -10$$

11.
$$3t + 8 = 6t - 13$$
$$3t - 3t + 8 = 6t - 3t - 13$$
$$8 = 3t - 13$$
$$8 + 13 = 3t - 13 + 13$$
$$21 = 3t$$
$$\frac{21}{3} = \frac{3t}{3}$$
$$7 = t$$

12.
$$\frac{x + 5}{6} = \frac{x - 3}{3}$$
$$3(x + 5) = 6(x - 3)$$
$$3x + 15 = 6x - 18$$
$$3x - 3x + 15 = 6x - 3x - 18$$
$$15 = 3x - 18$$
$$15 + 18 = 3x - 18 + 18$$
$$33 = 3x$$
$$\frac{33}{3} = \frac{3x}{3}$$
$$11 = x$$

13.
$$4(x - 2) = 3 + 5(x + 4)$$
$$4x - 8 = 3 + 5x + 20$$
$$4x - 8 = 5x + 23$$
$$4x - 4x - 8 = 5x - 4x + 23$$
$$-8 = x + 23$$
$$-8 - 23 = x + 23 - 23$$
$$-31 = x$$

14.
$$\frac{x}{4} + \frac{3}{5} = 7$$
$$20\left(\frac{x}{4} + \frac{3}{5}\right) = 20(7)$$
$$5x + 12 = 140$$
$$5x + 12 - 12 = 140 - 12$$
$$5x = 128$$
$$\frac{5x}{5} = \frac{128}{5}$$
$$x = \frac{128}{5}$$

15. $\dfrac{\frac{2}{1}}{\frac{1}{3}}=\dfrac{3}{x}$

$2x=3\left(\dfrac{1}{3}\right)$

$2x=1$

$\dfrac{2x}{2}=\dfrac{1}{2}$

$x=\dfrac{1}{2}$ cup

16. 1 hr 40 min = 60 min + 40 min = 100 min

$\dfrac{120}{100}=\dfrac{300}{x}$

$120x=100(300)$

$120x=30{,}000$

$\dfrac{120x}{120}=\dfrac{30{,}000}{120}$

$x=250$ min, or 4 hr 10 min

17. $A=bh$

$A=12(4)=48$

18. $V=2\pi R^2 r^2$

$V=2(3.14)(3)^2(1.75)^2$

$V=2(3.14)(9)(3.0625)$

$V=173.0925\approx173.1$

19. $z=\dfrac{\overline{x}-\mu}{\frac{\sigma}{\sqrt{n}}}$

$2=\dfrac{\overline{x}-100}{\frac{3}{\sqrt{16}}}$

$\dfrac{2}{1}=\dfrac{\overline{x}-100}{\frac{3}{4}}$

$2\left(\dfrac{3}{4}\right)=1(\overline{x}-100)$

$\dfrac{3}{2}=\overline{x}-100$

$\dfrac{3}{2}+100=\overline{x}-100+100$

$\dfrac{3}{2}+\dfrac{200}{2}=\overline{x}$

$\dfrac{203}{2}=\overline{x}$

$101.5=\overline{x}$

20. $K=\dfrac{1}{2}mv^2$

$4500=\dfrac{1}{2}m(30)^2$

$4500=450m$

$\dfrac{4500}{450}=\dfrac{450m}{450}$

$10=m$

21. $3x - 9y = 18$

$3x - 3x - 9y = -3x + 18$

$-9y = -3x + 18$

$\dfrac{-9y}{-9} = \dfrac{-3x + 18}{-9}$

$y = \dfrac{-3x + 18}{-9} = \dfrac{-(-3x + 18)}{9}$

$= \dfrac{3x - 18}{9} = \dfrac{3x}{9} - \dfrac{18}{9} = \dfrac{1}{3}x - 2$

22. $2x + 5y = 12$

$2x - 2x + 5y = -2x + 12$

$5y = -2x + 12$

$\dfrac{5y}{5} = \dfrac{-2x + 12}{5} = -\dfrac{2}{5}x + \dfrac{12}{5}$

23. $2x - 3y + 52 = 30$

$2x - 2x - 3y + 52 = -2x + 30$

$-3y + 52 = -2x + 30$

$-3y + 52 - 52 = -2x + 30 - 52$

$-3y = -2x - 22$

$\dfrac{-3y}{-3} = \dfrac{-2x - 22}{-3}$

$y = \dfrac{-2x - 22}{-3} = \dfrac{2x + 22}{3} = \dfrac{2}{3}x + \dfrac{22}{3}$

24. $-3x - 4y + 5z = 4$

$-3x + 3x - 4y + 5z = 3x + 4$

$-4y + 5z = 3x + 4$

$-4y + 5z - 5z = 3x - 5z + 4$

$-4y = 3x - 5z + 4$

$\dfrac{-4y}{-4} = \dfrac{3x - 5z + 4}{-4}$

$y = \dfrac{3x - 5z + 4}{-4}$

$= \dfrac{-(3x - 5z + 4)}{4}$

$= \dfrac{-3x + 5z - 4}{4}$

$= -\dfrac{3}{4}x + \dfrac{5}{4}z - 1$

25. $A = lw$

$\dfrac{A}{l} = \dfrac{lw}{l}$

$\dfrac{A}{l} = w$

26. $P = 2l + 2w$

$P - 2l = 2l - 2l + 2w$

$P - 2l = 2w$

$\dfrac{P - 2l}{2} = \dfrac{2w}{2}$

$\dfrac{P - 2l}{2} = w$

27. $L = 2(wh + lh)$

$L = 2wh + 2lh$

$L - 2wh = 2wh - 2wh + 2lh$

$L - 2wh = 2lh$

$\dfrac{L - 2wh}{2h} = \dfrac{2lh}{2h}$

$\dfrac{L - 2wh}{2h} = l$ or $l = \dfrac{L}{2h} - \dfrac{2wh}{2h} = \dfrac{L}{2h} - w$

28. $a_n = a_1 + (n - 1)d$

$a_n - a_1 = a_1 - a_1 + (n - 1)d$

$a_n - a_1 = (n - 1)d$

$\dfrac{a_n - a_1}{n - 1} = \dfrac{(n - 1)d}{n - 1}$

$\dfrac{a_n - a_1}{n - 1} = d$

29. $8+2x$

30. $3y-7$

31. $10+3r$

32. $\dfrac{8}{q}-11$

33.

$$\text{Let } x = \text{the number}$$
$$3x = 3 \text{ times the number}$$
$$4+3x = 4 \text{ increased by } 3 \text{ times the number}$$
$$4+3x = 22$$
$$4-4+3x = 22-4$$
$$3x = 18$$
$$\frac{3x}{3} = \frac{18}{3}$$
$$x = 6$$

34. Let $x =$ the number

$$3x = \text{the product of } 3 \text{ and a number}$$
$$3x+8 = \text{the product of } 3 \text{ and a number}$$
$$\text{increased by } 8$$
$$x-6 = 6 \text{ less than the number}$$
$$3x+8 = x-6$$
$$3x-x+8 = x-x-6$$
$$2x+8 = -6$$
$$2x+8-8 = -6-8$$
$$2x = -14$$
$$\frac{2x}{2} = \frac{-14}{2}$$
$$x = -7$$

35. Let $x =$ the number

$$x-4 = \text{the difference of a number and } 4$$
$$5(x-4) = 5 \text{ times the difference of a number}$$
$$\text{and } 4$$
$$5(x-4) = 45$$
$$5x-20 = 45$$
$$5x-20+20 = 45+20$$
$$5x = 65$$
$$\frac{5x}{5} = \frac{65}{5}$$
$$x = 13$$

36. Let $x =$ the number

$$10x = 10 \text{ times a number}$$
$$10x+14 = 14 \text{ more than } 10 \text{ times a number}$$
$$x+12 = \text{the sum of a number and } 12$$
$$8(x+12) = 8 \text{ times the sum of a number and } 12$$
$$10x+14 = 8(x+12)$$
$$10x+14 = 8x+96$$
$$10x-8x+14 = 8x-8x+96$$
$$2x+14 = 96$$
$$2x+14-14 = 96-14$$
$$2x = 82$$
$$\frac{2x}{2} = \frac{82}{2}$$
$$x = 41$$

37. Let $x =$ the amount invested in bonds

$$2x = \text{the amount invested in mutual funds}$$
$$x+2x = 15,000$$
$$3x = 15,000$$
$$\frac{3x}{3} = \frac{15,000}{3}$$
$$x = \$5000 \text{ in bonds}$$
$$2x = 2(5000) = \$10,000 \text{ in mutual funds}$$

38. Let $x =$ number of lawn chairs

$$9.50x = \text{variable cost per lawn chair}$$
$$9.50x+15,000 = 95,000$$
$$9.50x+15,000-15,000 = 95,000-15,000$$
$$9.50x = 80,000$$
$$\frac{9.50x}{9.50} = \frac{80,000}{9.50}$$
$$x = 8421.052632 \approx 8421 \text{ lawn chairs}$$

39. Let $x =$ the number of species at the
     Philadelphia Zoo

$2x + 140 =$ the number of species at the
     San Diego Zoo

$$x + 2x + 140 = 1130$$
$$3x + 140 = 1130$$
$$3x + 140 - 140 = 1130 - 140$$
$$3x = 990$$
$$\frac{3x}{3} = \frac{990}{3}$$
$$x = 330 \text{ species at the}$$
     Philadelphia Zoo

$$2x + 140 = 2(330) + 140 = 660 + 140$$
$$= 800 \text{ species at the San Diego Zoo}$$

40. Let $x =$ profit at restaurant B

$x + 12{,}000 =$ profit at restaurant A

$$x + (x + 12{,}000) = 68{,}000$$
$$2x + 12{,}000 = 68{,}000$$
$$2x + 12{,}000 - 12{,}000 = 68{,}000 - 12{,}000$$
$$2x = 56{,}000$$
$$\frac{2x}{2} = \frac{56{,}000}{2}$$
$$x = \$28{,}000 \text{ for restaurant B}$$

$x + 12{,}000 = 28{,}000 + 12{,}000 = \$40{,}000$ for restaurant A

41. $$s = \frac{k}{t}$$
$$10 = \frac{k}{3}$$
$$k = 10(3) = 30$$
$$s = \frac{30}{t}$$
$$s = \frac{30}{5} = 6$$

42. $$J = kA^2$$
$$32 = k(4)^2$$
$$32 = 16k$$
$$\frac{32}{16} = \frac{16k}{16}$$
$$k = 2$$
$$J = 2A^2$$
$$J = 2(7)^2 = 2(49) = 98$$

43. $$W = \frac{kL}{A}$$
$$80 = \frac{k(100)}{20}$$
$$100k = 1600$$
$$\frac{100k}{100} = \frac{1600}{100}$$
$$k = 16$$
$$W = \frac{16L}{A}$$
$$W = \frac{16(50)}{40} = \frac{800}{40} = 20$$

44. $$z = \frac{kxy}{r^2}$$
$$12 = \frac{k(20)(8)}{(8)^2}$$
$$160k = 768$$
$$\frac{160k}{160} = \frac{768}{160}$$
$$k = 4.8$$
$$z = \frac{4.8xy}{r^2}$$
$$z = \frac{4.8(10)(80)}{(3)^2} = \frac{3840}{9} = 426.\overline{6} \approx 426.7$$

45. a)
$$\frac{30 \text{ lb}}{2500 \text{ ft}^2} = \frac{x \text{ lb}}{12,500 \text{ ft}^2}$$
$$30(12,500) = 2500x$$
$$375,000 = 2500x$$
$$\frac{375,000}{2500} = \frac{2500x}{2500}$$
$$x = 150 \text{ lb}$$

b) $\dfrac{150}{30} = 5$ bags

46.
$$\frac{1 \text{ in.}}{30 \text{ mi}} = \frac{x \text{ in.}}{120 \text{ mi}}$$
$$30x = 120$$
$$\frac{30x}{30} = \frac{120}{30}$$
$$x = 4 \text{ in.}$$

47. $\dfrac{1 \; kWh}{\$0.162} = \dfrac{740 \; kWh}{x}$

$x = \$119.88$

48. $\quad d = kt^2$

$16 = k(1)^2$

$k = 16$

$d = 16t^2$

$d = 16(5)^2 = 16(25) = 400 \text{ ft}$

49.
$$5 + 9x \le 7x - 7$$
$$5 - 5 + 9x \le 7x - 7 - 5$$
$$9x \le 7x - 12$$
$$9x - 7x \le 7x - 7x - 12$$
$$2x \le -12$$
$$\frac{2x}{2} \le \frac{-12}{2}$$
$$x \le -6$$

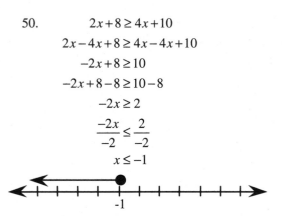

50.
$$2x + 8 \ge 4x + 10$$
$$2x - 4x + 8 \ge 4x - 4x + 10$$
$$-2x + 8 \ge 10$$
$$-2x + 8 - 8 \ge 10 - 8$$
$$-2x \ge 2$$
$$\frac{-2x}{-2} \le \frac{2}{-2}$$
$$x \le -1$$

51.
$$3(x + 9) \le 4x + 11$$
$$3x + 27 \le 4x + 11$$
$$3x - 4x + 27 \le 4x - 4x + 11$$
$$-x + 27 \le 11$$
$$-x + 27 - 27 \le 11 - 27$$
$$-x \le -16$$
$$\frac{-x}{-1} \ge \frac{-16}{-1}$$
$$x \ge 16$$

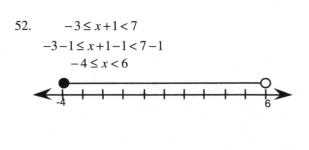

52.
$$-3 \le x + 1 < 7$$
$$-3 - 1 \le x + 1 - 1 < 7 - 1$$
$$-4 \le x < 6$$

53. $2 + 5x > -8$

$2 - 2 + 5x > -8 - 2$

$5x > -10$

$\dfrac{5x}{5} > \dfrac{-10}{5}$

$x > -2$

54. $5x + 13 \geq -22$

$5x + 13 - 13 \geq -22 - 13$

$5x \geq -35$

$\dfrac{5x}{5} \geq \dfrac{-35}{5}$

$x \geq -7$

55. $-1 < x \leq 9$

56. $-8 \leq x + 2 \leq 7$

$-8 - 2 \leq x + 2 - 2 \leq 7 - 2$

$-10 \leq x \leq 5$

57. - 60.

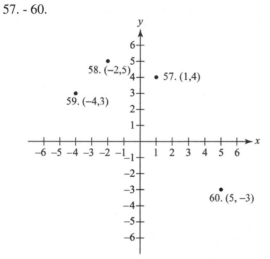

61.

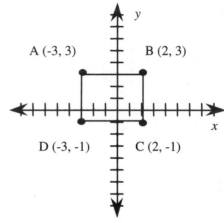

Area $= lw = 5(4) = 20$ square units

62.

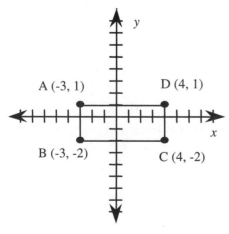

Area $= lw = 7(3) = 21$ square units

63.

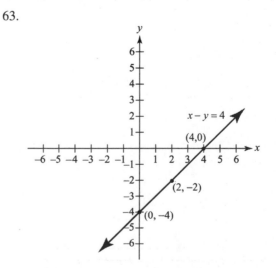

64.

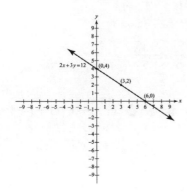

65.

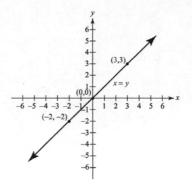

66.

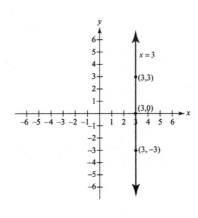

67.

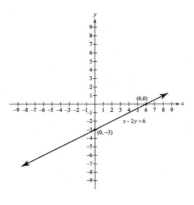

68.

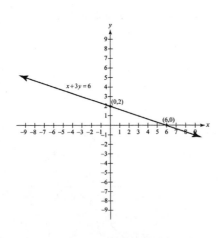

69.

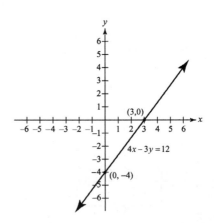

70.

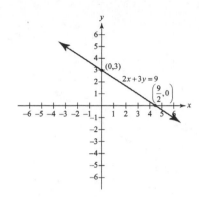

71. $m = \dfrac{5-3}{6-1} = \dfrac{2}{5}$

72. $m = \dfrac{-4-(-1)}{5-3} = \dfrac{-4+1}{5-3} = -\dfrac{3}{2}$

73. $m = \dfrac{3-(-4)}{2-(-1)} = \dfrac{3+4}{2+1} = \dfrac{7}{3}$

74. $m = \dfrac{-2-2}{6-6} = \dfrac{-4}{0}$ Undefined

75.

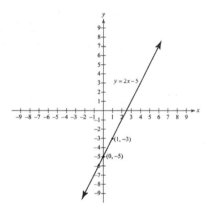

76.

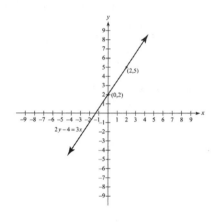

77.

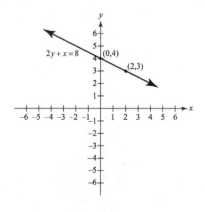

78.

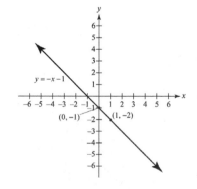

79. The y-intercept is 4, thus $b = 4$. Since the graph rises from left to right, the slope is positive. The change in y is 4 units while the change in x is 2. Thus, m, the slope is $\dfrac{4}{2}$ or 2. The equation is $y = 2x + 4$.

80. The y-intercept is 1, thus $b = 1$. Since the graph falls from left to right, the slope is negative. The change in y is 3 units while the change in x is 3. Thus, m, the slope is $\dfrac{-3}{3}$ or -1. The equation is $y = -x + 1$.

81. a)

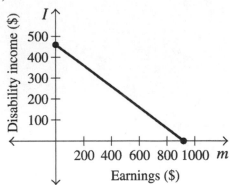

    b)   About $160

    c)   About $160

82. a)

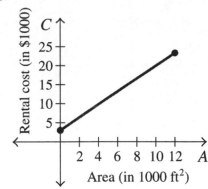

    b)   About $6400

    c)   About 4120 ft^2

83. Graph $4x + 3y = 12$. Since the original inequality is less than or equal to, a solid line is drawn. Since the point (0, 0) satisfies the inequality $4x + 3y \le 12$, all points on the line and in the half-plane below the line $4x + 3y = 12$ are in the solution set.

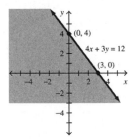

84. Graph $3x + 2y = 12$. Since the original inequality is greater than or equal to, a solid line is drawn. Since the point (0, 0) does not satisfy the inequality $3x + 2y \ge 12$, all points in the half plane above the line $3x + 2y = 12$ are in the solution set.

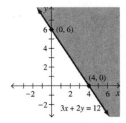

85. Graph $2x - 3y = 12$. Since the original inequality is strictly greater than, a dashed line is drawn. Since the point (0, 0) does not satisfy the inequality $2x - 3y > 12$, all points in the half-plane below the line $2x - 3y = 12$ are in the solution set.

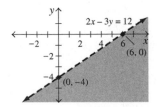

86. Graph $-7x - 2y = 14$. Since the original inequality is strictly less than, a dashed line is drawn. Since the point (0, 0) satisfies the inequality $-7x - 2y < 14$, all points in the half-plane to the right of the line $-7x - 2y = 14$ are in the solution set.

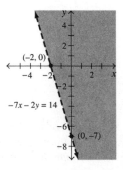

87. $x^2 + 9x + 18 = (x + 3)(x + 6)$

88. $x^2 + x - 20 = (x + 5)(x - 4)$

89. $x^2 - 10x + 24 = (x - 6)(x - 4)$

90. $x^2 - 9x + 20 = (x - 5)(x - 4)$

91. $6x^2 + 7x - 3 = (3x - 1)(2x + 3)$

92. $2x^2 + 13x - 7 = (2x - 1)(x + 7)$

93. $x^2 + 3x + 2 = 0$

$(x+1)(x+2) = 0$

$x+1 = 0$ or $x+2 = 0$

$x = -1 \qquad x = -2$

94. $x^2 - 5x = -4$

$x^2 - 5x + 4 = 0$

$(x-1)(x-4) = 0$

$x-1 = 0$ or $x-4 = 0$

$x = 1 \qquad x = 4$

95. $3x^2 - 17x + 10 = 0$

$(3x-2)(x-5) = 0$

$3x-2 = 0$ or $x-5 = 0$

$3x = 2 \qquad x = 5$

$x = \dfrac{2}{3}$

96. $3x^2 = -7x - 2$

$3x^2 + 7x + 2 = 0$

$(x+2)(3x+1) = 0$

$x+2 = 0$ or $3x+1 = 0$

$x = -2 \qquad 3x = -1$

$\qquad\qquad x = -\dfrac{1}{3}$

97. $x^2 - 4x - 1 = 0$

$a = 1, \ b = -4, \ c = -1$

$x = \dfrac{-(-4) \pm \sqrt{(-4)^2 - 4(1)(-1)}}{2(1)}$

$x = \dfrac{4 \pm \sqrt{16+4}}{2} = \dfrac{4 \pm \sqrt{20}}{2} = \dfrac{4 \pm 2\sqrt{5}}{2} = 2 \pm \sqrt{5}$

98. $x^2 - 3x + 2 = 0$

$a = 1, \ b = -3, \ c = 2$

$x = \dfrac{-(-3) \pm \sqrt{(-3)^2 - 4(1)(2)}}{2(1)}$

$x = \dfrac{3 \pm \sqrt{9-8}}{2} = \dfrac{3 \pm \sqrt{1}}{2} = \dfrac{3 \pm 1}{2}$

$x = \dfrac{4}{2} = 2$ or $x = \dfrac{2}{2} = 1$

99. $2x^2 - 3x + 4 = 0$

$a = 2, \ b = -3, \ c = 4$

$x = \dfrac{-(-3) \pm \sqrt{(-3)^2 - 4(2)(4)}}{2(2)}$

$x = \dfrac{3 \pm \sqrt{9-32}}{4} = \dfrac{3 \pm \sqrt{-23}}{4}$

No real solution

100. $2x^2 - x - 3 = 0$

$a = 2, \ b = -1, \ c = -3$

$x = \dfrac{-(-1) \pm \sqrt{(-1)^2 - 4(2)(-3)}}{2(2)}$

$x = \dfrac{1 \pm \sqrt{1+24}}{4} = \dfrac{1 \pm \sqrt{25}}{4} = \dfrac{1 \pm 5}{4}$

$x = \dfrac{6}{4} = \dfrac{3}{2}$ or $x = \dfrac{-4}{4} = -1$

101. Function since each value of x is paired with a unique value of y.

D: $x = $ -2, -1, 2, 3 \qquad R: $y = $ -1, 0, 2

102. Not a function since it is possible to draw a vertical line that intersects the graph at more than one point.

103. Not a function since it is possible to draw a vertical line that intersects the graph at more than one point.

104. Function since each vertical line intersects the graph at only one point.

D: all real numbers \qquad R: all real numbers

105. $f(x) = 5x - 2, \ x = 4$

$f(4) = 5(4) - 2 = 20 - 2 = 18$

106. $f(x) = -2x + 7, \ x = -3$

$f(-3) = -2(-3) + 7 = 6 + 7 = 13$

107. $f(x) = 2x^2 - 3x + 4$, $x = 5$

$f(5) = 2(5)^2 - 3(5) + 4 = 50 - 15 + 4 = 39$

108. $f(x) = -4x^2 + 7x + 9$, $x = 4$

$f(4) = -4(4)^2 + 7(4) + 9 = -64 + 28 + 9 = -27$

109. $y = -x^2 - 4x + 21$

a) $a = -1 < 0$, opens downward

b) $x = -2$ c) $(-2, 25)$ d) $(0, 21)$

e) $(-7, 0), (3, 0)$

f)

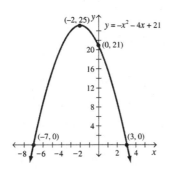

g) D: all real numbers   R: $y \le 25$

110. $f(x) = 2x^2 - 8x + 10$

a) $a = 2 > 0$, opens upward

b) $x = 2$ c) $(2, 2)$ d) $(0, 10)$

e) no x-intercepts

f)

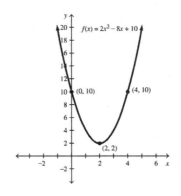

g) D: all real numbers   R: $y \ge 2$

111.

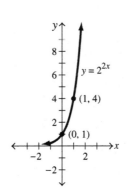

D: all real numbers   R: $y > 0$

112.

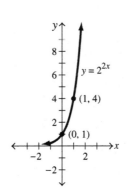

D: all real numbers   R: $y > 0$

113. $m = 30 - 0.002n^2$, $n = 60$

$m = 30 - 0.002(60)^2 = 30 - 0.002(3600)$

$\quad = 30 - 7.2 = 22.8$ mpg

114. $n = 2a^2 - 80a + 5000$

a) $a = 18$

$n = 2(18)^2 - 80(18) + 5000$

$\quad = 648 - 1440 + 5000 = 4208$

b) $a = 25$

$n = 2(25)^2 - 80(25) + 5000$

$\quad = 1250 - 2000 + 5000 = 4250$

115. $P = 100(0.92)^x$, $x = 4.5$

$P = 100(0.92)^{4.5}$

$\quad = 100(0.6871399881) = 68.71399881 \approx 68.7\%$

Chapter Test

1. $3x^2 + 4x - 1, \quad x = -2$

    $3(-2)^2 + 4(-2) - 1 = 12 - 8 - 1 = 3$

2. $3x + 5 = 2(4x - 7)$

    $3x + 5 = 8x - 14$

    $3x - 8x + 5 = 8x - 8x - 14$

    $-5x + 5 = -14$

    $-5x + 5 - 5 = -14 - 5$

    $-5x = -19$

    $\dfrac{-5x}{-5} = \dfrac{-19}{-5}$

    $x = \dfrac{19}{5}$

3. $-2(x - 3) + 6x = 2x + 3(x - 4)$

    $-2x + 6 + 6x = 2x + 3x - 12$

    $4x + 6 = 5x - 12$

    $4x - 5x + 6 = 5x - 5x - 12$

    $-x + 6 = -12$

    $-x + 6 - 6 = -12 - 6$

    $-x = -18$

    $\dfrac{-x}{-1} = \dfrac{-18}{-1}$

    $x = 18$

4. Let x = the number

    $2x$ = the product of the number and 2

    $2x + 7$ = the product of the number and 2, increased by 7

    $2x + 7 = 25$

    $2x + 7 - 7 = 25 - 7$

    $2x = 18$

    $\dfrac{2x}{2} = \dfrac{18}{2}$

    $x = 9$

5. Let x = the cost of the car before tax

    $0.07x$ = the amount of the sales tax

    $x + 0.07x = 26{,}750$

    $1.07x = 26{,}750$

    $\dfrac{1.07x}{1.07} = \dfrac{26{,}750}{1.07}$

    $x = \$25{,}000$

6. $L = ah + bh + ch; \ a = 3, \ b = 4, \ c = 5, \ h = 7$

    $L = 3(7) + 4(7) + 5(7)$

    $= 21 + 28 + 35 = 84$

7. $3x + 5y = 11$

    $3x - 3x + 5y = -3x + 11$

    $5y = -3x + 11$

    $\dfrac{5y}{5} = \dfrac{-3x + 11}{5}$

    $y = \dfrac{-3x + 11}{5} = -\dfrac{3}{5}x + \dfrac{11}{5}$

8. $L = \dfrac{kMN}{P}$

    $12 = \dfrac{k(8)(3)}{2}$

    $24k = 24$

    $k = \dfrac{24}{24} = 1$

    $L = \dfrac{(1)MN}{P}$

    $L = \dfrac{(1)(10)(5)}{15} = \dfrac{50}{15} = 3.\overline{3} = 3\dfrac{1}{3}$

9. $l = \dfrac{k}{w}$

    $15 = \dfrac{k}{9}$

    $k = 15(9) = 135$

    $l = \dfrac{135}{w}$

    $l = \dfrac{135}{20} = 6.75 \text{ ft}$

10. $-3x + 11 \le 5x + 35$

    $-3x - 5x + 11 \le 5x - 5x + 35$

    $-8x + 11 \le 35$

    $-8x + 11 - 11 \le 35 - 11$

    $-8x \le 24$

    $\dfrac{-8x}{-8} \ge \dfrac{24}{-8}$

    $x \ge -3$

11. $m = \dfrac{12 - 5}{7 - (-3)} = \dfrac{12 - 5}{7 + 3} = \dfrac{7}{10}$

12.

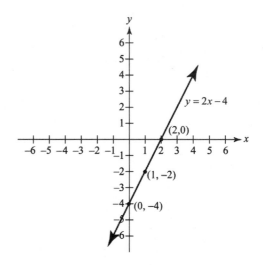

13.

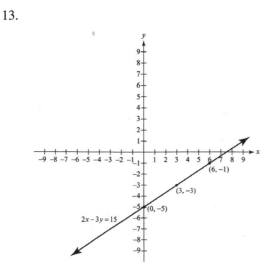

14. Graph $3y = 5x - 12$. Since the original statement is greater than or equal to, a solid line is drawn. Since the point (0, 0) satisfies the inequality $3y \ge 5x - 12$, all points on the line and in the half-plane above the line $3y = 5x - 12$ are in the solution set.

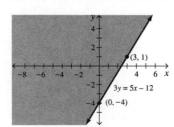

15. $x^2 - 3x = 28$

$x^2 - 3x - 28 = 0$

$(x - 7)(x + 4) = 0$

$x - 7 = 0$ or $x + 4 = 0$

$x = 7$. $x = -4$

16. $3x^2 + 2x = 8$

$3x^2 + 2x - 8 = 0$

$a = 3,\ b = 2,\ c = -8$

$x = \dfrac{-2 \pm \sqrt{(2)^2 - 4(3)(-8)}}{2(3)}$

$x = \dfrac{-2 \pm \sqrt{4 + 96}}{6} = \dfrac{-2 \pm \sqrt{100}}{6} = \dfrac{-2 \pm 10}{6}$

$x = \dfrac{8}{6} = \dfrac{4}{3}$ or $x = \dfrac{-12}{6} = -2$

17. Function since each vertical line intersects the graph at only one point.

18. $f(x) = -4x^2 - 11x + 5,\ x = -2$

$f(-2) = -4(-2)^2 - 11(-2) + 5$

$= -16 + 22 + 5 = 11$

19. $y = x^2 - 2x + 4$

a) $a = 1 > 0$, opens upward

b) $x = 1$ c) $(1, 3)$ d) $(0, 4)$

e) no x-intercepts

f)

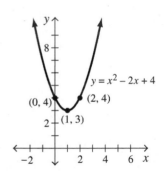

g) D: all real numbers   R: $y \geq 3$

CHAPTER SEVEN

SYSTEMS OF LINEAR EQUATIONS AND INEQUALITIES

Exercise Set 7.1

1. Two or more linear equations form a system of linear equations.

3. A consistent system of equations is a system that has a solution.

5. An inconsistent system of equations is a system that has no solution.

7. The graphs of the system of equations are parallel and do not intersect.

9. The graphs of the system of equations are in fact the same line.

11. $y = 2x - 6$ $y = -x + 3$ Therefore, $(3, 0)$ is a solution.

    $(0) = 2(3) - 6$ $(0) = -(3) + 3$

      $0 = 6 - 6$ $0 = 0$

      $0 = 0$

13.

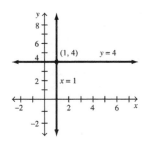

15.

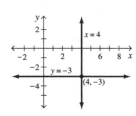

17.

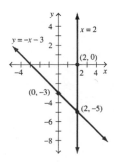

19.

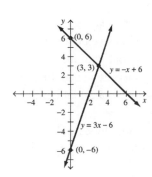

21.

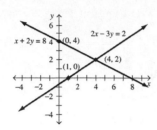

23.

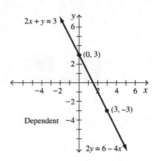

25.

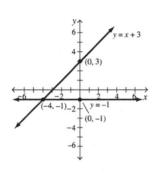

27.

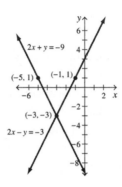

29.

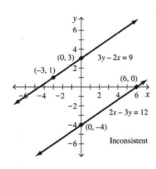

31.

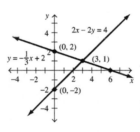

33. a) Two lines with different slopes are not parallel, and therefore have exactly one point of intersection giving one solution.

b) Two lines with the same slope and different y- intercepts are distinct parallel lines and have no solution.

c) Two lines with the same slopes and y-intercepts have infinitely many solutions, each point on the line.

35. $2x - y = 6$ $y = 2x - 6$
same slope, same y-intercept;
infinite number of solutions

37. $3x - 4y = 5$ $y = -3x + 8$
different slopes, different y-intercepts;
1 solution

39. $3x + y = 7$ $y = -3x + 9$
same slope, diff. y-intercepts; no solution

41. $2x - 3y = 6$ $x - (3/2)y = 3$
same slopes, same y-intercepts;
infinite number of solutions

43. $3x = 6y + 5$ $y = (1/2)x - 3$
same slope, diff. y-intercepts; no solution

45. $12x - 5y = 4$ $3x + 4y = 6$
diff. slopes, diff. y-intercepts; 1 solution

47. $5y - 2x = 15$ $2y - 5x = 2$
slopes are not negative reciprocals,
not perpendicular ($\searrow$)

49. $2x + y = 3$ $2y - x = 5$
slopes are negative reciprocals, $\searrow$

51. a) Let x = rate per hour
            y = cost
Cost for Tom's $y_T = 60x + 200$
Cost for Lawn Perfect $y_{LP} = 25x + 305$

b)

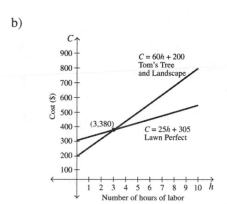

c) $60x + 200 = 25x + 305$
$\underline{-25x \quad -200 \quad -25x \quad -200}$
$35x \qquad = \qquad 105$

$\dfrac{35x}{35} = \dfrac{105}{35}$ →   x = 3 hours

53. a) Let C = cost , R = revenue
$C(x) = 15x + 400$
$R(x) = 25x$

b)

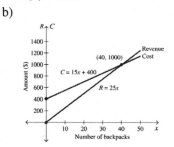

53. c) $25x = 15x + 400$
$\underline{-15x \quad -15x}$
$10x = 400$

$\dfrac{10x}{10} = \dfrac{400}{10}$ →   x = 40 backpacks

d) $P = R(x) - C(x) = 25x - (15x + 400)$
$P = 10x - 400$

e) $P = 10(30) - 400 = 300 - 400 = -\$100$ (loss)

f) $1000 = 10x - 400$ → $10x = 1400$
$x = 140$ BPs

55. a) Let C(x) = cost , R(x) = revenue
$C(x) = 155x + 8400$
$R(x) = 225x$

b)

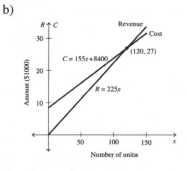

55. c) $225x = 155x + 8400$
$\underline{-155x \quad -155x}$
$70x = 8400$

$\dfrac{70x}{70} = \dfrac{8400}{70}$ →   x = 120 units

d) $P = R(x) - C(x) = 225x - (155x + 8400)$
$P = 70x - 8400$

e) $P = 70(100) - 8400 = 7000 - 8400$
$= -\$1400$ (loss)

f) $1260 = 70x - 8400$ → $70x = 9660$
$x = 138$ units

57. a) $P_1 = .15x + 300$

b)

   $P_2 = 450$ → $.15x + 300 = 450$

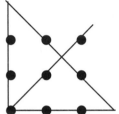

c) $.15x + 300 = 450$

   $\underline{\quad -300 \quad -300}$ → $\dfrac{.15x}{.15} = \dfrac{150}{.15}$

   $.15x \quad = 150$

   $x = \$1000$

59. a) 1 point   b) 3 pts.   c) 6 pts.

   d) 10 pts.

   e) For n lines, the maximum number of

      intersections is $\dfrac{n(n-1)}{2}$; for six lines,

      there are 15 points.

60.

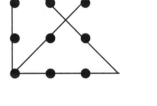

Exercise Set 7.2

1. Write the equations with the variables on one side and the constants on the other side. If necessary multiply one or both equations by a constant(s) so that when the equations are added one of the variables will be eliminated. Solve for the remaining variable and then substitute that value into one of the original equations to solve for the other variable.

3. The system is dependent if the result is of the form a = a.

5. Solve one equation for the variable that is most readily manipulated, then substitute into the other equation.

   $x + 3y = 3$

   $\underline{\quad -3y \quad -3y}$ → $3(3-3y) + 4y = 9$

   $x = 3 - 3y$

7. $y = x + 7$

   $y = -x + 5$

   Substitute $(x + 7)$ in place of y in the second equation.

   $x + 7 = -x + 5$ (solve for x)

   $\underline{+x \qquad +x}$

   $2x + 7 = 5$

   $\underline{\quad -7 \quad -7}$

   $2x \quad = -2$

   $\dfrac{2x}{2} = \dfrac{-2}{2}$ \qquad $x = -1$

   Now substitute -1 for x in an equation

   $y = x + 7$

   $y = (-1) + 7 = 6$

   The solution is $(-1, 6)$. Consistent

9. $2x + 4y = 8$ → $x = -2y + 4$

   $2x - y = -2$

   Substitute $(-2y + 4)$ in place of x in the second equation.

   $2(-2y + 4) - y = -2$ (solve for x)

   $-4y + 8 - y = -2$

   $-5y = -10$ \qquad $y = 2$

   Now substitute 2 for y in the 1st equation

   $2x + 4(2) = 8$

   $2x = 0$

   $x = 0$

   The solution is $(0, 2)$. Consistent

11. $y - x = 4$

$x - y = 3$

Solve the first equation for y.

$y - x + x = x + 4$

$y = x + 4$

Substitute $(x + 4)$ for y in the second equation.

$x - (x + 4) = 3$ (combine like terms)

$-4 \quad 3 \quad$ False

Since -4 does not equal 3, there is no solution to this system. The equations are inconsistent.

13. $3y + 2x = 4$

$y = 6 - x$

Solve the second equation for x.

$3y = 6 - x$

$3y - 6 = 6 - 6 - x$

$3y - 6 = -x$

$-3y + 6 \qquad = x$

Now substitute $(-3y + 6)$ for x in the 1st eq'n.

$3y + 2(6 - 3y) = 4$ (solve for y)

$3y + 12 - 6y = 4$

$-3y = -8$ (div. by -3) $y = 8/3$

Substitute 8/3 for y in the 2nd eq'n.

$3(8/3) n = 6 - x$

$8 = 6 - x \qquad x = -2$

The solution is $(-2, 8/3)$.   Consistent

15. $y - 2x = 3$

$2y = 4x + 6$

Solve the first equation for y.

$y - 2x + 2x = 2x + 3$

$y = 2x + 3$

Now substitute $(2x + 3)$ for y in the 2nd eq'n.

$2(2x + 3) = 4x + 6$

$4x + 6 = 4x + 6$

$4x - 4x + 6 = 4x - 4x + 6$

$6 = 6$

This statement is true for all values of x. The system is dependent.

17. $x = y + 3$

$x = -3$

Substitute -3 in place of x in the first equation.

$-3 = y + 3$

$-3 - 3 = y + 3 - 3$

$-6 = y$

The solution is $(-3, -6)$.   Consistent

19. $y + 3x - 4 = 0$

$2x - y = 7$

Solve the first equation for y.

$y + 3x - 4 = 0$

$y = 4 - 3x$

Substitute $4 - 3x$ for y in the second eq.

$2x - (4 - 3x) = 7$ (solve for x)

$2x - 4 + 3x = 7$

$5x = 11 \qquad x = 11/5$

Substitute 11/5 for x in the second eq'n.

$2(11/5) - y = 7$ (solve for y)

$22/5 - y = 7$

$-y = 13/5 \qquad y = -13/5$

The solution is $(11/5, -13/5)$.   Consistent

21. $x = 2y + 3$

$y = 3x - 1$

Substitute $(3x - 1)$ for y in the first equation.

$x = 2(3x - 1) + 3$

$x = 6x - 2 + 3$

$x = 6x + 1$

$x - 6x = 6x - 6x + 1$

$-5x = 1$

$\dfrac{-5x}{-5} = \dfrac{1}{-5} \qquad x = -1/5$

Substitute $-1/5$ for x in the second equation.

$y = 3(-1/5) - 1 = -3/5 - 5/5 = -8/5$

The solution is $(-1/5, -8/5)$.   Consistent

23. $y = -2x + 3$
     $4x + 2y = 12$
     Substitute $-2x + 3$ for y in the 2nd equation.
     $4x + 2(-2x + 3) = 12$
     $4x - 4x + 6 = 12$
     6   12   False
     Since 6 does not equal 12, there is no solution.
     The equations are inconsistent.

25. $3x + y = 10$
     $4x - y = 4$
     Add the equations to eliminate y.
     $7x = 14$ $x = 2$
     Substitute 2 for x in either eq'n.
     $3(2) + y = 10$ (solve for y)
     $6 + y = 10$ $y = 4$

     The solution is (2, 4)   Consistent

27. $x + y = 10$
     $x - 2y = -2$
     Multiply the 1st eq'n. by 2, then add the eq'ns.
     To eliminate y.
     $2x + 2y = 20$
     $x - 2y = -2$
     $3x = 18$ $x = 6$
     Substitute 6 for x in either eq'n.
     $(6) + y = 10$ (solve for y) $y = 4$

     The solution is (6, 4)   Consistent

29. $2x - y = -4$
     $-3x - y = 6$
     Multiply the second equation by -1,
     $2x - y = -4$
     $3x + y = -6$ add the equations to eliminate y
     $5x = -10$ $x = -2$
     Substitute -2 in place of x in the first equation.
     $2(-2) - y = -4$
     $-4 - y = -4$
     $-y = 0$ $y = 0$

     The solution is (-2, 0).   Consistent

31. $4x + 3y = -1$
     $2x - y = -13$
     Multiply the second equation by 3,
     $4x + 3y = -1$
     $6x - 3y = -39$ add the equations to eliminate
     y
     $10x = -40$ $x = -4$
     Substitute -4 for x in the 2nd equation.
     $2(-4) - y = -13$
     $-8 - y = -13$ $y = 5$
     The solution is (-4, 5).   Consistent

33. $2x + y = 11$
     $x + 3y = 18$
     Multiply the second equation by -2,
     $2x + y = 11$
     $-2x - 6y = -36$ add the equations to elim. x
     $-5y = -25$ $y = 5$
     Substitute 5 for y in the 2nd equation.
     $x + 3(5) = 18$
     $x + 15 = 18$ $x = 3$
     The solution is (3, 5).

35. $3x - 4y = 11$
     $3x + 5y = -7$
     Multiply the first equation by (-1),
     $-3x + 4y = -11$
     $3x + 5y = -7$ add the equations to elim. x
     $9y = -18$ $y = -2$
     Substitute -2 for y in the first equation.
     $3x - 4(-2) = 11$
     $3x = 3$ $x = 1$
     The solution is (1,- 2).   Consistent

37. $4x + y = 6$
     $-8x - 2y = 13$
     Multiply the first equation by 2,
     $8x + 2y = 12$
     $-8x - 2y = 13$ add the equations to elim. y
     0   25   False
     Since this statement is not true for any values
     of x and y, the equations are inconsistent.

39. $3x - 4y = 10$
$5x + 3y = 7$
Multiply the first equation by 3, and the
second equation by 4,
$9x - 12y = 30$
$20x + 12y = 28$ add the equations to elim. y
$29x = 58$ $x = 2$
Substitute 2 for x in the second equation.
$5(2) + 3y = 7$
$10 + 3y = 7$
$3y = -3$ $y = -1$
The solution is (2, -1).   Consistent

41. $S_1 = .15p + 12000$
$S_2 = .05p + 27000$

$.15p + 12000 = .05p + 27000$
$-.05p -12000 \quad -.05p -12000$
$.10p = 15000$
$p = \$150,000.00$

43. Let x = # of medium pizzas
$50 - x$ = # of large pizzas

$10.95x + 14.95(50-x) = 663.50$
$10.95x + 747.50 - 14.95x = 663.50$
$-4.00x = -84.00$ $x = 21$
Substitute 21 for x in 2nd let statement
$50 - x = 50 - (21) = 29$

21 medium pizzas and 29 large pizzas

45. Let x = # of liters at 25%
$10 - x$ = # of liters at 50%

$.25x + .50(10 - x) = .40(10)$
$.25x + 5 - .50x = 4$
$-.25x = -1$ $x = 4$
Substitute 4 for x in 2nd let statement
$10 - x = 10 - (4) = 6$

4 liters of 25% solution and
6 liters of 50% solution

47. Let c = monthly cost
x = number of copies

Eco. Sales: $c = 18 + 0.02x$
Office Sup.: $c = 24 + 0.015x$ set eq'ns.
equal
$18 + 0.02x = 24 + 0.015x$
$0.005x = 6$ $x = 1200$

1200 copies per month

49. Let x = no. of pounds of nuts
y = no. of pounds of pretzels

$x + y = 20$ $y = -x + 20$
$3x + 1y = 30$
Substitute $(20 - x)$ for y in the 2nd equation.
$3x + (20 - x) = 30$
$3x + 20 - x = 30$
$2x = 10$ $x = 5$ Solve for y
$y = 20 - 5 = 15$

Mix 5 lbs. of nuts with 15 lbs. of pretzels

51. Let x = no. of students
y = no. of adults

$x + y = 250$ $x = -y + 250$
$2x + 5y = 950$
Substitute $(-y + 250)$ for x in the 2nd equation.
$2(250 - y) + 5y = 950$
$500 - 2y + 5y = 950$
$3y = 450$ $y = 150$
Substitute 150 for y in the 1st eq'n.
$x + (150) = 250$ $x = 100$

100 students and 150 adults

53. $y_1 = -.58x + 31$
$y_5 = .32x + 7$

$-.58x + 31 = .32x + 7$
$\underline{.58x \quad -7 \quad .58x \quad -7}$
$24 = .90x$

$\dfrac{.90x}{.90} = \dfrac{24}{.90}$ $x = 26.666...$ 27 years
$1981 + 27 = 2008$ During 2007

55. $(1/u) + (2/v) = 8$

$(3/u) - (1/v) = 3$

Substitute x for $\frac{1}{u}$ and y for $\frac{1}{v}$.

(1) $x + 2y = 8$

(2) $3x - y = 3$

Multiply eq'n. (2) by 2,

$x + 2y = 8$

$6x - 2y = 6$ add to eliminate y

$7x = 14$ x= 2, thus u = ½

Substitute 2 for x in eq. (1).

$(2) + 2y = 8$

$2y = 6$ y = 3, thus v = 1/3

Answer: (1/2, 1/3)

57. a) $(2) + (1) + (4) = 7$ $(2) - (1) + 2(4) = 9$

$7 = 7$ $9 = 9$

$-(2) + 2(1) + (4) = 4$

$4 = 4$ (2,1,4) is a solution.

b) Add eq'ns. 1 and 2 to yield eq'n. 4

Multiply eq'n. 2 by 2, then add eq'ns. 2 and 3

to yield eq'n. 5

Combine eq'ns. 4 and 5 to find one variable.

Substitute back into various equations to find

the other 2 variables.

59. $y = 3x + 3$

$(1/3)y = x + 1$

If we multiply the 2nd eq'n. by 3, we get the

eq'n. $y = 3x + 1$, the same as eq'n. # 1.

Two lines that lie on top of one another have

an infinite number of solutions.

60. a) (0, 0)   b) (1, 0)   c) (0, 1)   d) (1, 1)

Exercise Set 7.3

1. A matrix is a rectangular array of elements.

3. A square matrix contains the same number of rows as columns.

5. A 3 x 2 matrix has 2 columns.

7. a) Add numbers in the same positions to produce an entry in that position.

b) $\begin{bmatrix} 1 & 4 & -1 \\ 3 & 2 & 5 \end{bmatrix} + \begin{bmatrix} 3 & 5 & -6 \\ -1 & 2 & 4 \end{bmatrix} = \begin{bmatrix} 1+3 & 4+5 & -1+(-6) \\ 3+(-1) & 2+2 & 5+4 \end{bmatrix} = \begin{bmatrix} 4 & 9 & -7 \\ 2 & 4 & 9 \end{bmatrix}$

9. a) The number of columns of the first matrix must be the same as the number of rows of the second matrix.

b) The dimensions of the resulting matrix will have the same number of rows as the first matrix and the same number of columns as the second matrix. The product of a 2 x 2 with a 2 x 3 matrix will yield a 2 x 3 matrix.

11. a) Identity matrix for 2x2 $\begin{bmatrix} 1 & 0 \\ 0 & 1 \end{bmatrix}$ b) Identity matrix for 3x3 $\begin{bmatrix} 1 & 0 & 0 \\ 0 & 1 & 0 \\ 0 & 0 & 1 \end{bmatrix}$

13. $A = \begin{bmatrix} 1 & 3 \\ 5 & 7 \end{bmatrix}$ $B = \begin{bmatrix} -5 & -1 \\ 7 & 2 \end{bmatrix}$ $A + B = \begin{bmatrix} 1+(-5) & 3+(-1) \\ 5+7 & 7+2 \end{bmatrix} = \begin{bmatrix} -4 & 2 \\ 12 & 9 \end{bmatrix}$

15. $A + B = \begin{bmatrix} 3 & 1 \\ 0 & 4 \\ 6 & 0 \end{bmatrix} + \begin{bmatrix} -3 & 3 \\ 4 & 0 \\ -1 & -1 \end{bmatrix} = \begin{bmatrix} 3+(-3) & 1+3 \\ 0+4 & 4+0 \\ 6+(-1) & 0+(-1) \end{bmatrix} = \begin{bmatrix} 0 & 4 \\ 4 & 4 \\ 5 & -1 \end{bmatrix}$

17. $A - B = \begin{bmatrix} 4 & -2 \\ -3 & 5 \end{bmatrix} - \begin{bmatrix} -2 & 5 \\ 9 & 1 \end{bmatrix} = \begin{bmatrix} 4-(-2) & -2-(-5) \\ -3-(9) & 5-1 \end{bmatrix} = \begin{bmatrix} 6 & -7 \\ -12 & 4 \end{bmatrix}$

19. $A - B = \begin{bmatrix} -4 & 3 \\ 6 & 2 \\ 1 & -5 \end{bmatrix} - \begin{bmatrix} -6 & -8 \\ -10 & -11 \\ 3 & -7 \end{bmatrix} = \begin{bmatrix} -4+6 & 3+8 \\ 6+10 & 2+11 \\ 1-3 & -5+7 \end{bmatrix} = \begin{bmatrix} 2 & 11 \\ 16 & 13 \\ -2 & 2 \end{bmatrix}$

21. $2B = 2\begin{bmatrix} 3 & 2 \\ 5 & 0 \end{bmatrix} = \begin{bmatrix} 2(3) & 2(2) \\ 2(5) & 2(0) \end{bmatrix} = \begin{bmatrix} 6 & 4 \\ 10 & 0 \end{bmatrix}$

23. $2B + 3C = 2\begin{bmatrix} 3 & 2 \\ 5 & 0 \end{bmatrix} + 3\begin{bmatrix} -2 & 3 \\ 4 & 0 \end{bmatrix} = \begin{bmatrix} 6 & 4 \\ 10 & 0 \end{bmatrix} + \begin{bmatrix} -6 & 9 \\ 12 & 0 \end{bmatrix} = \begin{bmatrix} 6-6 & 4+9 \\ 10+12 & 0+0 \end{bmatrix} = \begin{bmatrix} 0 & 13 \\ 22 & 0 \end{bmatrix}$

25. $3B - 2C = 3\begin{bmatrix} 3 & 2 \\ 5 & 0 \end{bmatrix} - 2\begin{bmatrix} -2 & 3 \\ 4 & 0 \end{bmatrix} = \begin{bmatrix} 9 & 6 \\ 15 & 0 \end{bmatrix} - \begin{bmatrix} -4 & 6 \\ 8 & 0 \end{bmatrix} = \begin{bmatrix} 9+4 & 6-6 \\ 15-8 & 0-0 \end{bmatrix} = \begin{bmatrix} 13 & 0 \\ 7 & 0 \end{bmatrix}$

27. $A \times B = \begin{bmatrix} 2 & 0 \\ 3 & 1 \end{bmatrix}\begin{bmatrix} 2 & 6 \\ 8 & 4 \end{bmatrix} = \begin{bmatrix} 2(2)+0(8) & 2(6)+0(4) \\ 3(2)+1(8) & 3(6)+1(4) \end{bmatrix} = \begin{bmatrix} 4 & 12 \\ 14 & 22 \end{bmatrix}$

29. $A \times B = \begin{bmatrix} 2 & 3 & -1 \\ 0 & 4 & 6 \end{bmatrix}\begin{bmatrix} 2 \\ 4 \\ 1 \end{bmatrix} = \begin{bmatrix} 2(2)+3(4)-1(1) \\ 0(2)+4(4)+6(1) \end{bmatrix} = \begin{bmatrix} 15 \\ 22 \end{bmatrix}$

31. $A \times B = \begin{bmatrix} 4 & 7 & 6 \\ -2 & 3 & 1 \\ 5 & 1 & 2 \end{bmatrix}\begin{bmatrix} 1 & 0 & 0 \\ 0 & 1 & 0 \\ 0 & 0 & 1 \end{bmatrix} = \begin{bmatrix} 4+0+0 & 0+7+0 & 0+0+6 \\ -2+0+0 & 0+3+0 & 0+0+1 \\ 5+0+0 & 0+1+0 & 0+0+2 \end{bmatrix} = \begin{bmatrix} 4 & 7 & 6 \\ -2 & 3 & 1 \\ 5 & 1 & 2 \end{bmatrix}$

33. $A + B = \begin{bmatrix} 1 & 3 & -2 \\ 4 & 0 & 3 \end{bmatrix} + \begin{bmatrix} 5 & -1 & 3 \\ 2 & -2 & 1 \end{bmatrix} = \begin{bmatrix} 1+5 & 3+(-1) & -2+3 \\ 4+2 & 0+(-2) & 3+1 \end{bmatrix} = \begin{bmatrix} 6 & 2 & 1 \\ 6 & -2 & 4 \end{bmatrix}$

$A \times B = \begin{bmatrix} 1 & 3 & -2 \\ 4 & 0 & 3 \end{bmatrix}\begin{bmatrix} 5 & -1 & 3 \\ 2 & -2 & 1 \end{bmatrix} =$ Operation cannot be performed because # of columns ≠ # of rows

35. Matrices A and B cannot be added because they do not have the same dimensions.

$A \times B = \begin{bmatrix} 4 & 5 & 3 \\ 6 & 2 & 1 \end{bmatrix} \times \begin{bmatrix} 3 & 2 \\ 4 & 6 \\ -2 & 0 \end{bmatrix} = \begin{bmatrix} 4(3)+5(4)+3(-2) & 4(2)+5(6)+3(0) \\ 6(3)+2(4)+1(-2) & 6(2)+2(6)+1(0) \end{bmatrix} = \begin{bmatrix} 26 & 38 \\ 24 & 24 \end{bmatrix}$

37. A and B cannot be added because they do not have the same dimensions.

$$A \times B = \begin{bmatrix} 1 & 2 \\ 3 & 4 \end{bmatrix}\begin{bmatrix} -3 \\ 2 \end{bmatrix} = \begin{bmatrix} 1(-3)+2(2) \\ 3(-3)+4(2) \end{bmatrix} = \begin{bmatrix} 1 \\ -1 \end{bmatrix}$$

39. $A + B = \begin{bmatrix} 1 & 2 \\ 2 & -3 \end{bmatrix} + \begin{bmatrix} 4 & 5 \\ 6 & 7 \end{bmatrix} = \begin{bmatrix} 1+4 & 2+5 \\ 2+6 & -3+7 \end{bmatrix} = \begin{bmatrix} 5 & 7 \\ 8 & 4 \end{bmatrix}$

$B + A = \begin{bmatrix} 4 & 5 \\ 6 & 7 \end{bmatrix} + \begin{bmatrix} 1 & 2 \\ 2 & -3 \end{bmatrix} = \begin{bmatrix} 4+1 & 5+2 \\ 6+2 & 7+(-3) \end{bmatrix} = \begin{bmatrix} 5 & 7 \\ 8 & 4 \end{bmatrix}$ Thus $A + B = B + A$.

41. $A + B = \begin{bmatrix} 0 & -1 \\ 3 & -4 \end{bmatrix} + \begin{bmatrix} 8 & 1 \\ 3 & -4 \end{bmatrix} = \begin{bmatrix} 0+8 & -1+1 \\ 3+3 & -4+(-4) \end{bmatrix} = \begin{bmatrix} 8 & 0 \\ 6 & -8 \end{bmatrix}$

$B + A = \begin{bmatrix} 8 & 1 \\ 3 & -4 \end{bmatrix} + \begin{bmatrix} 0 & -1 \\ 3 & -4 \end{bmatrix} = \begin{bmatrix} 8+0 & 1+(-1) \\ 3+3 & -4+(-4) \end{bmatrix} = \begin{bmatrix} 8 & 0 \\ 6 & -8 \end{bmatrix}$ Thus $A + B = B + A$.

43. $(A + B) + C = \left(\begin{bmatrix} 5 & 2 \\ 3 & 6 \end{bmatrix} + \begin{bmatrix} 3 & 4 \\ -2 & 7 \end{bmatrix} \right) + \begin{bmatrix} -1 & 4 \\ 5 & 0 \end{bmatrix} = \begin{bmatrix} 8 & 6 \\ 1 & 13 \end{bmatrix} + \begin{bmatrix} -1 & 4 \\ 5 & 0 \end{bmatrix} = \begin{bmatrix} 7 & 10 \\ 6 & 13 \end{bmatrix}$

$A + (B + C) = \begin{bmatrix} 5 & 2 \\ 3 & 6 \end{bmatrix} + \left(\begin{bmatrix} 3 & 4 \\ -2 & 7 \end{bmatrix} + \begin{bmatrix} -1 & 4 \\ 5 & 0 \end{bmatrix} \right) = \begin{bmatrix} 5 & 2 \\ 3 & 6 \end{bmatrix} + \begin{bmatrix} 2 & 8 \\ 3 & 7 \end{bmatrix} = \begin{bmatrix} 7 & 10 \\ 6 & 13 \end{bmatrix}$

Thus, $(A + B) + C = A + (B + C)$.

45. $(A + B) + C = \left(\begin{bmatrix} 7 & 4 \\ 9 & -36 \end{bmatrix} + \begin{bmatrix} 5 & 6 \\ -1 & -4 \end{bmatrix} \right) + \begin{bmatrix} -7 & -5 \\ -1 & 3 \end{bmatrix} = \begin{bmatrix} 12 & 10 \\ 8 & -40 \end{bmatrix} + \begin{bmatrix} -7 & -5 \\ -1 & 3 \end{bmatrix} = \begin{bmatrix} 5 & 5 \\ 7 & -37 \end{bmatrix}$

$A + (B + C) = \begin{bmatrix} 7 & 4 \\ 9 & -36 \end{bmatrix} + \left(\begin{bmatrix} 5 & 6 \\ -1 & -4 \end{bmatrix} + \begin{bmatrix} -7 & -5 \\ -1 & 3 \end{bmatrix} \right) = \begin{bmatrix} 7 & 4 \\ 9 & -36 \end{bmatrix} + \begin{bmatrix} -2 & 1 \\ -2 & -1 \end{bmatrix} = \begin{bmatrix} 5 & 5 \\ 7 & -37 \end{bmatrix}$

Thus, $(A + B) + C = A + (B + C)$.

47. $A \times B = \begin{bmatrix} 1 & -2 \\ 4 & -3 \end{bmatrix}\begin{bmatrix} -1 & -3 \\ 2 & 4 \end{bmatrix} = \begin{bmatrix} 1(-1)-2(2) & 1(-3)-2(4) \\ 4(-1)+(-3)(2) & 4(-3)+(-3)(4) \end{bmatrix} = \begin{bmatrix} -5 & -11 \\ -10 & -24 \end{bmatrix}$

$B \times A = \begin{bmatrix} -1 & -3 \\ 2 & 4 \end{bmatrix}\begin{bmatrix} 1 & -2 \\ 4 & -3 \end{bmatrix} = \begin{bmatrix} -1(1)+(-3)4 & -1(-2)+(-3)(-3) \\ 2(1)+4(4) & 2(-2)+4(-3) \end{bmatrix} = \begin{bmatrix} -13 & 11 \\ 18 & -16 \end{bmatrix}$

Thus, $A \times B \neq B \times A$.

49. $A \times B = \begin{bmatrix} 4 & 2 \\ 1 & -3 \end{bmatrix}\begin{bmatrix} 2 & 4 \\ -3 & 1 \end{bmatrix} = \begin{bmatrix} 4(2)+2(-3) & 4(4)+2(1) \\ 1(2)+(-3)(-3) & 1(4)+(-3)(1) \end{bmatrix} = \begin{bmatrix} 2 & 18 \\ 11 & 1 \end{bmatrix}$

$B \times A = \begin{bmatrix} 2 & 4 \\ -3 & 1 \end{bmatrix}\begin{bmatrix} 4 & 2 \\ 1 & -3 \end{bmatrix} = \begin{bmatrix} 2(4)+4(1) & 2(2)+4(-3) \\ -3(4)+1(1) & -3(2)+1(-3) \end{bmatrix} = \begin{bmatrix} 12 & -8 \\ -11 & -9 \end{bmatrix}$ Thus, $A \times B \neq B \times A$.

51. Since B = I, (the identity matrix), and $A \times I = I \times A = A$, we can conclude that $A \times B = B \times A$.

53. $(A \times B) \times C = \left(\begin{bmatrix} 1 & 3 \\ 4 & 0 \end{bmatrix}\begin{bmatrix} 4 & 2 \\ 3 & 1 \end{bmatrix}\right)\begin{bmatrix} 2 & 1 \\ 3 & 0 \end{bmatrix} = \begin{bmatrix} 13 & 5 \\ 16 & 8 \end{bmatrix}\begin{bmatrix} 2 & 1 \\ 3 & 0 \end{bmatrix} = \begin{bmatrix} 41 & 13 \\ 56 & 16 \end{bmatrix}$

$A \times (B \times C) = \begin{bmatrix} 1 & 3 \\ 4 & 0 \end{bmatrix}\left(\begin{bmatrix} 4 & 2 \\ 3 & 1 \end{bmatrix}\begin{bmatrix} 2 & 1 \\ 3 & 0 \end{bmatrix}\right) = \begin{bmatrix} 1 & 3 \\ 4 & 0 \end{bmatrix}\begin{bmatrix} 14 & 4 \\ 9 & 3 \end{bmatrix} = \begin{bmatrix} 41 & 13 \\ 56 & 16 \end{bmatrix}$ Thus, $(A \times B) \times C = A \times (B \times C)$.

55. $(A \times B) \times C = \left(\begin{bmatrix} 4 & 3 \\ -6 & 2 \end{bmatrix}\begin{bmatrix} 1 & 2 \\ 0 & 1 \end{bmatrix}\right)\begin{bmatrix} 4 & 3 \\ 0 & -2 \end{bmatrix} = \begin{bmatrix} 4 & 11 \\ -6 & -10 \end{bmatrix}\begin{bmatrix} 4 & 3 \\ 0 & -2 \end{bmatrix} = \begin{bmatrix} 16 & -10 \\ -24 & 2 \end{bmatrix}$

$A \times (B \times C) = \begin{bmatrix} 4 & 3 \\ -6 & 2 \end{bmatrix}\left(\begin{bmatrix} 1 & 2 \\ 0 & 1 \end{bmatrix}\begin{bmatrix} 4 & 3 \\ 0 & -2 \end{bmatrix}\right) = \begin{bmatrix} 4 & 3 \\ -6 & 2 \end{bmatrix}\begin{bmatrix} 4 & -1 \\ 0 & -2 \end{bmatrix} = \begin{bmatrix} 16 & -10 \\ -24 & 2 \end{bmatrix}$

Thus, $(A \times B) \times C = A \times (B \times C)$.

57. $(A \times B) \times C = \left(\begin{bmatrix} 3 & 4 \\ -1 & -2 \end{bmatrix}\begin{bmatrix} 0 & 1 \\ 1 & 0 \end{bmatrix}\right)\begin{bmatrix} 2 & 0 \\ 3 & 0 \end{bmatrix} = \begin{bmatrix} 4 & 3 \\ -2 & -1 \end{bmatrix}\begin{bmatrix} 2 & 0 \\ 3 & 0 \end{bmatrix} = \begin{bmatrix} 17 & 0 \\ -7 & 0 \end{bmatrix}$

$A \times (B \times C) = \begin{bmatrix} 3 & 4 \\ -1 & -2 \end{bmatrix}\left(\begin{bmatrix} 0 & 1 \\ 1 & 0 \end{bmatrix}\begin{bmatrix} 2 & 0 \\ 3 & 0 \end{bmatrix}\right) = \begin{bmatrix} 3 & 4 \\ -1 & -2 \end{bmatrix}\begin{bmatrix} 3 & 0 \\ 2 & 0 \end{bmatrix} = \begin{bmatrix} 17 & 0 \\ -7 & 0 \end{bmatrix}$

Thus, $(A \times B) \times C = A \times (B \times C)$.

59. $A \times B = \begin{bmatrix} 2 & 2 & .5 & 1 \\ 3 & 2 & 1 & 2 \\ 0 & 1 & 0 & 3 \\ .5 & 1 & 0 & 0 \end{bmatrix}\begin{bmatrix} 10 & 12 \\ 5 & 8 \\ 8 & 8 \\ 4 & 6 \end{bmatrix} = \begin{bmatrix} 2\cdot10+2\cdot5+.5\cdot8+1\cdot4 & 2\cdot12+2\cdot8+.5\cdot8+1\cdot6 \\ 3\cdot10+2\cdot5+1\cdot8+2\cdot4 & 3\cdot12+2\cdot8+1\cdot8+2\cdot6 \\ 0\cdot10+1\cdot5+0\cdot8+3\cdot4 & 0\cdot12+1\cdot8+0\cdot8+3\cdot6 \\ .5\cdot10+1\cdot5+0\cdot8+0\cdot4 & .5\cdot12+1\cdot8+0\cdot8+0\cdot6 \end{bmatrix} = \begin{bmatrix} 38 & 50 \\ 56 & 72 \\ 17 & 26 \\ 10 & 14 \end{bmatrix}$

61. $C(A \times B) = [40 \quad 30 \quad 12 \quad 20]\begin{bmatrix} 38 & 50 \\ 56 & 72 \\ 17 & 26 \\ 10 & 14 \end{bmatrix} = [36.04 \quad 47.52]$ cents   small $36.04, large $47.52

63. A must have 3 rows and B must have one column.

$A = \begin{bmatrix} 1 & -2 \\ 0 & 1 \\ 1 & 3 \end{bmatrix}$ $B = \begin{bmatrix} 6 \\ 7 \end{bmatrix}$ $A \times B = \begin{bmatrix} -8 \\ 7 \\ 27 \end{bmatrix}$

65. $A \times B = \begin{bmatrix} 5 & -2 \\ -2 & 1 \end{bmatrix}\begin{bmatrix} 1 & 2 \\ 2 & 5 \end{bmatrix} = \begin{bmatrix} 5(1)-2(2) & 5(2)-2(5) \\ -2(1)+1(2) & -2(2)+1(5) \end{bmatrix} = \begin{bmatrix} 1 & 0 \\ 0 & 1 \end{bmatrix}$

$B \times A = \begin{bmatrix} 1 & 2 \\ 2 & 5 \end{bmatrix}\begin{bmatrix} 5 & -2 \\ -2 & 1 \end{bmatrix} = \begin{bmatrix} 1(5)+2(-2) & 1(-2)+2(1) \\ 2(5)+5(-2) & 2(-2)+5(1) \end{bmatrix} = \begin{bmatrix} 1 & 0 \\ 0 & 1 \end{bmatrix}$

Thus, A and B are multiplicative inverses.

67. False. Let $A = [1 \quad 3]$ and $B = [2 \quad 1]$. Then $A - B = [-1, 2]$ and $B - A = [1, -2]$ $A - B \neq B - A$.

69. a) $1.4(14) + 0.7(10) + 0.3(7) = \$28.70$

b) $2.7(12) + 2.8(9) + 0.5(5) = \$60.10$

$$\text{c)   } L \times C = \begin{array}{c} \\ \\ \\ \\ \end{array} \begin{bmatrix} \overset{\text{Ames}}{28.7} & \overset{\text{Bay}}{24.6} \\ 41.3 & 35.7 \\ 69.3 & 60.1 \end{bmatrix} \begin{array}{l} \text{small} \\ \text{medium} \\ \text{large} \end{array}$$

This array shows the total cost of each sofa at each plant.

71. A + B cannot be calculated because the # of columns of A $\neq$ # of rows of B and the # of rows of A $\neq$ # of rows of B.

$$A \times B = \begin{bmatrix} 1 & 2 & 3 \\ 3 & 2 & 1 \end{bmatrix} \begin{bmatrix} 0 & 4 \\ 1 & 5 \\ 2 & 1 \end{bmatrix} = \begin{bmatrix} 1(0)+2(1)+3(2) & 1(4)+2(5)+3(1) \\ 3(0)+2(1)+1(2) & 3(4)+2(5)+1(1) \end{bmatrix} = \begin{bmatrix} 8 & 17 \\ 4 & 23 \end{bmatrix}$$

72. Answers will vary; an example is $A = \begin{bmatrix} 1 & 0 \\ 0 & 0 \end{bmatrix}$ and $B = \begin{bmatrix} 0 & 0 \\ 0 & 1 \end{bmatrix}$.

Exercise Set 7.4

1. a) An augmented matrix is a matrix formed with the coefficients of the variables and the constants. The coefficients are separated from the constants by a vertical bar.

b) $\begin{bmatrix} 1 & 3 & | & 7 \\ 2 & -1 & | & 4 \end{bmatrix}$

3. If you obtain an augmented matrix in which one row of numbers on the left side of the vertical line are all zeroes but a zero does not appear in the same row on the other side of the vertical line, the system is inconsistent.

5. 1) Multiply the 2nd row by -1/2; 2) add -3 times the 2nd row to the first row; and 3) identify the values of x and y.

1) $\begin{bmatrix} 1 & 3 & | & 5 \\ 0 & 1 & | & (-1/2) \end{bmatrix}$ 2) $\begin{bmatrix} 1+0 & 3+(-3) & | & 5+(3/2) \\ 0 & 1 & | & (-1/2) \end{bmatrix} = \begin{bmatrix} 1 & 0 & | & 13/2 \\ 0 & 1 & | & -1/2 \end{bmatrix}$ 3) $(x, y) = \left(\dfrac{13}{2}, \dfrac{-1}{2} \right)$

7. $x + 3y = 3 \qquad -x + y = -3$

$\begin{bmatrix} 1 & 3 & | & 3 \\ -1 & 1 & | & -3 \end{bmatrix} \rightarrow \begin{bmatrix} 1 & 3 & | & 3 \\ -1+1 & 1+3 & | & -3+3 \end{bmatrix} = \begin{bmatrix} 1 & 3 & | & 3 \\ 0 & 4 & | & 0 \end{bmatrix} \rightarrow \begin{bmatrix} 1+0 & 3+(-3) & | & 3+0 \\ 0 & 1 & | & 0 \end{bmatrix} = \begin{bmatrix} 1 & 0 & | & 3 \\ 0 & 1 & | & 0 \end{bmatrix} \rightarrow (3, 0)$

9. $x - 2y = -1 \qquad 2x + y = 8$

$\begin{bmatrix} 1 & -2 & | & -1 \\ 2 & 1 & | & 8 \end{bmatrix} \rightarrow \begin{bmatrix} 1 & -2 & | & -1 \\ 2-2 & 1+4 & | & 8+2 \end{bmatrix} = \begin{bmatrix} 1 & -2 & | & -1 \\ 0 & 5 & | & 10 \end{bmatrix} \rightarrow \begin{bmatrix} 1+0 & -2+2 & | & -1+4 \\ 0 & 5 & | & 10 \end{bmatrix} \rightarrow$

$\begin{bmatrix} 1 & 0 & | & 3 \\ 0 & 5 & | & 10 \end{bmatrix} = \begin{bmatrix} 1 & 0 & | & 3 \\ 0 & 1 & | & 2 \end{bmatrix} \rightarrow (3, 2)$

11. $\begin{bmatrix} 2 & -5 & | & -6 \\ -4 & 10 & | & 12 \end{bmatrix} \underset{(r_2 + 2r_1)}{=} \begin{bmatrix} 2 & -5 & | & -6 \\ 0 & 0 & | & 0 \end{bmatrix} \Rightarrow$ Dependent system

The solution is all points on the line $2x - 5y = -6$.

13. $\begin{bmatrix} 2 & -3 & | & 10 \\ 2 & 2 & | & 5 \end{bmatrix} \begin{matrix} (r_1 \div 2) \\ (r_2 - 2r_1) \end{matrix} \begin{bmatrix} 1 & \frac{-3}{2} & | & 5 \\ 0 & 5 & | & -5 \end{bmatrix} \begin{matrix} = \\ (r_2 \div (5)) \end{matrix} \begin{bmatrix} 1 & -\frac{3}{2} & | & 5 \\ 0 & 1 & | & -1 \end{bmatrix} \begin{matrix} (r_1 + \frac{3}{2}r_2) \\ = \end{matrix} \begin{bmatrix} 1 & 0 & | & \frac{7}{2} \\ 0 & 1 & | & -1 \end{bmatrix}$ The solution is $(7/2, -1)$.

15. $\begin{bmatrix} 4 & 2 & | & -10 \\ -2 & 1 & | & -7 \end{bmatrix} \begin{matrix} = \\ r_2 + \frac{1}{2}r_1 \end{matrix} \begin{bmatrix} 4 & 2 & | & -10 \\ 0 & 2 & | & -12 \end{bmatrix} \begin{matrix} (r_1 \div 4) \\ = \end{matrix} \begin{bmatrix} 1 & 1/2 & | & -10/4 \\ 0 & 2 & | & -12 \end{bmatrix} \begin{matrix} = \\ (r_2 \div 2) \end{matrix} \begin{bmatrix} 1 & 1/2 & | & -10/4 \\ 0 & 1 & | & -6 \end{bmatrix}$

$\begin{bmatrix} 1 & 1/2 & | & -10/4 \\ 0 & 1 & | & -6 \end{bmatrix} \begin{matrix} (r_1 - \frac{1}{2}r_2) \\ = \end{matrix} \begin{bmatrix} 1 & 0 & | & 1/2 \\ 0 & 1 & | & -6 \end{bmatrix}$ The solution is $(1/2, -6)$.

17. $\begin{bmatrix} -3 & 6 & | & 5 \\ 2 & -4 & | & 8 \end{bmatrix} \begin{matrix} (r_1 \div (-3)) \\ = \end{matrix} \begin{bmatrix} 1 & -2 & | & \frac{-5}{3} \\ 2 & -4 & | & 8 \end{bmatrix} \begin{matrix} = \\ (r_2 - 2r_1) \end{matrix} \begin{bmatrix} 1 & -2 & | & \frac{-5}{3} \\ 0 & 0 & | & \frac{34}{3} \end{bmatrix} \Rightarrow$ Inconsistent system   No solution.

19. $\begin{bmatrix} 2 & 1 & | & 11 \\ 1 & 3 & | & 18 \end{bmatrix} \begin{matrix} (r_1 \div 2) \\ = \end{matrix} \begin{bmatrix} 1 & \frac{1}{2} & | & \frac{11}{2} \\ 1 & 3 & | & 18 \end{bmatrix} \begin{matrix} = \\ (r_2 - r_1) \end{matrix} \begin{bmatrix} 1 & \frac{1}{2} & | & \frac{11}{2} \\ 0 & \frac{5}{2} & | & \frac{25}{2} \end{bmatrix} \begin{matrix} = \\ (\frac{2}{5}r_2) \end{matrix} \begin{bmatrix} 1 & \frac{1}{2} & | & \frac{11}{2} \\ 0 & 1 & | & 5 \end{bmatrix} \begin{matrix} (r_1 - \frac{1}{2}r_2) \\ = \end{matrix} \begin{bmatrix} 1 & 0 & | & 3 \\ 0 & 1 & | & 5 \end{bmatrix}$

The solution is $(3, 5)$

21. $S + L = 55$ $4S + 6L = 290$

$\begin{bmatrix} 1 & 1 & | & 55 \\ 4 & 6 & | & 290 \end{bmatrix} \begin{matrix} = \\ (r_2 - 4r_1) \end{matrix} \begin{bmatrix} 1 & 1 & | & 55 \\ 0 & 2 & | & 70 \end{bmatrix} \begin{matrix} = \\ (r_2 \div 2) \end{matrix} \begin{bmatrix} 1 & 1 & | & 55 \\ 0 & 1 & | & 35 \end{bmatrix} \begin{matrix} (r_1 - r_2) \\ = \end{matrix} \begin{bmatrix} 1 & 0 & | & 20 \\ 0 & 1 & | & 35 \end{bmatrix}$ The solution is $(20, 35)$;

20 small flags and 35 large ones

23. Let T = # of hours for truck driver L = # of hours for laborer

$10T + 8L = 144$ $L = T + 2$ → $T = L - 2$

$\begin{bmatrix} 10 & 8 & | & 144 \\ 1 & -1 & | & -2 \end{bmatrix} \begin{matrix} r_1 + 8r_2 \\ = \end{matrix} \begin{bmatrix} 18 & 0 & | & 128 \\ 1 & -1 & | & -2 \end{bmatrix} \begin{matrix} (r_1 \div 18) \\ = \end{matrix} \begin{bmatrix} 1 & 0 & | & 64/9 \\ 1 & -1 & | & -2 \end{bmatrix} \begin{matrix} = \\ (r_2 - r_1) \end{matrix} \begin{bmatrix} 1 & 0 & | & 64/9 \\ 0 & -1 & | & -82/9 \end{bmatrix}$

$\begin{bmatrix} 1 & 0 & | & 64/9 \\ 0 & -1 & | & -82/9 \end{bmatrix} \begin{matrix} = \\ (r_2 \bullet -1) \end{matrix} \begin{bmatrix} 1 & 0 & | & 64/9 \\ 0 & 1 & | & 82/9 \end{bmatrix}$ $(64/9, 82/9)$

7 1/9 hours for the truck driver and 9 1/9 hours for the laborer.

25. $1.5x + 2y = 337.5$ $x + y = 200$

$\begin{bmatrix} 1.5 & 2 & | & 337.5 \\ 1 & 1 & | & 200 \end{bmatrix} = \begin{bmatrix} 1 & 1.\overline{33} & | & 225 \\ 1 & 1 & | & 200 \end{bmatrix} = \begin{bmatrix} 1 & 1.\overline{33} & | & 225 \\ 1 & 1 & | & 200 \end{bmatrix} = \begin{bmatrix} 1 & 1.\overline{33} & | & 225 \\ 0 & -.\overline{33} & | & -25 \end{bmatrix} = \begin{bmatrix} 1 & 1.\overline{33} & | & 225 \\ 0 & 1 & | & 75 \end{bmatrix} = \begin{bmatrix} 1 & 0 & | & 125 \\ 0 & 1 & | & 75 \end{bmatrix}$

The solution is 125 non-refillable pencils @ $1.50 and 75 refillable pencils @ $2.00.

Exercise Set 7.5

1. The solution set of a system of linear inequalities is the set of points that satisfy all inequalities in the system.

3.

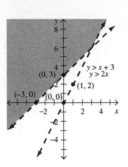

5.

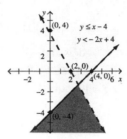

7.

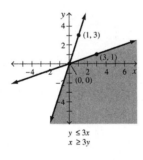

9.

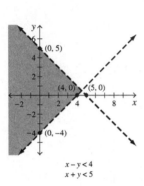

11.

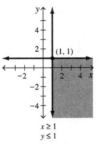

13.

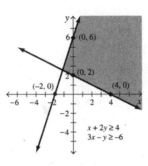

15.

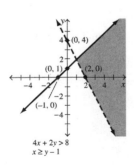

17.

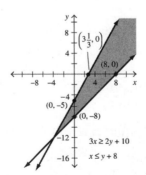

19. a) Let P = Panasonic,   S = Sony
        600P + 900S ≤ 18000
        P ≥ 2S   P ≥ 10
                                  S ≥ 5

    c) (15, 6) means 15 Panasonic models and 6
        Sony models.
        600 (15) + 900 (6) = 9000 + 5400 or
        $ 14,400

b)

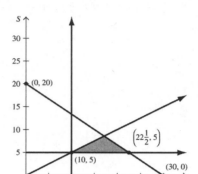

21. a) No, if the lines are parallel there may not be a
        solution to the system.

    b) Example: y ≥ x   y ≤ x – 2
        This system has no solution.

23. No. Every line divides the plane into two halves
    only one of which can be part of the solution.
    Therefore, the points in the other half cannot
    satisfy both inequalities and so do not solve
    the system.

    Example: y ≥ x   x ≥ 2

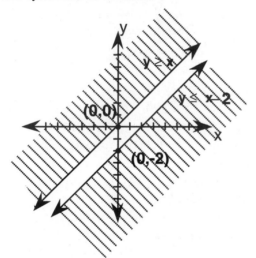

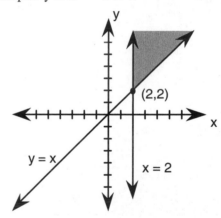

24. $y \leq x$, $y \geq x$, $y \leq 0$, $y \geq 0$

25. $y \leq x$, $y \geq x$

Exercise Set 7.6

1. Constraints are restrictions that are represented
    as linear inequalities.

3. Vertices

5. If a linear equation of the form K = Ax + By is evaluated at each point in a closed polygonal region, the
    maximum and minimum values of the equation occur at a corner.

7. At (0, 0), K= 6(0) + 4(0) = 0
    At (0, 4), K= 6(0) + 4(4) = 16
    At (2, 3), K= 6(2) + 4(3) = 24
    At (5, 0), K= 6(5) + 4(0) = 30
    The maximum value is 30 at (5, 0);
    the minimum value is 0 at (0, 0).

9. a)

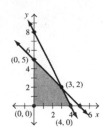

b) $x + y \le 5$ $2x + y \le 8$ $x \ge 0$ $y \ge 0$

   $P = 5x + 4y$

   At $(0,0)$, $P = 5(0) + 4(0) = 0$ min. at $(0, 0)$

   At $(0,4)$, $P = 5(0) + 4(4) = 16$

   At $(3, 2$, $P = 5(3) + 4(2) = 23$ max. at $(3, 2)$

   At $(0,5)$, $P = 5(0) + 4(5) = 20$

11. a)

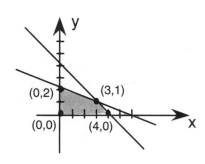

b) $P = 7x + 6y$

   At $(0,0)$, $P = 7(0) + 6(0) = 0$ min. at $(0, 0)$

   At $(0,2)$, $P = 7(0) + 6(2) = 12$

   At $(3,1)$, $P = 7(3) + 6(1) = 27$

   At $(4,0)$, $P = 7(4) + 6(0) = 28$ max. at $(4, 0)$

13. a)

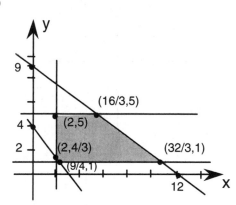

13 b) $P = 2.20x + 1.65y$

   At $(2,4/3)$, $P = 2.20(2)+1.65(4/3) = 6.60$

   At $(2, 5)$, $P = 2.20(2)+1.65(5) = 12.65$

   At $(16/3,5)$, $P=2.20(16/3)+1.65(5) = 19.98$

   At $(32/3,1)$, $P=2.20(32/3)+1.65(1) = 25.12$

   At $(9/4,1)$, $P=2.20(9/4)+1.65(1) = 6.60$

   Maximum of 25.12 at $(32/3, 1)$ and

   Minimum of 6.60 at $(2,4/3)$, $(9/4, 1)$

15. a) Let x = number of rolls of Kodak film

   y = number of rolls of Fuji film

   $x + y \le 24$ $x \ge 2y$

   $x \ge 0$ $y \ge 4$

b) $P = .35x + .50y$

c)

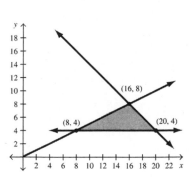

d) At $(8, 4)$, $P = .35(8) + .50(4) = .35$

   At $(16, 8)$, $P = .35(16) + .50(8) = 9.6$ max. at

   $(16, 8)$

   At $(20, 4)$, $P = .35(20) + .50(4) = 9$

e) 16 rolls of Kodak film and 8 rolls of Fuji film

f) Max. profit = \$9.60

17. Let x = gallons of indoor paint
    y = gallons of outdoor paint
    $x \geq 60$ $y \geq 100$
    (a) $3x + 4y \geq 60$ $x \geq 0$
        $10x + 5y \geq 100$ $y \geq 0$
    (b) $C = 28x + 33y$
    (c)

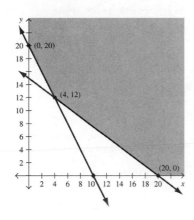

d) At (0, 20), C = 28(0) + 33(20) = 660
   At (20, 0), C = 28(20) + 33(0) = 560
   At (4, 12), C = 28(4) + 33(12) = 508
e) 4 hours on Mach. 1 and 12 hours on Mach. 2
f) Max. profit = $ 660.00

19. Let x = # of car seats
    y = # of strollers
    $x + 3y \leq 24$ $2x + y \leq 16$ $x + y \leq 10$

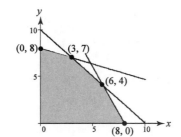

P = 25x + 35y
At (0, 8), P = 25(0) + 35(8) = 280
At (3, 7), P = 25(3) + 35(7) = 320
At (6, 4), P = 25(6) + 35(4) = 290
At (8, 0), P = 25(8) + 35(0) = 200
3 car seats and 7 strollers
 Max. profit = $ 320.00

Review Exercises

1.

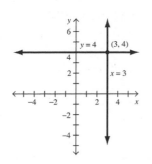

2.

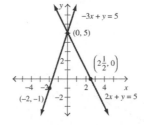

3.

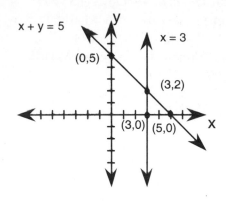

The solution is (3,2).

4.

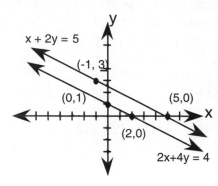

Inconsistent

5. $y = (2/3)x + 5$
    $y = (2/3)x + 5$
    Same slope and y-intercept. Infinite # of solutions.

6. $y = 2x + 6$
    $y = 2x + 7.5$
    Same slope but different y-intercepts. No solution.

7. $6y - 2x = 20$ becomes $y = (1/3)x + 10/3$
    $4y + 2x = 10$ becomes $y = - (1/2)x + 5/2$
    Different slopes. One solution.

8. $y = (1/2)x - 2$
    $y = 2x + 6$
    Different slopes. One solution.

9. (1) $- x + y = 12$
    (2) $\underline{x + 2y = - 3}$ (add)
    $\qquad\qquad 3y = 9 \qquad\qquad y = 3$
    Substitute 3 in place of y in the first equation.
    $- x + 3 = 12$
    $- x = 9 \qquad x = - 9$
    The solution is $(- 9, 3)$.

10. $x - 2y = 9$
    $y = 2x - 3$
    Substitute $(2x - 3)$ in place of y in the 1st equation.
    $x - 2(2x - 3) = - 11$ (solve for x)
    $x - 4x - 6 = - 11$
    $5x - 6 = - 1$
    $5x = - 5 \qquad x = - 1$
    Substitute $(- 1)$ in place of x in the 2nd equation.
    $y = 2(- 1) - 3 = - 2 - 3 = - 5$
    The solution is $(- 1, - 5)$.

11. $2x - y = 4 \qquad y = 2x - 4$
    $3x - y = 2$
    Substitute $2x - 4$ for y in the second equation.
    $3x - (2x - 4) = 2$ (solve for x)
    $3x - 2x + 4 = 2$
    $x + 4 = 2 \qquad x = -2$
    Substitute -2 for x in an equation.
    $2(-2) - y = 4$
    $-4 - y = 4 \qquad y = -8$
    The solution is $(-2, -8)$.

12. $3x + y = 1 \qquad y = -3x + 1$
    $3y = - 9x - 4$
    Substitute $-3x + 1$ for y in the second equation.
    $3(-3x + 1) = -9x - 4$ (solve for x)
    $-9x + 3 = -9x - 4$
    $3 \neq 4$ False   There is no solution to this system.
    The equations are inconsistent.

13. (1) $x - 2y = 1$
    (2) $2x + y = 7$
    Multiply the second equation by 2.
    $x - 2y = 1$
    $\underline{4x + 2y = 14}$
    $5x \qquad = 15$
    $\qquad x = 3$
    Substitute 3 in place of x in the equation (2).
    $2(3) + y = 7$
    $\qquad y = 7 - 6$
    $\qquad y = 1$
    The solution is (3,1).

14. (1) $2x + y = 2$
    (2) $\underline{-3x - y = 5}$ (add)
    $\quad -x = 7 \qquad\qquad x = -7$
    Substitute (-7) in place of x in the 1st equation.
    $2(-7) + y = 2$
    $-14 \, y = 2 \qquad\qquad y = 16$
    The solution is ($-7, 16$).

15. (1) $x + y = 2$
    (2) $x + 3y = -2$
    Multiply the first equation by -1.
    $-x - y = -2$
    $\underline{x + 3y = -2}$ (add)
    $2y = -4 \qquad\qquad y = -2$
    Substitute (-2) for y in equation (2).
    $x + 3(-2) = -2$
    $x - 6 = -2 \qquad x = 4$
    The solution is (4,-2).

16. (1) $4x - 8y = 16$
    (2) $x - 2y = 4 \qquad x = 2y + 4$
    Substitute $2y + 4$ for x in the first equation.
    $4(2y + 4) - 8y = 16$
    $8y + 16 - 8y = 16$
    $16 = 16$ True
    There are an infinite number of solutions.
    The system is dependent.

17. (1) $3x + 5y = 15$
    (2) $2x + 4y = 0$
    Multiply the first equation by 2, and the 2nd
    equation by (-3).
    $6x + 10y = 30$
    $\underline{-6x - 12y = 0}$ (add)
    $-2y = 30 \qquad\qquad y = -15$
    Substitute (-15) for y in the second equation.
    $2x + 4(-15) = 0$
    $2x - 60 = 0 \qquad x = 30$
    The solution is (30, -15).

18. (1) $3x + 4y = 6$
    (2) $2x - 3y = 4$
    Multiply the first equation by 2, and the second
    equation by -3.
    $6x + 8y = 12$
    $\underline{-6x + 9y = -12}$ (add)
    $17y = 0 \qquad\qquad y = 0$
    Substitute 0 for y in the first equation.
    $3x + 4(0) = 6$
    $3x = 6 \qquad\qquad x = 2$
    The solution is (2,0).

19. $A + B = \begin{bmatrix} 1 & -3 \\ 2 & 4 \end{bmatrix} + \begin{bmatrix} -2 & -5 \\ 6 & 3 \end{bmatrix} = \begin{bmatrix} 1+(-2) & -3+(-5) \\ 2+6 & 4+3 \end{bmatrix} = \begin{bmatrix} -1 & -8 \\ 8 & 7 \end{bmatrix}$

20. $A - B = \begin{bmatrix} 1 & -3 \\ 2 & 4 \end{bmatrix} - \begin{bmatrix} -2 & -5 \\ 6 & 3 \end{bmatrix} = \begin{bmatrix} 1-(-2) & -3-(-5) \\ 2-6 & 4-3 \end{bmatrix} = \begin{bmatrix} 3 & 2 \\ -4 & 1 \end{bmatrix}$

21. $2A = 2 \begin{bmatrix} 1 & -3 \\ 2 & 4 \end{bmatrix} = \begin{bmatrix} 2(1) & 2(-3) \\ 2(2) & 2(4) \end{bmatrix} = \begin{bmatrix} 2 & -6 \\ 4 & 8 \end{bmatrix}$

22. $2A - 3B = 2\begin{bmatrix} 1 & -3 \\ 2 & 4 \end{bmatrix} - 3\begin{bmatrix} -2 & -5 \\ 6 & 3 \end{bmatrix} = \begin{bmatrix} 2 & -6 \\ 4 & 8 \end{bmatrix} + \begin{bmatrix} 6 & 15 \\ -18 & -9 \end{bmatrix} = \begin{bmatrix} 2+6 & -6+15 \\ 4-18 & 8-9 \end{bmatrix} = \begin{bmatrix} 8 & 9 \\ -14 & -1 \end{bmatrix}$

23. $A \times B = \begin{bmatrix} 1 & -3 \\ 2 & 4 \end{bmatrix} \times \begin{bmatrix} -2 & -5 \\ 6 & 3 \end{bmatrix} = \begin{bmatrix} 1(-2)+(-3)6 & 1(-5)+(-3)3 \\ 2(-2)+4(6) & 2(-5)+4(3) \end{bmatrix} = \begin{bmatrix} -20 & -14 \\ 20 & 2 \end{bmatrix}$

24. $B \times A = \begin{bmatrix} -2 & -5 \\ 6 & 3 \end{bmatrix} \times \begin{bmatrix} 1 & -3 \\ 2 & 4 \end{bmatrix} = \begin{bmatrix} (-2)1+(-5)2 & (-2)(-3)+(-5)4 \\ 6(1)+3(2) & 6(-3)+3(4) \end{bmatrix} = \begin{bmatrix} -12 & -14 \\ 12 & -6 \end{bmatrix}$

25. $\begin{bmatrix} 1 & 2 & | & 4 \\ 1 & 1 & | & 2 \end{bmatrix} \overset{=}{\underset{(r_2 - r_1)}{}} \begin{bmatrix} 1 & 2 & | & 6 \\ 0 & -1 & | & -2 \end{bmatrix} \overset{(r_1 - 2r_2)}{\underset{2r_2}{}} \begin{bmatrix} 1 & 0 & | & -2 \\ 0 & 1 & | & 2 \end{bmatrix}$ The solution is (2, 2).

26. $\begin{bmatrix} -1 & 1 & | & 4 \\ 1 & 2 & | & 2 \end{bmatrix} = \begin{bmatrix} 1 & -1 & | & -4 \\ 0 & 3 & | & 6 \end{bmatrix} = \begin{bmatrix} 1 & 0 & | & -2 \\ 0 & 1 & | & 2 \end{bmatrix}$ The solution is (− 2, 2).

27. $\begin{bmatrix} 2 & 1 & | & 3 \\ 3 & -1 & | & 12 \end{bmatrix} \overset{(r_1 \div 2)}{\underset{=}{}} \begin{bmatrix} 1 & \frac{1}{2} & | & \frac{3}{2} \\ 3 & -1 & | & 12 \end{bmatrix} \overset{=}{\underset{(r_2 - 3r_1)}{}} \begin{bmatrix} 1 & \frac{1}{2} & | & \frac{3}{2} \\ 0 & -\frac{5}{2} & | & \frac{15}{2} \end{bmatrix} \overset{=}{\underset{(-\frac{2}{5}r_2)}{}} \begin{bmatrix} 1 & \frac{1}{2} & | & \frac{3}{2} \\ 0 & 1 & | & -3 \end{bmatrix} \overset{(-\frac{1}{2}r_2 + r_1)}{\underset{=}{}} \begin{bmatrix} 1 & 0 & | & 3 \\ 0 & 1 & | & -3 \end{bmatrix}$
The solution is (3,− 3).

28. $\begin{bmatrix} 2 & 3 & | & 2 \\ 4 & -9 & | & 4 \end{bmatrix} = \begin{bmatrix} 1 & \frac{3}{2} & | & 1 \\ 0 & -15 & | & 0 \end{bmatrix} = \begin{bmatrix} 1 & \frac{3}{2} & | & 1 \\ 0 & 1 & | & 0 \end{bmatrix} = \begin{bmatrix} 1 & 0 & | & 1 \\ 0 & 1 & | & 0 \end{bmatrix}$ The solution is (1,0)

29. $\begin{bmatrix} 1 & 3 & | & 3 \\ 3 & -2 & | & 2 \end{bmatrix} = \begin{bmatrix} 1 & 3 & | & 3 \\ 0 & -11 & | & -7 \end{bmatrix} = \begin{bmatrix} 1 & 3 & | & 3 \\ 0 & 1 & | & \frac{7}{11} \end{bmatrix} = \begin{bmatrix} 1 & 0 & | & \frac{12}{11} \\ 0 & 1 & | & \frac{7}{11} \end{bmatrix}$ The solution is $\left(\frac{12}{11}, \frac{7}{11} \right)$

30. $\begin{bmatrix} 3 & -6 & | & -9 \\ 4 & 5 & | & 14 \end{bmatrix} \overset{(r_1 \bullet -1)}{\longrightarrow} \begin{bmatrix} -3 & 6 & | & 9 \\ 4 & 5 & | & 14 \end{bmatrix} \overset{(r_2 + r_1)}{\longrightarrow} \begin{bmatrix} 1 & 11 & | & 23 \\ 4 & 5 & | & 14 \end{bmatrix} \overset{(r_1 \bullet -4)}{\underset{(r_2 + r_1)}{}} \begin{bmatrix} 1 & 11 & | & 23 \\ 0 & -39 & | & -78 \end{bmatrix}$

$\begin{bmatrix} 1 & 11 & | & 23 \\ 0 & -39 & | & -78 \end{bmatrix} \overset{(r_1 \bullet 11/39)}{\underset{(r_2 + r_1)}{}} \begin{bmatrix} 1 & 0 & | & 1 \\ 0 & -39 & | & -78 \end{bmatrix} \overset{(r_2 \div -2)}{\longrightarrow} \begin{bmatrix} 1 & 0 & | & 1 \\ 0 & 1 & | & 2 \end{bmatrix}$ The solution is (1, 2).

31. Let x = amount borrowed at 8%   y = amount borrowed at 10%
    .08x + .10y = 53000   x + y = 600000
    $\begin{bmatrix} .08 & .10 & | & 53000 \\ 1 & 1 & | & 600000 \end{bmatrix} \overset{(r_2 \bullet -10)}{\underset{(r_2 + r)}{}} \begin{bmatrix} -2 & 0 & | & -700000 \\ 1 & 1 & | & 600000 \end{bmatrix} \overset{(r_1 \bullet -1)}{\underset{(r_2 + r)}{}} \begin{bmatrix} 1 & 0 & | & 350000 \\ 0 & 1 & | & 250000 \end{bmatrix}$
    x = $350,000 and y = $ 250,000

32. Let s = liters of 80% acid solution
    w = liters of 50% acid solution
    s + w = 100
    0.80s + 0.50w = 100(0.75)
    0.80s + 0.50w = 75
    s = 100 − w
    0.80(100 − w) + 0.50w = 75
    80 − 0.80w + 0.50w = 75
    − 0.30w = − 5
    w = − 5/(− 0.30) = 16 2/3 liters
    s = 100 − 16 2/3 = 83 1/3 liters

33. Let s = salary   r = commission rate
    (1) s + 4000r = 660
    (2) s + 6000r = 740 (subtract 1 from 2)
          2000r = 80
    r = 80/2000 = 0.04
    Substitute 0.04 for r in eq'n. 1.
    s = 660 − 4000(.04)   s = 500
    His salary is 500 per week and his commission
    rate is 4%.

34. Let c = total cost   x = no. of months to
operate
    a) model 1600A: c_A = 950 + 32x
        model 6070B: c_B = 1275 + 22x
        950 + 32x = 1275 + 22x
        10x = 325   x = 32.5 months
        After 32.5 months of operation the total cost
        of the units will be equal.
    b) After 32.5 months or 2.7 years, the most cost
        effective unit is the unit with the lower per
        month to operate cost. Thus, model 6070B is
        the better deal in the long run.

35. a) Let C = total cost for parking
            x = number of additional hours
        All-Day: C = 5 + 0.50x
        Sav-A-Lot: C = 4.25 + 0.75x
        5 + 0.50x = 4.25 + 0.75x
        0.75 = 0.25x   3 = x
        The total cost will be the same after 3
        additional hours or 4 hours total.
    b) After 5 hours or x = 4 additional hours:
        All-Day: C = 5 + 0.50(4) = $7.00
        Sav-A-Lot: C = 4.25 + 0.75(4) = $7.25
        All-Day would be less expensive.

36.

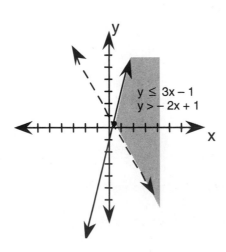

37.

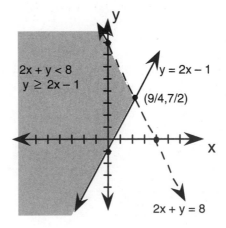

38.

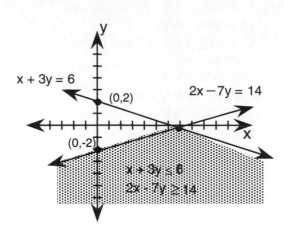

39.

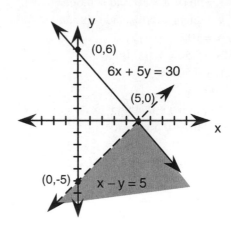

40. $P = 6x + 3y$

At $(0,0)$, $P = 6(0) + 3(0) = 0$

At $(0,10)$, $P = 6(0) + 3(10) = 30$

At $(9,0)$, $P = 6(9) + 3(0) = 54$

The maximum profit is $54 at $(9,0)$.

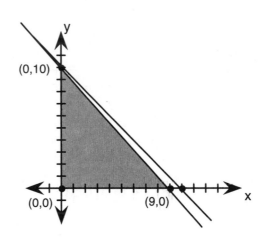

Chapter Test

1. If the lines do not intersect (parallel) the system of equations is inconsistent. The system of equations is consistent if the lines intersect only once. If both equations represent the same line then the system of equations is dependent.

2.

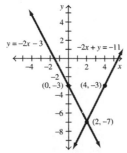

The solution is $(2, -7)$.

3. Write each equation in slope intercept form, then

   compare slopes and intercepts.

   $4x + 5y = 6$ $-3x + 5y = 13$

   $5y = -4x + 6$ $5y = 3x + 13$

   $y = -(4/5)x + 6/5$

   $y = (3/5)x + 13/5$

   The slopes are different so there is only
   one solution.

4. $x - y = 5 \qquad x = y + 5$

$2x + 3y \qquad\quad = -5$

Substitute $(y + 5)$ for x in the second equation.

$2(y + 5) + 3y = -5$ (solve for y)

$2y + 10 + 3y = -5$

$5y + 10 = -5$

$5y = -15 \qquad y = -3$

Substitute (-3) for y in the equation $x = y + 5$.

$x = -3 + 5 = 2 \qquad$ The solution is $(2, -3)$.

5. $y = 5x + 7 \qquad y = 2x + 1$

Substitute $(5x + 7)$ for y in the second equation.

$5x + 7 = 2x + 1$ (solve for x)

$3x = -6 \qquad x = -2$

Substitute -2 for x in the first equation.

$y = 5(-2) + 7 = -10 + 7 = -3$

The solution is $(-2, -3)$.

6. $x - y = 4$

$\underline{2x + y = 5}$ (add)

$3x = 9 \qquad\qquad x = 3$

Substitute 3 for x in the 2nd equation.

$2(3) + y = 5$

$6 + y = 5 \qquad y = -1 \qquad$ The solution is $(3, -1)$.

7. $4x + 3y = 5$

$2x + 4y = 10$

Multiply the second equation by (-2).

$4x + 3y = 5$

$\underline{-4x - 8y = -20}$ (add)

$-5y = -15 \qquad\qquad y = 3$

Substitute 3 for y in the first equation.

$4x + 3(3) = 5$

$4x + 9 = 5$

$4x = -4 \qquad\qquad x = -1$

The solution is $(-1, 3)$.

8. $3x + 4y = 6$

$2x - 3y = 4$

Multiply the 1st eq'n. by 3 and the 2nd eq'n. by 4.

$9x + 12y = 18$

$\underline{8x - 12y = 16}$

$17x = 34 \qquad\qquad x = 2$

8. Substitute 2 for x in an equation.

$2(2) - 3y = 4$ (solve for y)

$-3y = 0 \qquad\qquad y = 0$

The solution is $(2, 0)$.

9. $\begin{bmatrix} 1 & 3 & | & 4 \\ 5 & 7 & | & 4 \end{bmatrix} \underset{(-5r_1 + r_2)}{=} \begin{bmatrix} 1 & 3 & | & 4 \\ 0 & -8 & | & -16 \end{bmatrix} \underset{(r_2 \div (-8))}{=} \begin{bmatrix} 1 & 3 & | & 4 \\ 0 & 1 & | & 2 \end{bmatrix} \overset{(r_1 - 3r_2)}{=} \begin{bmatrix} 1 & 0 & | & -2 \\ 0 & 1 & | & 2 \end{bmatrix}$

The solution is $(-2, 2)$.

10. $A + B = \begin{bmatrix} 2 & -5 \\ 1 & 3 \end{bmatrix} + \begin{bmatrix} -1 & -3 \\ 5 & 2 \end{bmatrix} = \begin{bmatrix} 2 + (-1) & -5 - 3 \\ 1 + 5 & 3 + 2 \end{bmatrix} = \begin{bmatrix} 1 & -8 \\ 6 & 5 \end{bmatrix}$

11. $3A - B = 3\begin{bmatrix} 2 & -5 \\ 1 & 3 \end{bmatrix} - \begin{bmatrix} -1 & -3 \\ 5 & 2 \end{bmatrix} = \begin{bmatrix} 3(2) - (-1) & 3(-5) - (-3) \\ 3(1) - 5 & 3(3) - 2 \end{bmatrix} = \begin{bmatrix} 7 & -12 \\ -2 & 7 \end{bmatrix}$

12. $A \times B = \begin{bmatrix} 2 & -5 \\ 1 & 3 \end{bmatrix}\begin{bmatrix} -1 & -3 \\ 5 & 2 \end{bmatrix} = \begin{bmatrix} 2(-1) + (-5)(5) & 2(-3) + (-5)(2) \\ 1(-1) + (3)(5) & 1(-3) + 3(2) \end{bmatrix} = \begin{bmatrix} -27 & -16 \\ 14 & 3 \end{bmatrix}$

13. $y < -2x + 2$ $y > 3x + 2$

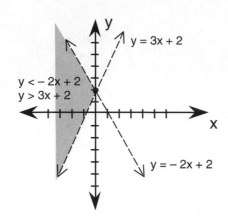

14. Let x = lb of $6.00 coffee
    y = lb of $7.50 coffee
    $x + y = 30$ $y = 30 - x$
    $6x + 7.5y = 7.00(30)$
    Substitute $(30 - x)$ for y in the 2nd equation.
    $6x + 7.5(30 - x) = 210$
    $6x + 225 - 7.5x = 210$
    $-1.5x = -15$ $x = 10$
    Substitute 10 for x in the equation
    $y = 30 - x$.
    $y = 30 - 10 = 20$
    Mix 10 lb of the $6.00 coffee with 20 lb of
    the $7.50 coffee.

15. (a) Let x = no. of checks written in one month.
        Cost at Union Bank: $6 + .10x$
        Cost at Citrus Bank: $2 + .20x$
        These are equal when:
          $2 + .2x = 6 + .1x$
            $.1x = 4$
              $x = 40$
    (b) Since $14 < 40$, the bank with the lower
        monthly fee is better, which is Citrus Bank.

16. a)

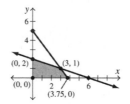

   b) $P = 5x + 3y$
      At $(0, 0)$ $P = 5(0) + 3(0) = 0$
      At $(0, 2)$ $P = 5(0) + 3(2) = 6$
      At $(3, 1)$ $P = 5(3) + 3(1) = 18$
      At $(3.75, 0)$ $P = 5(3.75) + 3(0) = 18.75$
      Max. at $(3.75, 0)$ and Min. at $(0, 0)$

CHAPTER EIGHT

THE METRIC SYSTEM

Exercise Set 8.1

1. The metric system.

3. It is the worldwide accepted standard of measurement. There is only 1 basic unit of measurement for each quantity. It is based on the number 10, which makes many calculations easier.

5. a) Move the decimal point one place for each change in unit of measure.

   b) $714.6 \text{ cm} = \dfrac{714.6}{10^5} \text{ km} = 714.6 \times 10^{-5} \text{ km} = 0.007146 \text{ km}$

   c) $30.8 \text{ hm} = (30.8)(1000) \text{ dm} = 30800 \text{ dm}$

7.
kilo	1000 times the base unit	k
hecto	100 times the base unit	h
deka	10 times the base unit	da
deci	1/10 times the base unit	d
centi	1/100 times the base unit	c
milli	1/1000 times the base unit	m

9. a) 100 times greater
   b) 1 dam = 100 dm
   c) 1dm = 0.01 dam

11. 2 pounds

13. 5 grams

15. 22° C

17. kilo d

19. hecto c

21. deci f

23. a) 10 liters
    b) 1/100 liter
    c) 1/1000 liter
    d) 1/10 liter
    e) 1000 liters
    f) 100 liters

25. mg   1/1000 gm

27. dg   1/10 gm

29. hg   100 gm

31. Max. load 320 kg = (320 x 1,000) g = 320 000 g

33. 2 m = (2 x 100) cm = 200 cm

35. 0.095 hℓ = (0.095)(100) = 9.5 ℓ

37. 242.6 cm = (242.6)(0.0001) hm = 0.02426 hm

39. 4036 mg = (4036)(0.00001) hg = 0.04036 hg

41. 1.34 hm = (1.34)(10000) cm = 13,400 cm

43. 92.5 kg = 92,500 g

45. 895 ℓ = 895,000 mℓ

47. 240 cm = 0.0240 hm

49. 40,302 mℓ = 4.0302 daℓ

51. 590 cm, 5.1 dam, 0.47 km

53. 2.2 kg, 2,400 g, 24,300 dg

55. 203,000 mm, 2.6 km, 52.6 hm

57. Jim, since a meter is longer than a yard.

59. The pump that removes 1 daℓ of water per min.
     1 dekaliter > 1 deciliter

61. a) Perimeter= 2l + 2w= 2(74) + 2(99)= 346 cm
     b) 346 cm = (346 x 10) mm = 3,460 mm

63. a) (4)(27 m) = 108 m   b) 108 m = 0.108 km
     c) 108 m = 108 000 mm

65. 8 (400) m = 3,200 m;   3,200 m = 3.2 km

67. a) 6(360) ml = 2,160 mℓ
     b) 2160(1000) = 2.16 ℓ
     c) 2.45 / 2.16 = $1.13 per liter

69. a) (6.9)(1000) = 6,900 gm
     b) 6,900 / 3 = 2300 gm   2300 gm = 23,000 dg

71. 1 gigameter = 1000 megameters

73. 1 teraliter = 1 x 10^{24} picoliters

75. 0.8/.027 = 29.6 ≈ 30 eggs

77. 195 mg = 0.195 g
     0.8/0.195 = 4.1 cups

79. 5000 cm = 5 dam

81. 0.00006 hg = 6 mg

83. 0.02 kℓ = 2 daℓ

85. magr   gram

86. migradec   decigram

87. rteli   liter

88. raktileed   dekaliter

89. terem   meter

90. leritililm   milliliter

91. reketolim   kilometer

92. timenceret   centimeter

93. greeed sulesic degree celsius

94. togmeharc   hectogram

Exercise Set 8.2

1. volume

3. area

5. volume

7. volume

9. area

11. length

13. Answers will vary.

15. Answers will vary.

17. Answers will vary.

19. 1 cubic decimeter

21. 1 cubic centimeter

23. area

25. centimeters

27. cm or mm

29. centimeters

31. millimeters

33. cm or mm

35. kilometers

37. c   27 m

39. c   5 km

41. a   2 cm

43. b   1000 m

45. mm (AWV)

47. cm or m   (AWV)

49. mm or cm   (AWV)

51. cm, km

53. m

55. cm

57. sq. cm.

59. sq. m.

61. sq. m. or hectares

63. sq. mm. or sq. cm.

65. hectares or sq. km.

67. b   2.2 sq .m.

69. a   800 sq. m.

71. c   360 sq. cm.

73. c   1200 sq. mm.

75. AWV

77. AWV

79. AWV

81. kiloliters

83. milliliters

85. liters

87. cubic meters

89. liters

91. c   7780 cu. cm.

93. c   55 kℓ

95. a   550 cu. m.

97. a   30 cu. m.

99. a) AWV

99. b) 152,561 cc

101. a) AWV   b) v ≈ (3.14)(0.25)^2 (1) = 0.20 m^3

103. a) AWV   b) A = lw = (4)(2.2) = 8.8 cm^2

105. (82)(62) - (50)(42) = 5084 - 2100 = 2984 cm^2

107. a) (3.75)(1.4) = 5.25 km
      b) (5.25)(100 ha) = 525 ha

109. a) (18)(10)(2.5) = 450 m^3
      b) 450 m^3 = 450 kℓ

111. a) V = lwh = (70)(40)(20) = 56,000 cm^3 b) 56,000 cm^3 = 56,000 mℓ c) 56 000 mℓ = $\left(\dfrac{56000}{1000}\right)\ell$ = 56 ℓ

113. $10^2 = 100$ times larger   115. $10^3 = 1000$ times larger   117. 1,000,000 mm^2

119. 100 hm^2 121. 0.001 cm^3 123. 1,000,000 cm^3

125. 435 cm^3 = 435 mℓ 127. 76 kℓ = 76 m^3 129. $(6.0 \times 10^4)(10) = 600{,}000$ dℓ

131. AWV

133. 6.7 kl = 6.7 m^3 = (6.7×10^3) dm^3 = 6,700 dm^3 135. a) 1 sq mi = $(1\ \text{mi}^2)(5280)^2 \dfrac{\text{ft}^2}{\text{mi}^2} = 27{,}878{,}400\ \text{ft}^2$

$27{,}878{,}400\ \text{ft}^2 \times (12)^2 \dfrac{\text{in}^2}{\text{ft}^2} = 4{,}014{,}489{,}600\ \text{in}^2$

     b) It is easier to convert in the metric system because it is a base 10 system.

137. Answers will vary.   138. a) 1.5 m = 150 cm

                                                 150 – 50 = 100 cm.

                                   b) 150/50 = 3   3 times longer

                                   c) No.

139. a) 5150.7 liters / day

     b) 493.2 liters / day

Exercise Set 8.3

1. kilogram   3. 2 lb   5. approx. 35° C AWV   7. Answers will vary

9. grams   11. grams   13. grams   15. kilograms

17. grams   19. b   2.26 kg   21. b   1.4 kg   23. b   2800 kg

25. AWV   27. AWV   29. c   0° C   31. b   27° C

33. b   5° C   35. c   177° C   37. b   -7° C

39. $F = \dfrac{9}{5}(30) + 32 = 54 + 32 = 86°\,F$ 41. $C = \dfrac{5}{9}(92 - 32) = \dfrac{5}{9}(60) = 33.3°\,C$

43. $C = \dfrac{5}{9}(180 - 32) = \dfrac{5}{9}(148) = 82.2°\,C$ 45. $F = \dfrac{9}{5}(37) + 32 = 66.6 + 32 = 98.6°\,F$

47. $C = \dfrac{5}{9}(13 - 32) = \dfrac{5}{9}(-19) = -10.6°\,C$ 49. $F = \dfrac{9}{5}(45) + 32 = 81 + 32 = 113°\,F$

51. $C = \dfrac{5}{9}(-20 - 32) = \dfrac{5}{9}(-52) = -28.9°\,C$ 53. $F = \dfrac{9}{5}(22) + 32 = 39.6 + 32 = 71.6°\,F$

53. $F = \dfrac{9}{5}(22) + 32 = 39.6 + 32 = 71.6°\,F$ 55. $F = \dfrac{9}{5}(35.1) + 32 = 63.2 + 32 = 95.2°\,F$

57. low: $F = \dfrac{9}{5}(17.8) + 32 = 32 + 32 = 64.04°\,F$ 59. cost = (6.2)(.70) = \$ 4.34

     high: $F = \dfrac{9}{5}(23.5) + 32 = 42.3 + 32 = 74.3°\,F$

     Range = 74.30 – 64.04 = 10.26° F

59. cost = (6.2)(.70) = \$ 4.34   61. total mass = 45 g + 29 g + 370 mℓ =

                                                     45 g + 29 g + 370 g = 444 g

63. a) V = lwh,   l = 16 m, w = 12 m, h = 12 m   65. Yes, $78°\,F = \dfrac{5}{9}(78 - 32) \approx 25.6°\,C$, not 20° C

       V = (16)(12)(12) = 2304 m^3

     b) 2304 m^3 = 2304 kℓ

c) 2304 kℓ= 2304 t

67. $4.2 \text{ kg} = (4.2 \text{ kg})\left(\dfrac{1 \text{ t}}{1000 \text{ kg}}\right) = 0.0042 \text{ t}$ 69. $17.4 \text{ t} = (17.4 \text{ t})\left(\dfrac{1000 \text{ kg}}{1 \text{ t}}\right) = 17,400 \text{ kg} =$

    $17,400,000 \text{ g}$

71. $1.2 \, \ell = 1200 \text{ m}\ell$ a) 1200 g   b) 1200 cm^3 74. 3 kg _____ x

73. a) $V = lwh$ $l = 1 \text{ yd} = 3 \text{ ft}$ $w = 15 \text{ in} = 1.25 \text{ ft}$

            $h = 1.5 \text{ ft}$

        $V = (3)(1.25)(1.5) = 5.625 \text{ cubic feet}$

    $(3)(2) = 6 = 4x$

    $x = 6/4 = 3/2 = 1.5$ 1.5 kg

    1.5 kg = 1500 g

   b) $(5.625 \text{ ft}^3)\left(62.5 \, \dfrac{\text{lbs}}{\text{ft}^3}\right) = 351.6 \text{ lb}$

   c) $(351.6 \text{ lb})\left(\dfrac{1 \text{ gal}}{8.3 \text{ lb}}\right) = 42.4 \text{ gal}$

75. a) -62.11° C $F = \dfrac{9}{5}(-62.11) + 32 = -111.798 + 32 = -79.8°\text{ F}$

   b) 2.5° C $F = \dfrac{9}{5}(2.5) + 32 = 4.5 + 32 = 36.5°\text{ F}$

   c) 918,000,000° F $C = \dfrac{5}{9}(918,000,000 - 32) = \dfrac{5}{9}(917,999,968) = 509,999,982.2 \approx 510,000,000°\text{ C}$

Exercise Set 8.4

1. **Dimensional analysis** is a procedure used to convert from one unit of measurement to a different unit of measurement.

3. $\dfrac{60 \text{ seconds}}{1 \text{ minute}}$ or $\dfrac{1 \text{ minute}}{60 \text{ seconds}}$ because 60 seconds = 1 minute

5. $\dfrac{1 \text{ ft}}{30 \text{ cm}}$ Since we need to eliminate centimeters, cm must appear in the denominator. Since we need to convert to feet, ft must appear in the numerator.

7. $\dfrac{3.8 \, \ell}{1 \text{ gal}}$ Since we need to eliminate gallons, gal must appear in the denominator. Since we need to convert to liters, l must appear in the numerator.

9. $52 \text{ in.} = (52 \text{ in.})\left(\dfrac{2.54 \text{ cm}}{1 \text{ in.}}\right) = 132.08 \text{ cm}$

11. $4.2 \text{ ft} = (4.2 \text{ ft})\left(\dfrac{30 \text{ cm}}{1 \text{ ft}}\right)\left(\dfrac{1 \text{ m}}{100 \text{ cm}}\right) = 1.26 \text{ m}$

13. $15 \text{ yd}^2 = \left(15 \text{ yd}^2\right)\left(\dfrac{0.8 \text{ m}^2}{1 \text{ yd}^2}\right) = 12 \text{ m}^2$

15. $39 \text{ mi} = \left(39 \text{ mi}\right)\left(\dfrac{1.6 \text{ km}}{1 \text{ mi}}\right) = 62.4 \text{ km}$

17. $675 \text{ ha} = \left(675 \text{ ha}\right)\left(\dfrac{1 \text{ acre}}{0.4 \text{ ha}}\right) = 1687.5 \text{ acres}$

19. $15.6 \; \ell = \left(15.6 \; \ell\right)\left(\dfrac{1 \text{ pt}}{0.47 \; \ell}\right) = 33.19148936 \approx 33.19 \text{ pints}$

21. $45.6 \text{ m}\ell = \left(45.6 \text{ m}\ell\right)\left(\dfrac{1 \text{ fl oz}}{30 \text{ m}\ell}\right) = 1.52 \text{ fl oz}$

23. $120 \text{ lb} = \left(120 \text{ lb}\right)\left(\dfrac{0.45 \text{ kg}}{1 \text{ lb}}\right) = 54 \text{ kg}$

25. 28 grams

27. 0.45 kilogram

29. 2.54 centimeters, 1.6 kilometers

31. $10 \text{ yd} = \left(10 \text{ yd}\right)\left(\dfrac{0.9 \text{ m}}{1 \text{ yd}}\right) = 9 \text{ meters}$

33. $505 \text{ m} = \left(505 \text{ m}\right)\left(\dfrac{1 \text{ yd}}{0.9 \text{ m}}\right) = 561.\overline{1} \approx 561.11 \text{ yd}$

35. $344 \text{ m} = \left(344 \text{ m}\right)\left(\dfrac{100 \text{ cm}}{1 \text{ m}}\right)\left(\dfrac{1 \text{ ft}}{30 \text{ cm}}\right) = 1146.\overline{6} \approx 1146.67 \text{ ft}$

37. $85 \text{ km} = \left(85 \text{ km}\right)\left(\dfrac{1 \text{ mi}}{1.6 \text{ km}}\right) = 53.125 \approx 53.13 \text{ mph}$

39. $\left(6 \text{ yd}\right)\left(9 \text{ yd}\right) = 54 \text{ yd}^2$

$54 \text{ yd}^2 = \left(54 \text{ yd}^2\right)\left(\dfrac{0.8 \text{ m}^2}{1 \text{ yd}^2}\right) = 43.2 \text{ m}^2$

41. $400 \text{ g} = \left(400 \text{ g}\right)\left(\dfrac{1 \text{ oz}}{28 \text{ g}}\right) = 14.28571429 \approx 14.29 \text{ oz}$

43. $8 \text{ fl oz} = \left(8 \text{ fl oz}\right)\left(\dfrac{30 \text{ m}\ell}{1 \text{ fl oz}}\right) = 240 \text{ m}\ell$

45. $\left(50 \text{ ft}\right)\left(30 \text{ ft}\right)\left(8 \text{ ft}\right) = 12,000 \text{ ft}^3$

$12,000 \text{ ft}^3 = \left(12,000 \text{ ft}^3\right)\left(\dfrac{0.03 \text{ m}^3}{1 \text{ ft}^3}\right) = 360 \text{ m}^3$

47. $1 \text{ kg} = \left(1 \text{ kg}\right)\left(\dfrac{1 \text{ lb}}{0.45 \text{ kg}}\right) = 2.\overline{2} \text{ lb}$

$\dfrac{\$1.10}{2.\overline{2}} = \$0.495 \text{ per pound}$

49. $34.5 \text{ k}\ell = \left(34.5 \text{ k}\ell\right)\left(\dfrac{1000 \; \ell}{1 \text{ k}\ell}\right)\left(\dfrac{1 \text{ gal}}{3.8 \; \ell}\right) = 9078.947368 \approx 9078.95 \text{ gal}$

51. a) $8 \text{ stones} = (8 \text{ stones})\left(\dfrac{70 \text{ kg}}{11 \text{ stones}}\right) = 50.\overline{90} \approx 50.91 \text{ kg}$

b) $50.\overline{90} \text{ kg} = (50.\overline{90} \text{ kg})\left(\dfrac{1 \text{ lb}}{0.45 \text{ kg}}\right) = 113.\overline{13} \approx 113.13 \text{ lb}$

53. a) $-282 \text{ ft} = (-282 \text{ ft})\left(\dfrac{30 \text{ cm}}{1 \text{ ft}}\right) = -8460 \text{ cm}$

b) $-8460 \text{ cm} = (-8460 \text{ cm})\left(\dfrac{1 \text{ m}}{100 \text{ cm}}\right) = -84.6 \text{ m}$

55. a) $1 \text{ m}^2 = (1 \text{ m}^2)\left(\dfrac{(3.3)^2 \text{ ft}^2}{1 \text{ m}^2}\right) = 10.89 \text{ ft}^2$

b) $1 \text{ m}^3 = (1 \text{ m}^3)\left(\dfrac{(3.3)^3 \text{ ft}^3}{1 \text{ m}^3}\right) = 35.937 \text{ ft}^3$

57. $56 \text{ lb} = (56 \text{ lb})\left(\dfrac{0.45 \text{ kg}}{1 \text{ lb}}\right)\left(\dfrac{1 \text{ mg}}{1 \text{ kg}}\right) = 25.2 \text{ mg}$

59. $76 \text{ lb} = (76 \text{ lb})\left(\dfrac{0.45 \text{ kg}}{1 \text{ lb}}\right)\left(\dfrac{200 \text{ mg}}{1 \text{ kg}}\right) = 6840 \text{ mg}$

$6840 \text{ mg} = (6840 \text{ mg})\left(\dfrac{1 \text{ g}}{1000 \text{ mg}}\right) = 6.84 \text{ g}$

61. a) $2 \text{ teaspoons} = (2 \text{ teaspoons})\left(\dfrac{12.5 \text{ mg}}{1 \text{ teaspoon}}\right) = 25 \text{ mg}$

b) $12 \text{ fl oz} = (12 \text{ fl oz})\left(\dfrac{30 \text{ m}\ell}{1 \text{ fl oz}}\right)\left(\dfrac{12.5 \text{ mg}}{5 \text{ m}\ell}\right) = 900 \text{ mg}$

63. a) $964 \text{ ft} = (964 \text{ ft})\left(\dfrac{30 \text{ cm}}{1 \text{ ft}}\right)\left(\dfrac{1 \text{ m}}{100 \text{ cm}}\right) = 289.2 \text{ m}$

b) $85,000 \text{ tons} = (85,000 \text{ tons})\left(\dfrac{0.9 \text{ tonne}}{1 \text{ ton}}\right) = 76\ 500 \text{ t}$

c) $28 \text{ mi} = (28 \text{ mi})\left(\dfrac{1.6 \text{ km}}{1 \text{ mi}}\right) = 44.8 \text{ kph}$

65. a) $(37 \text{ m})\left(\dfrac{1 \text{ yd}}{0.9 \text{ m}}\right) = 41.\overline{1} \approx 41.1 \text{ yd}$

b) $(370\ 140 \text{ km})\left(\dfrac{1 \text{ mi}}{1.6 \text{ km}}\right) = 231,337.5 \text{ mi}$

c) $(44 \text{ km})\left(\dfrac{1 \text{ mi}}{1.6 \text{ km}}\right) = 27.5 \text{ mi}$

d) $1260°\text{ C} = \dfrac{9}{5}(1260) + 32 = 2300° \text{ F}$

e) $(335 \text{ km})\left(\dfrac{1 \text{ mi}}{1.6 \text{ km}}\right) = 209.375 \text{ mph}$

f) $(29\ 484 \text{ kg})\left(\dfrac{1 \text{ lb}}{0.45 \text{ kg}}\right) = 65,520 \text{ lb}$

g) $(4.5 \text{ m})\left(\dfrac{1 \text{ yd}}{0.9 \text{ m}}\right) \times (18 \text{ m})\left(\dfrac{1 \text{ yd}}{0.9 \text{ m}}\right) = 5 \text{ yd} \times 20 \text{ yd}$

h) $(171\ 396\ \ell)\left(\dfrac{1\ gal}{3.8\ \ell}\right) = 45{,}104.21053 \approx 45{,}104.21$ gal/min

i) $(63\ 588\ \ell)\left(\dfrac{1\ gal}{3.8\ \ell}\right) = 16{,}733.68421 \approx 16{,}733.68$ gal/min

j) $(46.89\ m)\left(\dfrac{1\ yd}{0.9\ m}\right) = 52.1$ yd

k) $(8.4\ m)\left(\dfrac{1\ yd}{0.9\ m}\right) = 9.\overline{3} \approx 9.33$ yd

l) $(632\ \ 772\ kg)\left(\dfrac{1\ lb}{0.45\ kg}\right) = 1{,}406{,}160$ lb

m) $(106\ \ 142\ kg)\left(\dfrac{1\ lb}{0.45\ kg}\right) = 235{,}871.\overline{1} \approx 235{,}871.11$ lb

n) $-251°\ C = \dfrac{9}{5}(-251) + 32 = -419.8°\ F$

67. $15(130\ lb) = 1950$ lb

   $(1950\ lb)\left(\dfrac{0.18\ kg}{100\ lb}\right)\left(\dfrac{1\ lb}{0.45\ kg}\right) = 7.8$ lb

69. A meter

70. A kilogram

71. A hectare

72. A liter

73. A tonne

74. A decimeter

75. wonton

77. 1 kilohurtz

79. 1 megaphone

81. 2 kilomockingbird

83. 1 decaration

Review Exercises

1. $\dfrac{1}{100}$ of base unit

2. $1000\times$ base unit

3. $\dfrac{1}{1000}$ of base unit

4. $100\times$ base unit

5. 10 times base unit

6. $\dfrac{1}{10}$ of base unit

7. 20 cg = 0.20 g

8. 3.2 ℓ= 320 cℓ

9. 0.0004 cm = 0.004 mm

10. 1 000 000 mg = 1 kg

11. 4.62 kℓ= 4620 ℓ

12. 192.6 dag = 19 260 dg

13. 2.67 kℓ= 2 670 000 mℓ
     14 630 cℓ= 146 300 mℓ
     3000 mℓ 14 630 cℓ 2.67 kℓ

14. 0.047 km = 47 m
     47 000 cm = 470 m
     0.047 km, 47 000 cm,
     4700 m

15. Centimeters

16. Grams

17. Degrees Celsius

18. Millimeters or centimeters

19. Square meters

20. Milliliters or cubic centimeters

21. Millimeters

22. Kilograms or tonnes

23. Kilometers

24. Meters or centimeters

25. a) and b) Answers will vary.

26. a) and b) Answers will vary.

27. c

28. b

29. c

30. a

31. a

32. b

33. $2500 \text{ kg} = (2500 \text{ kg}) \left(\dfrac{1 \text{ lb}}{0.45 \text{ kg}} \right) \left(\dfrac{1 \text{ T}}{2000 \text{ lb}} \right) \left(\dfrac{0.9 \text{ t}}{1 \text{ T}} \right) = 2.5 \text{ t}$

34. $6.3 \text{ t} = (6.3 \text{ t}) \left(\dfrac{1 \text{ T}}{0.9 \text{ t}} \right) \left(\dfrac{2000 \text{ lb}}{1 \text{ T}} \right) \left(\dfrac{0.45 \text{ kg}}{1 \text{ lb}} \right) \left(\dfrac{1000 \text{ g}}{1 \text{ kg}} \right) = 6 \ 300 \ 000 \text{ g}$

35. $18° \text{ C} = \dfrac{9}{5}(18) + 32 = 64.4° \text{ F}$

36. $68° \text{ F} = \dfrac{5}{9}(68 - 32) = 20° \text{ C}$

37. $-6° \text{ F} = \dfrac{5}{9}(-6 - 32) = -21.\overline{1} \approx -21.1° \text{ C}$

38. $39° \text{ C} = \dfrac{9}{5}(39) + 32 = 102.2° \text{ F}$

39. $l = 4 \text{ cm}, \ w = 1.6 \text{ cm}$

    $A = lw = 4(1.6) = 6.4 \text{ cm}^2$

40. $r = 1.5 \text{ cm}$

    $A = \pi r^2 \approx 3.14(1.5)^2 = 7.065 \approx 7.07 \text{ cm}^2$

41. a) $V = lwh = (10)(4)(2) = 80 \text{ m}^3$

    b) $(80 \text{ m}^3) \left(\dfrac{1 \text{ kl}}{1 \text{ m}^3} \ell \right) \left(\dfrac{1000 \ \ell}{1 \text{ k}\ell} \right) \left(\dfrac{1 \text{ kg}}{1 \ \ell} \right) = 80 \ 000 \text{ kg}$

42. a) $A = lw = 30(22) = 660 \text{ m}^2$

    b) $660 \text{ m}^2 = (660 \text{ m}^2) \left(\dfrac{1 \text{ km}^2}{(1000)^2 \text{ m}^2} \right) = 0.000 \ 66 \text{ km}^2$

43. a) $V = lwh = (80)(40)(30) = 96 \ 000 \text{ cm}^3$

    b) $96 \ 000 \text{ cm}^3 = (96 \ 000 \text{ cm}^3) \left(\dfrac{1 \text{ m}^3}{(100)^3 \text{ cm}^3} \right) = 0.096 \text{ m}^3$

    c) $96 \ 000 \text{ cm}^3 = (96 \ 000 \text{ cm}^3) \left(\dfrac{1 \text{ m}\ell}{1 \text{ cm}^3} \right) = 96 \ 000 \text{ m}\ell$

    d) $0.096 \text{ m}^3 = (0.096 \text{ m}^3) \left(\dfrac{1 \text{ k}\ell}{1 \text{ m}^3} \right) = 0.096 \text{ k}\ell$

44. Since $1 \text{ km} = 100 \times 1 \text{ dam}, \ 1 \text{ km}^2 = 100^2 \times 1 \text{ dam}^2 = 10 \ 000 \text{ dam}^2$.

    Thus, 1 square kilometer is 10,000 times larger than a square dekameter.

45. $(20 \text{ cm}) \left(\dfrac{1 \text{ in.}}{2.54 \text{ cm}} \right) = 7.874015748 \approx 7.87 \text{ in.}$

46. $(105 \text{ kg}) \left(\dfrac{1 \text{ lb}}{0.45 \text{ kg}} \right) = 233.\overline{3} \approx 233.33 \text{ lb}$

47. $(83 \text{ yd}) \left(\dfrac{0.9 \text{ m}}{1 \text{ yd}} \right) = 74.7 \text{ m}$

48. $(100 \text{ m}) \left(\dfrac{1 \text{ yd}}{0.9 \text{ m}} \right) = 111.\overline{1} \approx 111.11 \text{ yd}$

49. $(45 \text{ mi}) \left(\dfrac{1.6 \text{ km}}{1 \text{ mi}} \right) = 72 \text{ kph}$

50. $(40 \ \ell) \left(\dfrac{1 \text{ qt}}{0.95 \ \ell} \right) = 42.10526316 \approx 42.11 \text{ qt}$

51. $(15 \text{ gal}) \left(\dfrac{3.8 \ \ell}{1 \text{ gal}} \right) = 57 \ \ell$

52. $(40 \text{ m}^3) \left(\dfrac{1 \text{ yd}^3}{0.729 \text{ m}^3} \right) = 54.8696845 \approx 54.87 \text{ yd}^3$

53. $\left(83 \text{ cm}^2\right)\left(\dfrac{1 \text{ in.}^2}{6.45 \text{ cm}^2}\right) = 12.86821705 \approx 12.87 \text{ in.}^2$

54. $\left(4 \text{ qt}\right)\left(\dfrac{0.95 \ \ell}{1 \text{ qt}}\right) = 3.8 \ \ell$

55. $\left(15 \text{ yd}^3\right)\left(\dfrac{0.729 \text{ m}^3}{1 \text{ yd}^3}\right) = 10.9 \text{ m}^3$

56. $\left(62 \text{ mi}\right)\left(\dfrac{1.6 \text{ km}}{1 \text{ mi}}\right) = 99.2 \text{ km}$

57. $\left(27 \text{ cm}\right)\left(\dfrac{1 \text{ ft}}{30 \text{ cm}}\right) = 0.9 \text{ ft}$

58. $\left(3.25 \text{ in.}\right)\left(\dfrac{2.54 \text{ cm}}{1 \text{ in.}}\right)\left(\dfrac{10 \text{ mm}}{1 \text{ cm}}\right) = 82.55 \text{ mm}$

59. a) $700(1.5 \text{ kg}) = 1050 \text{ kg}$

   b) $1050 \text{ kg} = \left(1050 \text{ kg}\right)\left(\dfrac{1 \text{ lb}}{0.45 \text{ kg}}\right) = 2333.\overline{3} \approx 2333.33 \text{ lb}$

60. $A = lw = (24)(15) = 360 \text{ ft}^2$

   $360 \text{ ft}^2 = \left(360 \text{ ft}^2\right)\left(\dfrac{0.09 \text{ m}^2}{1 \text{ ft}^2}\right) = 32.4 \text{ m}^2$

61. a) $\left(50{,}000 \text{ gal}\right)\left(\dfrac{3.8 \ \ell}{1 \text{ gal}}\right)\left(\dfrac{1 \text{ k}\ell}{1000 \ \ell}\right) = 190 \text{ k}\ell$

   b) $\left(190 \text{ k}\ell\right)\left(\dfrac{1000 \ \ell}{1 \text{ k}\ell}\right)\left(\dfrac{1 \text{ kg}}{1 \ \ell}\right) = 190\ 000 \text{ kg}$

62. a) $35 \text{ mi} = \left(35 \text{ mi}\right)\left(\dfrac{1.6 \text{ km}}{1 \text{ mi}}\right) = 56 \text{ kph}$

   b) $56 \text{ km} = \left(56 \text{ km}\right)\left(\dfrac{1000 \text{ m}}{1 \text{ km}}\right) = 56\ 000 \text{ meters per hour}$

63. a) $V = lwh = (90)(70)(40) = 252\ 000 \text{ cm}^3$

   $252\ 000 \text{ cm}^3 = \left(252\ 000 \text{ cm}^3\right)\left(\dfrac{1 \text{ m}\ell}{1 \text{ cm}^3}\right)\left(\dfrac{1 \ \ell}{1000 \text{ m}\ell}\right) = 252 \ \ell$

   b) $252 \ \ell = \left(252 \ \ell\right)\left(\dfrac{1 \text{ kg}}{1 \ \ell}\right) = 252 \text{ kg}$

64. $1 \text{ kg} = \left(1 \text{ kg}\right)\left(\dfrac{1 \text{ lb}}{0.45 \text{ kg}}\right) = 2.\overline{2} \text{ lb}$

   $\dfrac{\$3.50}{2.\overline{2}} = \$1.575 \approx \$1.58 \text{ per pound}$

Chapter Test

1. $204 \text{ c}\ell = 0.204 \text{ da}\ell$

2. $123 \text{ km} = 123\ 000\ 000 \text{ mm}$

3. $1 \text{ km} = \left(1 \text{ km}\right)\left(\dfrac{100 \text{ dam}}{1 \text{ km}}\right) = 100 \text{ dam or } 100 \text{ times greater}$

4. $400(6) = 2400 \text{ m}$

   $\left(2400 \text{ m}\right)\left(\dfrac{1 \text{ km}}{1000 \text{ m}}\right) = 2.4 \text{ km}$

5. b

6. a

7. c

8. c

9. b

10. $1 \text{ m}^2 = \left(1 \text{ m}^2\right)\left(\dfrac{100^2 \text{ cm}^2}{1 \text{ m}^2}\right) = 10\ 000 \text{ cm}^2$ or 10,000 times greater

11. $1 \text{ m}^3 = \left(1 \text{ m}^3\right)\left(\dfrac{1000^3 \text{ mm}^3}{1 \text{ m}^3}\right) = 1\ 000\ 000\ 000 \text{ mm}^3$ or 1,000,000,000 times greater

12. $452 \text{ in.} = \left(452 \text{ in.}\right)\left(\dfrac{2.54 \text{ cm}}{1 \text{ in.}}\right) = 1148.08 \text{ cm}$

13. $150 \text{ m} = \left(150 \text{ m}\right)\left(\dfrac{1 \text{ yd}}{0.9 \text{ m}}\right) = 166.\overline{6} \approx 166.67 \text{ yd}$

14. $-10^\circ \text{ F} = \dfrac{5}{9}(-10 - 32) = -23.\overline{3} \approx -23.33^\circ \text{ C}$

15. $20^\circ \text{ C} = \dfrac{9}{5}(20) + 32 = 68^\circ \text{ F}$

16. $12 \text{ ft} = \left(12 \text{ ft}\right)\left(\dfrac{30 \text{ cm}}{1 \text{ ft}}\right) = 360 \text{ cm}$ or $12 \text{ ft} = \left(12 \text{ ft}\right)\left(\dfrac{12 \text{ in.}}{1 \text{ ft}}\right)\left(\dfrac{2.54 \text{ cm}}{1 \text{ in.}}\right) = 365.76 \text{ cm}$

17. a) $V = lwh = 20(20)(8) = 3200 \text{ m}^3$

    b) $3200 \text{ m}^3 = \left(3200 \text{ m}^3\right)\left(\dfrac{1000 \text{ k}\ell}{1 \text{ m}^3}\right) = 3200 \text{ k}\ell$

    c) $3200 \text{ k}\ell = (3200 \text{ k}\ell)\left(\dfrac{1000 \text{ }\ell}{1 \text{ k}\ell}\right)\left(\dfrac{1 \text{ kg}}{1 \text{ }\ell}\right) = 3\ 200\ 000 \text{ kg}$

18. Total surface area: $2lh + 2wh = 2(20)(6) + 2(15)(6) = 420 \text{ m}^2$

    Liters needed for first coat: $\left(420 \text{ m}^2\right)\left(\dfrac{1 \text{ }\ell}{10 \text{ m}^2}\right) = 42 \text{ }\ell$

    Liters needed for second coat: $\left(420 \text{ m}^2\right)\left(\dfrac{1 \text{ }\ell}{15 \text{ m}^2}\right) = 28 \text{ }\ell$

    Total liters needed: $42 + 28 = 70 \text{ }\ell$

    Total cost: $\left(70 \text{ }\ell\right)\left(\dfrac{\$3.50}{1 \text{ }\ell}\right) = \$245$

CHAPTER NINE

GEOMETRY

Exercise Set 9.1

1. a) Undefined terms, definitions, postulates (axioms), and theorems
    b) First, Euclid introduced **undefined terms**. Second, he introduced certain **definitions**. Third, he stated primitive propositions called **postulates (axioms)** about the undefined terms and definitions. Fourth, he proved, using deductive reasoning, other propositions called **theorems**.

3. Two lines in the same plane that do not intersect are **parallel lines**.

5. Two angles in the same plane are **adjacent angles** when they have a common vertex and a common side but no common interior points.

7. Two angles the sum of whose measure is 90° are called **complementary angles**.

9. An angle whose measure is greater than 90° but less than 180° is an **obtuse angle**.

11. An angle whose measure is 90° is a **right angle**.

13. Half line, $\overset{\circ\!\!\rightarrow}{AB}$

15. Line segment, $\overline{AB}$

17. Line, $\overleftrightarrow{AB}$

19. Open line segment, $\overset{\circ\!\!-\!\!\circ}{AB}$

21. $\overrightarrow{BD}$

23. $\overset{\circ\!\!\rightarrow}{BD}$

25. $\{B, F\}$

27. $\{C\}$

29. $\overline{BC}$

31. $\overrightarrow{BC}$

33. $\varnothing$

35. $\overline{BC}$

37. $\angle ABE$

39. $\angle EBC$

41. $\overset{\circ\!\!-\!\!\circ}{AC}$

43. $\overset{\circ-}{BE}$

45. Obtuse

47. Straight

49. Right

51. None of these

53. $90° - 19° = 71°$

55. $90° - 32\frac{3}{4}° = 57\frac{1}{4}°$

57. $90° - 64.7° = 25.3°$

59. $180° - 91° = 89°$

61. $180° - 20.5° = 159.5°$

63. $180° - 43\frac{5}{7}° = 136\frac{2}{7}°$

65. d

67. c

69. e

71. Let x = measure of $\measuredangle 2$

$x+4$ = measure of $\measuredangle 1$

$x + x + 4 = 90$

$2x + 4 = 90$

$2x = 86$

$x = \dfrac{86}{2} = 43°, m\measuredangle 2$

$x + 4 = 43 + 4 = 47°, m\measuredangle 1$

73. Let x = measure of $\measuredangle 1$

$180 - x$ = measure of $\measuredangle 2$

$x - (180 - x) = 88$

$x - 180 + x = 88$

$2x - 180 = 88$

$2x = 268$

$x = \dfrac{268}{2} = 134°, m\measuredangle 1$

$180 - x = 180 - 134 = 46°, m\measuredangle 2$

75. $m\measuredangle 1 + 125° = 180°$

$m\measuredangle 1 = 55°$

$m\measuredangle 2 = m\measuredangle 1$ (vertical angles)

$m\measuredangle 3 = 125°$ (vertical angles)

$m\measuredangle 5 = m\measuredangle 2$ (alternate interior angles)

$m\measuredangle 4 = m\measuredangle 3$ (alternate interior angles)

$m\measuredangle 7 = m\measuredangle 4$ (vertical angles)

$m\measuredangle 6 = m\measuredangle 5$ (vertical angles)

Measures of angles 3, 4, and 7 are each 125°.

Measures of angles 1, 2, 5, and 6 are each 55°.

77. $m\measuredangle 1 + 25° = 180°$

$m\measuredangle 1 = 155°$

$m\measuredangle 3 = m\measuredangle 1$ (vertical angles)

$m\measuredangle 2 = 25°$ (vertical angles)

$m\measuredangle 4 = m\measuredangle 3$ (alternate interior angles)

$m\measuredangle 7 = m\measuredangle 4$ (vertical angles)

$m\measuredangle 5 = m\measuredangle 2$ (corresponding angles)

$m\measuredangle 6 = m\measuredangle 5$ (vertical angles)

Measures of angles 2, 5, and 6 are each 25°.

Measures of angles 1, 3, 4, and 7 are each 155°.

79. $x + 3x + 10 = 90$

$4x + 10 = 90$

$4x = 80$

$x = \dfrac{80}{4} = 20°, m\measuredangle 2$

$3x + 10 = 3(20) + 10 = 70°, m\measuredangle 1$

81. $x + 2x - 9 = 90$

$3x - 9 = 90$

$3x = 99$

$x = \dfrac{99}{3} = 33°, m\measuredangle 1$

$2x - 9 = 2(33) - 9 = 57°, m\measuredangle 2$

83. $x + 2x - 15 = 180$

$3x - 15 = 180$

$3x = 195$

$x = \dfrac{195}{3} = 65°, m\measuredangle 2$

$2x - 15 = 2(65) - 15 = 115°, m\measuredangle 1$

85. $x + 5x + 6 = 180$

$6x + 6 = 180$

$6x = 174$

$x = \dfrac{174}{6} = 29°, m\measuredangle 1$

$5x + 6 = 5(29) + 6 = 151°, m\measuredangle 2$

87. a) An infinite number of lines can be drawn through a given point.

b) An infinite number of planes can be drawn through a given point.

89. An infinite number of planes can be drawn through a given line.

For Exercises 91 - 97, the answers given are one of many possible answers.

91. Plane ABG and plane JCD

93. $\overleftrightarrow{BG}$ and $\overleftrightarrow{DG}$

95. Plane $AGB \cap$ plane $ABC \cap$ plane $BCD = \{B\}$

97. $\overleftrightarrow{BC} \cap$ plane $ABG = \{B\}$

99. Always true. If any two lines are parallel to a third line, then they must be parallel to each other.

101. Sometimes true. Vertical angles are only complementary when each is equal to 45°.

103. Sometimes true. Alternate interior angles are only complementary when each is equal to 45°.

105. No. Line *m* and line *n* may intersect.

107.

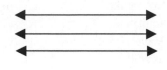

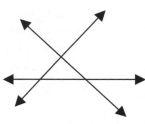

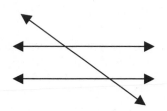

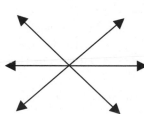

108.   360°

109. a)

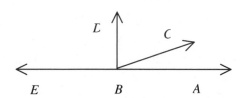

Other answers are possible.

b) Let $m\angle ABC = x$ and $m\angle CBD = y$.

$x + y = 90°$ and $y = 2x$

Substitute $y = 2x$ into $x + y = 90°$.

$x + 2x = 90°$

$3x = 90°$

$\dfrac{3x}{3} = \dfrac{90°}{3}$

$x = 30° = m\angle ABC$

c) $m\angle CBD = y$

$y = 2x = 2(30°) = 60°$

d) $m\angle ABD + m\angle DBE = 180°$

$m\angle ABD = x + y = 30° + 60° = 90°$.

$90° + m\angle DBE = 180°$

$m\angle DBE = 180° - 90° = 90°$.

Exercise Set 9.2
1. A **polygon** is a closed figure in a plane determined by three or more straight line segments.
3. The different types of triangles are acute, obtuse, right, isosceles, equilateral, and scalene. Descriptions will vary.
5. If the corresponding sides of two similar figures are the same length, the figures are **congruent figures**.

7. a) Rectangle   9. a) Hexagon   11. a) Rhombus   13. a) Octagon
   b) Not regular   b) Regular   b) Not regular   b) Not regular
15. a) Scalene   17. a) Isosceles   19. a) Equilateral   21. a) Scalene
    b) Right   b) Obtuse   b) Acute   b) Obtuse
23. Parallelogram   25. Rhombus   27. Trapezoid

29. The measures of the other two angles of the triangle are 138° and 25° (by vertical angles). Therefore, the measure of angle x is 180° - 138° - 25° = 17°.
31. The measure of one angle of the triangle is 27° (by vertical angles). The measure of another angle of the triangle is 180° - 57° = 123°. The measure of the third angle of the triangle is 180° - 27° - 123° = 30°. The measure of angle x is 180° - 30° = 150° (The 30° angle and angle x form a straight angle.).

33.

Angle	Measure	Reason
1	50°	$\angle$ 1 and $\angle$ 5 are vertical angles
2	63°	Vertical angle with the given 63° angle
3	67°	$\angle$ 1, $\angle$ 2, and $\angle$ 3 form a straight angle
4	67°	$\angle$ 3 and $\angle$ 4 are vertical angles
5	50°	$\angle$ 5 and $\angle$ 12 are corresponding angles
6	113°	$\angle$ 6 and the given 67° angle form a straight angle
7	50°	The sum of the measures of the interior angles of a triangle is 180°
8	130°	$\angle$ 8 and $\angle$ 12 form a straight angle
9	67°	$\angle$ 4 and $\angle$ 9 are corresponding angles
10	113°	$\angle$ 6 and $\angle$ 10 are vertical angles
11	130°	$\angle$ 8 and $\angle$ 11 are vertical angles
12	50°	$\angle$ 7 and $\angle$ 12 are vertical angles

35. $n = 5$
    $(5 - 2) \times 180° = 3 \times 180° = 540°$

37. $n = 6$
    $(6 - 2) \times 180° = 4 \times 180° = 720°$

39. $n = 20$
    $(20 - 2) \times 180° = 18 \times 180° = 3240°$

41. a) The sum of the measures of the interior angles of a triangle is 180°. Dividing by 3, the number of angles, each interior angle measures 60°.
    b) Each exterior angle measures 180° - 60° = 120°.

43. a) The sum of the measures of the interior angles of an octagon is (8 - 2) × 180° = 6 × 180° = 1080°. Dividing by 8, the number of angles, each interior angle measures 135°.
    b) Each exterior angle measures 180° - 135° = 45°.

45. a) The sum of the measures of the interior angles of a dodecagon is (12 - 2) × 180° = 10 × 180° = 1800°. Dividing by 12, the number of angles, each interior angle measures 150°.
    b) Each exterior angle measures 180° - 150° = 30°.

47. Let $x = BC$

$$\frac{BC}{B'C'} = \frac{AB}{A'B'}$$

$$\frac{x}{2.4} = \frac{10}{4}$$

$$4x = 24$$

$$x = 6$$

Let $y = A'C'$

$$\frac{A'C'}{AC} = \frac{A'B'}{AB}$$

$$\frac{y}{8} = \frac{4}{10}$$

$$10y = 32$$

$$y = \frac{32}{10} = \frac{16}{5}$$

49. Let $x = DC$

$$\frac{DC}{D'C'} = \frac{AB}{A'B'}$$

$$\frac{x}{6} = \frac{4}{10}$$

$$10x = 24$$

$$x = \frac{24}{10} = \frac{12}{5}$$

Let $y = B'C'$

$$\frac{B'C'}{BC} = \frac{A'B'}{AB}$$

$$\frac{y}{3} = \frac{10}{4}$$

$$4y = 30$$

$$y = \frac{30}{4} = \frac{15}{2}$$

51. Let $x = AC$

$$\frac{AC}{A'C'} = \frac{BC}{B'C'}$$

$$\frac{x}{0.75} = \frac{2}{1.25}$$

$$1.25x = 1.5$$

$$x = 1.2$$

Let $y = A'B'$

$$\frac{A'B'}{AB} = \frac{B'C'}{BC}$$

$$\frac{y}{1} = \frac{1.25}{2}$$

$$2y = 1.25$$

$$y = 0.625$$

53. Let $x = BC$

$$\frac{BC}{EC} = \frac{AB}{DE}$$

$$\frac{x}{2} = \frac{6}{2}$$

$$2x = 12$$

$$x = 6$$

55. $AD = AC - DC = 10 - \dfrac{10}{3} = \dfrac{30}{3} - \dfrac{10}{3} = \dfrac{20}{3}$

57. $A'B' = AB = 14$

59. $AC = A'C' = 28$

61. $m\angle ACB = m\angle A'C'B' = 28°$

63. $A'B' = AB = 8$

65. $B'C' = BC = 16$

67. $m\angle A'D'C' = m\angle ADC = 70°$

69. $180° - 125° = 55°$

71. $180° - 90° - 55° = 35°$

73. Let $x =$ height of silo

$$\frac{x}{6} = \frac{105}{9}$$

$$9x = 630$$

$$x = 70 \text{ ft}$$

75. a) $197 \text{ mi} = (197 \text{ mi})\left(\frac{5280 \text{ ft}}{1 \text{ mi}}\right)\left(\frac{12 \text{ in.}}{1 \text{ ft}}\right)$

$= 12,481,920 \text{ in.}$

Let $x =$ the actual distance from Dallas to Houston

$$\frac{x}{3.75} = \frac{12,481,920}{3}$$

$$3x = 46,807,200$$

$$x = 15,602,400 \text{ in.}$$

$15,602,400 \text{ in.} = (15,602,400 \text{ in.})\left(\frac{1 \text{ ft}}{12 \text{ in.}}\right)\left(\frac{1 \text{ mi}}{5280 \text{ ft}}\right)$

$= 246.25 \text{ mi}$

b) Let $x =$ the actual distance from Dallas to San Antonio

$$\frac{x}{4.125} = \frac{12,481,920}{3}$$

$$3x = 51,487,920$$

$$x = 17,162,640 \text{ in.}$$

$17,162,640 \text{ in.} = (17,162,640 \text{ in.})\left(\frac{1 \text{ ft}}{12 \text{ in.}}\right)\left(\frac{1 \text{ mi}}{5280 \text{ ft}}\right)$

$= 270.875 \text{ mi}$

77.

$$\frac{DE}{D'E'} = 3 \qquad\qquad \frac{EF}{E'F'} = 3 \qquad\qquad \frac{DF}{D'F'} = 3$$

$$\frac{12}{D'E'} = 3 \qquad\qquad \frac{15}{E'F'} = 3 \qquad\qquad \frac{9}{D'F'} = 3$$

$$3D'E' = 12 \qquad\qquad 3E'F' = 15 \qquad\qquad 3D'F' = 9$$

$$\overline{D'E'} = 4 \qquad\qquad \overline{E'F'} = 5 \qquad\qquad \overline{D'F'} = 3$$

79. a) $m\angle HMF = m\angle TMB$, $m\angle HFM = m\angle TBM$, $m\angle MHF = m\angle MTB$

b) Let $x =$ height of the wall

$$\frac{x}{20} = \frac{5.5}{2.5}$$

$$2.5x = 110$$

$$x = \frac{110}{2.5} = 44 \text{ ft}$$

80. a) $m\angle CED = m\angle ABC$; $m\angle ACB = m\angle DCE$ (vertical angles); $m\angle BAC = m\angle CDE$ (alternate interior angles)

b) Let $x = DE$

$$\frac{x}{AB} = \frac{CE}{BC}$$

$$\frac{x}{543} = \frac{1404}{356}$$

$$356x = 762,372$$

$$x = 2141.494382 \approx 2141.49 \text{ ft}$$

Exercise Set 9.3

Throughout this section, on exercises involving π, we used the π key on a scientific calculator to determine the answer. If you use 3.14 for π, your answers may vary slightly.

1. a) The **perimeter** of a two-dimensional figure is the sum of the lengths of the sides of the figure.
    b) The **area** of a two-dimensional figure is the region within the boundaries of the figure.
    c)

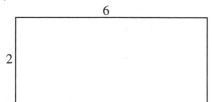

$A = lw = 6(2) = 12$ square units

$P = 2l + 2w = 2(6) + 2(2) = 12 + 4 = 16$ units

3. a) To determine the number of square inches, multiply the number of square feet by $12 \times 12 = 144$.
    b) To determine the number of square feet, divide the number of square inches by $12 \times 12 = 144$.

5. $A = \dfrac{1}{2}bh = \dfrac{1}{2}(10)(7) = 35$ in.^{2}

7. $A = \dfrac{1}{2}bh = \dfrac{1}{2}(7)(5) = 17.5$ cm^2

9. $A = lw = (15)(7) = 105$ ft^2

    $P = 2l + 2w = 2(15) + 2(7) = 44$ ft

11. $3 \text{ m} = 3(100) = 300$ cm

    $A = bh = 300(20) = 6000$ cm^2

    $P = 2l + 2w = 2(300) + 2(27) = 654$ cm

13. $2 \text{ ft} = 2(12) = 24$ in.

    $A = \dfrac{1}{2}h(b_1 + b_2) = \dfrac{1}{2}(24)(5 + 19)$

    $= \dfrac{1}{2}(24)(24) = 288$ in.^{2}

    $P = s_1 + s_2 + b_1 + b_2 = 25 + 25 + 5 + 19 = 74$ in.

15. $A = \pi r^2 = \pi(7)^2 = 49\pi = 153.93804 \approx 153.94$ in.^{2}

    $C = 2\pi r = 2\pi(7) = 14\pi = 43.98229715 \approx 43.98$ in.

17. $r = \dfrac{9}{2} = 4.5$ ft

    $A = \pi r^2 = \pi(4.5)^2 = 20.25\pi = 63.61725124$

    ≈ 63.62 ft^2

    $C = 2\pi r = 2\pi(4.5) = 9\pi = 28.27433388$

    ≈ 28.27 ft

19. a) $a^2 + 12^2 = 15^2$

    $a^2 + 144 = 225$

    $a^2 = 81$

    $a = \sqrt{81} = 9$ in.

    b) $P = s_1 + s_2 + s_3 = 9 + 12 + 15 = 36$ in.

    c) $A = \dfrac{1}{2}bh = \dfrac{1}{2}(9)(12) = 54$ in.^{2}

21. a) $c^2 = 10^2 + 24^2$

    $c^2 = 100 + 576$

    $c^2 = 676$

    $c = \sqrt{676} = 26$ cm

    b) $P = s_1 + s_2 + s_3 = 10 + 24 + 26 = 60$ cm

    c) $A = \dfrac{1}{2}bh = \dfrac{1}{2}(10)(24) = 120$ cm^2

23. Area of larger circle:

    $\pi(4)^2 = 16\pi = 50.26548246$ cm^2

    Area of smaller circle:

    $\pi(3)^2 = 9\pi = 28.27433388$ cm^2

    Shaded area:

    $50.26548246 - 28.27433388 = 21.99114858$

    ≈ 21.99 cm^2

25. Use the Pythagorean Theorem to find the length of a side of the shaded square.

$$x^2 = 2^2 + 2^2$$

$$x^2 = 4 + 4$$

$$x^2 = 8$$

$$x = \sqrt{8}$$

Shaded area: $\sqrt{8}\left(\sqrt{8}\right) = 8$ in.^{2}

27. Area of trapezoid:

$$\frac{1}{2}(8)(9+20) = \frac{1}{2}(8)(29) = 116 \text{ in.}^2$$

Area of circle: $\pi(4)^2 = 16\pi = 50.26548246$ in.^{2}

Shaded area:

$116 - 50.26548246 = 65.73451754 \approx 65.73$ in.^{2}

29. Area of small rectangle on the right side:

$$12(6) = 72 \text{ ft}^2$$

Area of semi-circle on the right side:

$$\frac{1}{2}\pi(6)^2 = 18\pi = 56.54866776 \text{ ft}^2$$

Area of shaded region on the right side:

$72 - 56.54866776 = 15.45133224 \text{ ft}^2$

Area of shaded region on the left side:

15.45133224 ft^2

Area of triangle: $\frac{1}{2}(14)(12) = 84 \text{ ft}^2$

Shaded area:

$15.45133224 + 15.45133224 + 84$

$= 114.9026645 \approx 114.90 \text{ ft}^2$

31. Length of rectangle: $3(8) = 24$ in.

Area of rectangle: $24(8) = 192$ in.^{2}

Radius of each circle: $\frac{8}{2} = 4$ in.

Area of each circle: $\pi(4)^2 = 16\pi = 50.26548246$

Shaded area:

$192 - 50.26548246 - 50.26548246 - 50.26548246$

$= 41.20355262 \approx 41.20$ in.^{2}

33. $\dfrac{1}{x} = \dfrac{9}{107}$

$9x = 107$

$x = \dfrac{107}{9} = 11.\overline{8} \approx 11.89 \text{ yd}^2$

35. $\dfrac{1}{14.7} = \dfrac{9}{x}$

$x = 14.7(9) = 132.3 \text{ ft}^2$

37. $\dfrac{1}{23.4} = \dfrac{10,000}{x}$

$x = 23.4(10,000) = 234,000 \text{ cm}^2$

39. $\dfrac{1}{x} = \dfrac{10,000}{1075}$

$10,000x = 1075$

$x = \dfrac{1075}{10,000} = 0.1075 \text{ m}^2$

41. Area of living/dining room: $25(22) = 550 \text{ ft}^2$

a) $550(5.89) = \$3239.50$

b) $550(8.89) = \$4889.50$

43. Area of kitchen: $12(14) = 168 \text{ ft}^2$

Area of first floor bathroom: $6(10) = 60 \text{ ft}^2$

Area of second floor bathroom: $8(14) = 112 \text{ ft}^2$

Total area: $168 + 60 + 112 = 340 \text{ ft}^2$

Cost: $340(\$5) = \$1700$

45. Area of bedroom 1: $10(14) = 140 \text{ ft}^2$

    Area of bedroom 2: $10(20) = 200 \text{ ft}^2$

    Area of bedroom 3: $10(14) = 140 \text{ ft}^2$

    Total area: $140 + 200 + 140 = 480 \text{ ft}^2$

    Cost: $480(\$6.06) = \$2908.80$

47. Area of entire lawn if all grass:
    $200(100) = 20{,}000 \text{ ft}^2$

    Area of patio: $40(10) = 400 \text{ ft}^2$

    Area of shed: $10(8) = 80 \text{ ft}^2$

    Area of house: $50(25) = 1250 \text{ ft}^2$

    Area of drive: $30(10) = 300 \text{ ft}^2$

    Area of pool: $\pi(12)^2 = 144\pi = 452.3893421 \text{ ft}^2$

    Area of lawn:
    $20{,}000 - 400 - 80 - 1250 - 300 - 452.3893421$

    $= 17{,}517.61066 \text{ ft}^2 = \dfrac{17{,}517.61066}{9}$

    $= 1946.401184 \text{ yd}^2$

    Cost:
    $1946.401184(\$0.02) = \$38.92802368 \approx \$38.93$

49. a) $A = 11.5(15.4) = 177.1 \text{ m}^2$

    b) $\dfrac{1}{x} = \dfrac{10{,}000}{177.1}$

    $10{,}000x = 177.1$

    $x = \dfrac{177.1}{10{,}000} = 0.01771 \text{ hectare}$

51. Let c = length of guy wire

    $90^2 + 52^2 = c^2$

    $8100 + 2704 = c^2$

    $10{,}804 = c^2$

    $c = \sqrt{10{,}804} = 103.9422917 \approx 103.94 \text{ ft}$

53. Let a = horizontal distance from dock to boat

    $a^2 + 9^2 = 41^2$

    $a^2 + 81 = 1681$

    $a^2 = 1600$

    $a = \sqrt{1600} = 40 \text{ ft}$

55. a) $A = s^2$

    b) $A = (2s)^2 = 4s^2$

    c) The area of the square in part b) is four times larger than the area of the square in part a).

57. $s = \dfrac{1}{2}(a + b + c) = \dfrac{1}{2}(8 + 6 + 10) = 12$

    $A = \sqrt{12(12 - 8)(12 - 6)(12 - 10)}$

    $= \sqrt{12(4)(6)(2)} = \sqrt{576} = 24 \text{ cm}^2$

59. Answers will vary.

Exercise Set 9.4

In this section, we use the π key on the calculator to determine answers in calculations involving π. If you use 3.14 for π, your answers may vary slightly.

1. **Volume** is a measure of the capacity of a figure.
3. A **polyhedron** is a closed surface formed by the union of polygonal regions.
   A **regular polyhedron** is one whose faces are all regular polygons of the same size and shape.
5. A **prism** and a **pyramid** are both polyhedrons, but a prism has a top and a bottom base while a pyramid only has one base.

7. $V = s^3 = (3)^3 = 27 \text{ ft}^3$

9. $1 \text{ ft} = 12 \text{ in.}$

$V = \pi r^2 h = \pi(2)^2(12) = 48\pi$

$= 150.7964474 \approx 150.80 \text{ in.}^3$

11. $V = \frac{1}{3}\pi r^2 h = \frac{1}{3}\pi(3)^2(14) = 42\pi$

$= 131.9468915 \approx 131.95 \text{ cm}^3$

13. Area of the base:

$B = \frac{1}{2}h(b_1 + b_2) = \frac{1}{2}(10)(8+12) = 100 \text{ in.}^2$

$V = Bh = 100(24) = 2400 \text{ in.}^3$

15. $r = \frac{9}{2} = 4.5 \text{ cm}$

$V = \frac{4}{3}\pi r^3 = \frac{4}{3}\pi(4.5)^3 = 121.5\pi$

$= 381.7035074 \approx 381.70 \text{ cm}^3$

17. Area of the base: $B = s^2 = (11)^2 = 121 \text{ cm}^2$

$V = \frac{1}{3}Bh = \frac{1}{3}(121)(13) = 524.\overline{3} \approx 524.33 \text{ cm}^3$

19. Area of the base:

$B = \frac{1}{2}h(b_1 + b_2) = \frac{1}{2}(5)(7+9) = 40 \text{ in.}^2$

$V = \frac{1}{3}Bh = \frac{1}{3}(40)(8) = 106.\overline{6} \approx 106.67 \text{ in.}^3$

21. $V = $ volume of rect. solid - volume of cylinder

$= 6(4)(3) - \pi(1)^2(4) = 72 - 4\pi$

$= 72 - 12.56637061$

$= 59.43362939 \approx 59.43 \text{ m}^3$

23. $V = $ volume of rect. solid - volume of sphere

$= 4(4)(4) - \frac{4}{3}\pi(2)^3 = 64 - 33.51032164$

$= 30.48967836 \approx 30.49 \text{ ft}^3$

25. $V = $ vol. of large cylinder - vol. of small cylinder

$= \pi(1.5)^2(5) - \pi(0.5)^2(5) = 11.25\pi - 1.25\pi = 10\pi$

$= 31.41592654 \approx 31.42 \text{ m}^3$

27. $V = $ volume of rect. solid - volume of pyramid

$= 3(3)(4) - \frac{1}{3}(3)^2(4) = 36 - 12 = 24 \text{ ft}^3$

29. $7 \text{ yd}^3 = 7(27) = 189 \text{ ft}^3$

31. $153 \text{ ft}^3 = \frac{153}{27} = 5.\overline{6} \approx 5.67 \text{ yd}^3$

33. $5.9 \text{ m}^3 = 5.9(1,000,000) = 5,900,000 \text{ cm}^3$

35. $3,000,000 \text{ cm}^3 = \frac{3,000,000}{1,000,000} = 3 \text{ m}^3$

37. a) $V = 46(25)(25) = 28,750 \text{ in.}^3$

b) $(1 \text{ ft})^3 = (12 \text{ in.})(12 \text{ in.})(12 \text{ in.}) = 1728 \text{ in.}^3$

$28,750 \text{ in.}^3 = \frac{28,750}{1728} = 16.63773148 \approx 16.64 \text{ ft}^3$

39. $V = 12(4)(3) = 144 \text{ in.}^3$

$144 \text{ in.}^3 = 144(0.01736) = 2.49984 \approx 2.50 \text{ qt}$

41. a) Cylinder 1:

$V = \pi\left(\frac{10}{2}\right)^2(12) = 300\pi = 942.4777961 \approx 942.48 \text{ in.}^3$

Cylinder 2:

$V = \pi\left(\frac{12}{2}\right)^2(10) = 360\pi$

$= 1130.973355 \approx 1130.97 \text{ in.}^3$

The container with the larger diameter holds more.

b) $1130.97 - 942.48 = 188.49 \approx 188.50 \text{ in.}^3$

43. $V = \frac{1}{3}Bh = \frac{1}{3}(720)^2(480) = 82,944,000 \text{ ft}^3$

45. $r = \frac{3.875}{2} = 1.9375 \text{ in.}$

Volume of each cylinder:

$\pi r^2 h = \pi(1.9375)^2(3)$

$= 11.26171875\pi = 35.37973289$

Total volume:

$8(35.37973289) = 283.0378631 \approx 283.04 \text{ in.}^3$

47. a) $5.5 \text{ ft} = 5.5(12) = 66 \text{ in.}$

$r = \frac{2.5}{2} = 1.25 \text{ in.}$

$V = \pi r^2 h = \pi(1.25)^2(66) = 103.125\pi$

$= 323.9767424 \approx 323.98 \text{ in.}^3$

b) $\frac{323.98}{1728} = 0.187488426 \approx 0.19 \text{ ft}^3$

49. a) Round pan:

$A = \pi r^2 = \pi\left(\frac{9}{2}\right)^2 = 20.25\pi$

$= 63.61725124 \approx 63.62 \text{ in.}^2$

Rectangular pan: $A = lw = 7(9) = 63 \text{ in.}^2$

b) Round pan:

$V = \pi r^2 h \approx 63.62(2) = 127.24 \text{ in.}^3$

Rectangular pan: $V = lwh = 7(9)(2) = 126 \text{ in.}^3$

c) Round pan

51. a) $B = \text{area of trapezoid} = \frac{1}{2}(9)(8+12) = 90 \text{ in.}^2$

$4 \text{ ft} = 4(12) = 48 \text{ in.}$

$V = Bh = 90(48) = 4320 \text{ in.}^3$

b) $1 \text{ ft}^3 = (12)(12)(12) = 1728 \text{ in.}^3$

$4320 \text{ in.}^3 = \frac{4320}{1728} = 2.5 \text{ ft}^3$

53. $8 - x + 3 = 2$
$11 - x = 2$
$-x = -9$
$x = 9 \text{ edges}$

55. $x - 8 + 4 = 2$
$x - 4 = 2$
$x = 6 \text{ vertices}$

57. $11 - x + 5 = 2$
$16 - x = 2$
$-x = -14$
$x = 14 \text{ edges}$

59. Let $r = $ the radius of one of the cans of orange juice

The length of the box $= 6r$ and the width of the box $= 4r$

Volume of box - volume of cans:

$lwh - 6(\pi r^2 h) = (6r)(4r)h - 6\pi r^2 h = 24r^2 h - 6\pi r^2 h = 6r^2 h(4-\pi)$

Percent of the volume of the interior of the box that is not occupied by the cans:

$\frac{6r^2 h(4-\pi)}{lwh} = \frac{6r^2(4-\pi)}{(6r)(4r)} = \frac{4-\pi}{4} = 0.2146018366 \approx 21.46\%$

61. a) – e) Answers will vary.

f) If we double the radius of a sphere, the new volume will be eight times the original volume.

63. a) Find the volume of each numbered region. Since the length of each side is $a+b$, the sum of the volumes of each region will equal $(a+b)^3$.

b) $V_1 = a(a)(a) = a^3$ $\qquad V_2 = a(a)(b) = a^2b$ $\qquad V_3 = a(a)(b) = a^2b$ $\qquad V_4 = a(b)(b) = ab^2$

$V_5 = a(a)(b) = a^2b$ $\qquad V_6 = a(b)(b) = ab^2$ $\qquad V_7 = b(b)(b) = b^3$

c) The volume of the piece not shown is ab^2.

64. a) $5.5 \text{ ft} = 5.5(12) = 66 \text{ in.}$

$V = Bh = 5(66) = 330 \text{ in.}^3$

b) Radius of cylinder: $\dfrac{0.75}{2} = 0.375 \text{ in.}$

Volume of cylinder: $\pi r^2 h = \pi (0.375)^2 (66) = 9.28125\pi = 29.15790682 \text{ in.}^3$

Volume of hollow noodle: $330 - 29.15790682 = 300.8420932 \approx 300.84 \text{ in.}^3$

Exercise Set 9.5

1. The act of moving a geometric figure from some starting position to some ending position without altering its shape or size is called **rigid motion**. The four main rigid motions studied in this section are reflections, translations, rotations, and glide reflections.

3. A **reflection** is a rigid motion that moves a figure to a new position that is a mirror image of the figure in the starting position.

5. A **translation** is a rigid motion that moves a figure by sliding it along a straight line segment in the plane.

7. A **rotation** is a rigid motion performed by rotating a figure in the plane about a specific point.

9. A **glide reflection** is a rigid motion formed by performing a translation (or glide) followed by a reflection.

11. A geometric figure is said to have **reflective symmetry** if the positions of a figure before and after a reflection are identical (except for vertex labels).

13. A **tessellation** is a pattern consisting of the repeated use of the same geometric figures to entirely cover a plane, leaving no gaps.

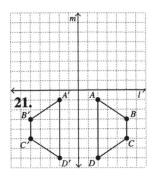

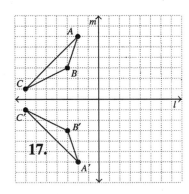

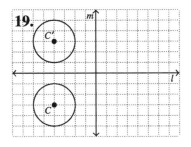

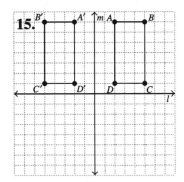

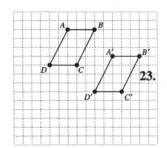

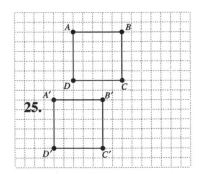

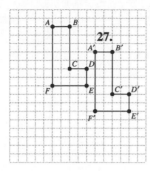

27.

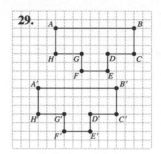

29.

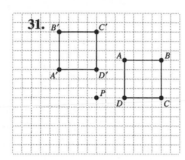

31.

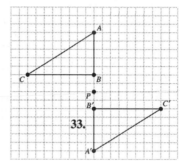

33.

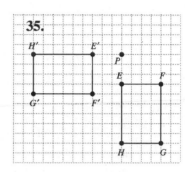

35.

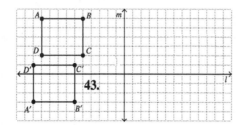

37.

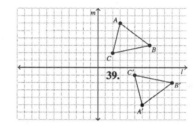

39.

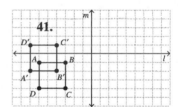

41.

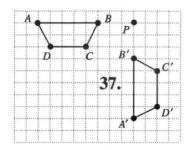

43.

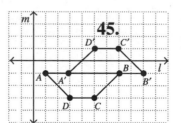

45.

47. a)

b) Yes

c) Yes

49. a)

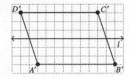

    b) No
    c) No

51. a)

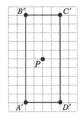

    b) No
    c) No
    d)

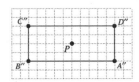

    e) Yes
    f) Yes

53. a) – c)

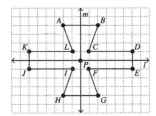

55. a) – b)

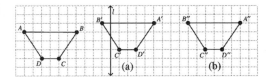

    c) No
    d) The order in which the translation and the reflection are performed is important. The figure obtained in part b) is the glide reflection.

57. Answers will vary..

59. a) Answers will vary.
    b) A regular pentagon cannot be used as a tessellating shape.

60. Although answers will vary depending on the font, the following capital letters have reflective symmetry about a horizontal line drawn through the center of the letter: B, C, D, E, H, I, K, O, X.

61. Although answers will vary depending on the font, the following capital letters have reflective symmetry about a vertical line drawn through the center of the letter: A, H, I, M, O, T, U, V, W, X, Y.

62 Although answers will vary depending on the font, the following capital letters have 180° rotational symmetry about a point in the center of the letter: H, I, O, S, X, Z.

Exercise Set 9.6

1. Topology is sometimes referred to as "rubber sheet geometry" because it deals with bending and stretching of geometric figures.

3. You can construct a Möbius strip by taking a strip of paper, giving one end a half twist, and taping the ends together.

5. Four

7. A **Jordan curve** is a topological object that can be thought of as a circle twisted out of shape.

9. The number of holes in the object determines the **genus** of an object.

11. 1, 4, 6 – Red; 2,3 – Yellow; 7 – Green; 5 – Blue

13. 1, 4, 6 – Red; 2, 5, 8 – Blue; 3, 7, 9 – Yellow

15. 1 – Red; 2, 5 – Yellow; 3, 6 – Blue; 4 – Green

17. YT, NU, AB, ON – Red
    NT, QC – Blue
    BC, SK – Green
    MB – Yellow

19. TX, KS, MS, KY, SC, FL – Red
    OK, LA, TN – Green
    MO, GA, VA – Blue
    AR, AL, NC – Yellow

21. Outside; a straight line from point *A* to a point clearly outside the curve crosses the curve an even number of times.

23. Outside; a straight line from point *A* to a point clearly outside the curve crosses the curve an even number of times.

25. Outside; a straight line from point *C* to a point clearly outside the curve crosses the curve an even number of times.

27. Inside; a straight line from point *E* to a point clearly outside the curve crosses the curve an odd number of times.

29. 1

31. 1

33. Larger than 5

35. 5

37. 0

39. 5

41. a) - d) Answers will vary.

43. One

45. Two

47. The smaller one is a Möbius strip; the larger one is not.

49. Yes. "Both sides" of the belt experience wear.

51. Ecuador, Brazil, Chile – Red
    Colombia, Guyana, French Guiana, Bolivia – Green
    Peru, Venezuela, Suriname, Paraguay, Uruguay – Yellow
    Argentina – Blue

53. a) 1
    b) 1
    c) Answers will vary.

Exercise Set 9.7

1. Girolamo Saccheri - proved many theorems of hyperbolic geometry

3. Carl Friedrich Gauss - discovered hyperbolic geometry

5. G.F. Bernhard Riemann - discovered elliptical geometry

7. a) Euclidean - Given a line and a point not on the line, one and only one line can be drawn parallel to the given line through the given point.

   b) Hyperbolic - Given a line and a point not on the line, two or more lines can be drawn through the given point parallel to the given line.

   c) Elliptical - Given a line and a point not on the line, no line can be drawn through the given point parallel to the given line.

9. A plane

11. A pseudosphere

13. Spherical - elliptical geometry; flat - Euclidean geometry; saddle-shaped - hyperbolic geometry

15.

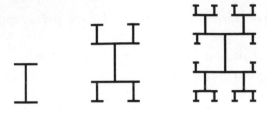

17.

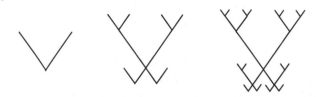

19. a)

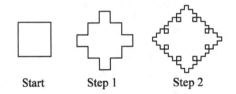

Start   Step 1   Step 2

b) Infinite.

c) Finite since it covers a finite or closed area.

Review Exercises

In the Review Exercises and Chapter Test questions, the π key on the calculator is used to determine answers in calculations involving π. If you use 3.14 for π, your answers may vary slightly.

1. $\{F\}$

2. $\triangle BFC$

3. $\overline{BC}$

4. $\overrightarrow{BH}$

5. $\{F\}$

6. $\{\ \}$

7. $90° - 51.2° = 38.8°$

8. $180° - 124.7° = 55.3°$

9. Let $x = BC$

$$\frac{BC}{B'C} = \frac{AC}{A'C}$$

$$\frac{x}{3.4} = \frac{12}{4}$$

$$4x = 40.8$$

$$x = \frac{40.8}{4} = 10.2 \text{ in.}$$

10. Let $x = A'B'$

$$\frac{A'B'}{AB} = \frac{A'C}{AC}$$

$$\frac{x}{6} = \frac{4}{12}$$

$$12x = 24$$

$$x = \frac{24}{12} = 2 \text{ in.}$$

11. $m\angle ABC = m\angle A'B'C$

$m\angle A'B'C = 180° - 88° = 92°$

Thus, $m\angle ABC = 92°$

$m\angle BAC = 180° - 30° - 92° = 58°$

12. $m\angle ABC = m\angle A'B'C$

$m\angle A'B'C = 180° - 88° = 92°$

Thus, $m\angle ABC = 92°$

13. $m\angle 1 = 180° - 110° = 70°$

$m\angle 6 = 70°$ (angle 1 and angle 6 are vertical angles)

The measure of the top angle of the triangle is $50°$, by vertical angles. The measure of the angle on the bottom right of the triangle is $180° - 70° - 50° = 60°$.

$m\angle 2 = 60°$ (angle 2 and the angle on the bottom right of the triangle are vertical angles)

The measure of the alternate interior angle of angle 2 is $60°$. Thus, $m\angle 3 = 180° - 60° = 120°$.

The measure of the alternate interior angle of angle 6 is $70°$. Thus, $m\angle 5 = 180° - 70° = 110°$.

$m\angle 4 = 180° - 110° = 70°$

14. $n = 6$

$(n-2)180° = (6-2)180° = 4(180°) = 720°$

15. $A = lw = 9(7) = 63 \text{ cm}^2$

16. $A = \dfrac{1}{2}bh = \dfrac{1}{2}(14)(5) = 35 \text{ in.}^2$

17. $A = \dfrac{1}{2}h(b_1 + b_2) = \dfrac{1}{2}(2)(4+9) = 13 \text{ in.}^2$

18. $A = bh = 12(7) = 84 \text{ in.}^2$

19. $A = \pi r^2 = \pi(13)^2 = 169\pi$

$= 530.9291585 \approx 530.93 \text{ cm}^2$

20. $A = lw = 14(16) = 224 \text{ ft}^2$

Cost: $224(\$2.75) = \$616$

21. $V = \pi r^2 h = \pi(5)^2(15) = 375\pi$

$= 1178.097245 \approx 1178.10 \text{ in.}^3$

22. $V = lwh = 10(3)(4) = 120 \text{ cm}^3$

23. If h represents the height of the triangle which is the base of the pyramid, then

$$h^2 + 3^2 = 5^2$$
$$h^2 + 9 = 25$$
$$h^2 = 16$$
$$h = \sqrt{16} = 4 \text{ ft}$$

$B = \dfrac{1}{2}bh = \dfrac{1}{2}(6)(4) = 12 \text{ ft}^2$

$V = \dfrac{1}{3}Bh = \dfrac{1}{3}(12)(7) = 28 \text{ ft}^3$

24. $B = \dfrac{1}{2}bh = \dfrac{1}{2}(9)(12) = 54 \text{ m}^2$

$V = Bh = 54(8) = 432 \text{ m}^3$

25. $r = \dfrac{12}{2} = 6 \text{ mm}$

$V = \dfrac{1}{3}\pi r^2 h = \dfrac{1}{3}\pi(6)^2(16) = 192\pi$

$= 603.1857895 \approx 603.19 \text{ mm}^3$

26. $V = \dfrac{4}{3}\pi r^3 = \dfrac{4}{3}\pi(7)^3 = 457.\overline{3}\pi$

     $= 1436.75504 \approx 1436.76 \text{ ft}^3$

27. $h^2 + 1^2 = 3^2$

                 $h^2 + 1 = 9$

                   $h^2 = 8$

                   $h = \sqrt{8}$

$A = \dfrac{1}{2}h(b_1 + b_2) = \dfrac{1}{2}(\sqrt{8})(2+4) = 8.485281374 \text{ ft}^2$

a) $V = Bh = 8.485281374(8)$

      $= 67.88225099 \approx 67.88 \text{ ft}^3$

b)   Weight:

    $67.88(62.5) + 375 = 4617.5 \text{ lb}$

Yes, it will support the trough filled with water.

c) $(4617.5 - 375) = 4242.5 \text{ lb of water}$

$\dfrac{4242.5}{8.3} = 511.1445783 \approx 511.14 \text{ gal}$

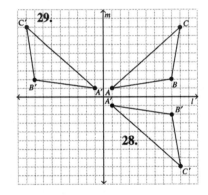

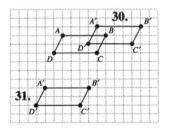

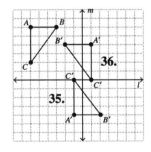

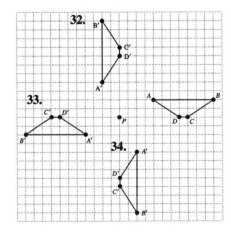

37. Yes   38. No   39. No   40. Yes   41. 1

42. Saarland, North Rhine-Westphalia, Bremen, Mecklenburg-Western Pomerania, Berlin, Thuringia, Baden-Württemberg, Hamburg – Red

Rhineland-Palatinate, Lower Saxony, Saxony – Green

Schleswig-Holstein, Hesse, Brandenburg – Yellow

Bavaria, Saxony-Anhalt – Blue

43. Outside; a straight line from point A to a point clearly outside the curve crosses the curve an even number of times.

44. Euclidean: Given a line and a point not on the line, one and only one line can be drawn parallel to the given line through the given point.

Elliptical: Given a line and a point not on the line, no line can be drawn through the given point parallel to the given line.

Hyperbolic: Given a line and a point not on the line, two or more lines can be drawn through the given point parallel to the given line.

45.

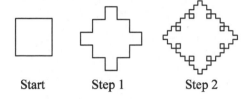

Start   Step 1   Step 2

Chapter Test

1. $\overleftrightarrow{EF}$

2. $\triangle BCD$

3. $\{D\}$

4. $\overline{AC}$

5. $90° - 36.9° = 53.1°$

6. $180° - 101.5° = 78.5°$

7. The other two angles of the triangle are 48° (by vertical angles) and 180° - 112° = 68°. Thus, the measure of angle $x = 180° - 48° - 68° = 64°$.

8. $n = 8$

$$(n-2)180° = (8-2)180° = 6(180°) = 1080°$$

9. Let $x = B'C'$

$$\frac{B'C'}{BC} = \frac{A'C'}{AC}$$

$$\frac{x}{7} = \frac{5}{13}$$

$$13x = 35$$

$$x = \frac{35}{13} = 2.692307692 \approx 2.69 \text{ cm}$$

10. a)

$$x^2 + 5^2 = 13^2$$

$$x^2 + 25 = 169$$

$$x^2 = 144$$

$$x = \sqrt{144} = 12 \text{ in.}$$

b) $P = 5 + 13 + 12 = 30 \text{ in.}$

c) $A = \frac{1}{2}bh = \frac{1}{2}(5)(12) = 30 \text{ in.}^2$

11. $r = \frac{16}{2} = 8 \text{ cm}$

$$V = \frac{4}{3}\pi r^3 = \frac{4}{3}\pi(8)^3 = 682.\overline{6}\pi$$

$$= 2144.660585 \approx 2144.66 \text{ cm}^3$$

12. $B = 9(14) + \pi(4.5)^2 = 126 + 20.25\pi = 189.6172512$

$V = Bh = 189.6172512(6) = 1137.703507 \text{ ft}^3$

$$1137.703507 \text{ ft}^3 = \frac{1137.703507}{27}$$

$$= 42.13716694 \approx 42.14 \text{ yd}^3$$

13. $B = lw = 4(7) = 28 \text{ ft}^2$

$$V = \frac{1}{3}Bh = \frac{1}{3}(28)(12) = 112 \text{ ft}^3$$

14.

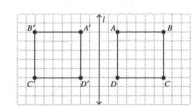

15.

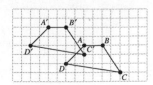

16.

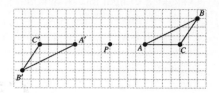

17.

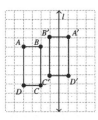

18. a) No
    b) Yes

19.   A **Möbius strip** is a surface with one side and
       one edge.

20.   a) and b) Answers will vary.

21. Euclidean: Given a line and a point not on the line,
     one and only one line can be drawn parallel to the
     given line through the given point.

     Elliptical: Given a line and a point not on the line, no
     line can be drawn through the given point parallel to
     the given line.

     Hyperbolic: Given a line and a point not on the line,
     two or more lines can be drawn through the given
     point parallel to the given line.

CHAPTER TEN

MATHEMATICAL SYSTEMS

Exercise Set 10.1

1. A binary operation is an operation that is performed on two elements, and the result is a single element.

3. Each of these operations can be performed on only two elements at a time and the result is always a single element.   a) $2 + 3 = 5$ b) $5 - 3 = 2$ c) $2 \times 3 = 6$ d) $6 \div 3 = 2$

5. Closure, identity, each element must have a unique inverse, associative property, commutative property.

7. If a binary operation is performed on any two elements of a set and the result is an element of the set, then that set is closed under the given binary operation. For all integers a and b, $a + b$ is an integer. Therefore, the set of integers is closed under the operation of addition.

9. When a binary operation is performed on two elements in a set and the result is the identity element for the binary operation, then each element is said to be the inverse of the other. The additive inverse of 2 is (-2) since $2 + (-2) = 0$, and the multiplicative inverse of 2 is $(1/2)$ since $2 \times 1/2 = 1$.

11. No. Every commutative group is also a group.

13. d   The Commutative property need not apply.

15. The associative property of addition states that $(a + b) + c = a + (b + c)$, for any elements a, b, and c.
Example: $(3 + 4) + 5 = 3 + (4 + 5)$

17. The commutative property of multiplication stated that $a \times b = b \times a$, for any real numbers a, b, and c.
Example: $3 \times 4 = 4 \times 3$

19. $8 \div 4 = 2$, but $4 \div 8 = \frac{1}{2}$

21. $(6 - 3) - 2 = 3 - 2 = 1$, but $6 - (3 - 2) = 6 - 1 = 5$

23. No. No identity element

25. Yes. Satisfies 5 properties needed

27. No. Not closed

29. No. No identity element

31. No. Not all elements have inverses

33. Yes. Satisfies 4 properties needed

35. No. Not closed   ie.: 1/0 is undefined

37. No. Does not satisfy Associative property

39. No; the system is not closed, $\pi + (-\pi) = 0$ which is not an irrational number.

41. Yes. Closure: The sum of any two real numbers is a real number. The identity element is zero.   Example: $5 + 0 = 0 + 5 = 5$
Each element has a unique inverse.
Example: $6 + (-6) = 0$
The associative property holds:
Example: $(2 + 3) + 4 = 2 + (3 + 4)$

43. Answers will vary.

44. 9^{9^9}

45. 9/19/29/39/49/59/69/79/89
90/91/92/93/94/95/96/97/98/99
The sign maker needs 20 nines.

Exercise Set 10.2

1. The clock addition table is formed by adding all pairs of integers between 1 and 12 using the 12 hour clock to determine the result. Example: If the clock is at 7 and we add 8, then the clock will read 3. Thus, $7 + 8 = 3$ in clock arithmetic.

3. a) First add $(4 + 10)$ on the clock, then add that result to 3 on the clock to obtain the final answer.

   b) $(4 + 10) + 3 = 2$ $(2) + 3 = 5$

5. a) Add 12 to 5 to get 17, which is larger than 9. Then subtract 9 from 17.

   b) $5 - 9$ $5 + 12 = 17$ $17 - 9 = 8$

   c) Since 12 is the identity element, you can add 12 to any number without changing the answer.

7. Yes, the sum of any two numbers in clock 12 arithmetic is a number in clock 12 arithmetic.

9. Yes. 1 and 11 are inverses, 2 and 10 are inverses, 3 and 9 are inverses, 4 and 8 are inverses, 5 and 7 are inverses, 6 is its own inverse, and 12 is its own inverse.

11. Yes. $6 + 9 = 3$ and $9 + 6 = 3$

13. a) Identity element = 5

   b) Additive inverse of 2 is 3. $2 + 3 = 5$

15. Yes. Commutative, symmetrical around main diagonal

17. Identity element = C, Row 3 is identical to top row and column 3 is identical to left column

19. The inverse of A is B, because A Ⓨ B = C

   and B Ⓨ A = C.

21. $4 + 7 = 11$

23. $9 + 8 = 5$

25. $4 + 12 = 4$

27. $3 + (8 + 9) = 3 + 5 = 8$

29. $(6 + 4) + 8 = 10 + 8 = 6$

31. $(7 + 8) + (9 + 6) = 3 + 3 = 6$

33. $7 - 4 = 3$

35. $4 - 12 = 4$

37. $5 - 10 = 7$

39. $1 - 12 = 1$

41. $5 - 5 = 12$

43. $12 - 12 = 12$

45.

+	1	2	3	4	5	6
1	2	3	4	5	6	1
2	3	4	5	6	1	2
3	4	5	6	1	2	3
4	5	6	1	2	3	4
5	6	1	2	3	4	5
6	1	2	3	4	5	6

55.

+	1	2	3	4	5	6	7
1	2	3	4	5	6	7	1
2	3	4	5	6	7	1	2
3	4	5	6	7	1	2	3
4	5	6	7	1	2	3	4
5	6	7	1	2	3	4	5
6	7	1	2	3	4	5	6
7	1	2	3	4	5	6	7

47. $1 + 6 = 1$

49. $5 - 2 = 3$

51. $2 - 6 = 2$

53. $4 - 6 = 4$

55. See above

57. $6 + 5 = 4$

59. $7 + 6 = 6$

61. $3 - 6 = 4$

63. $(4 - 5) - 6 = 6 - 6 = 7$

65. Yes. Satisfies 5 required properties

67. a) $\{0, 1, 2, 3\}$

   b) ✈ airplane

   c) Yes. All elements in the table are in the original set.

   d) Identity element is 0.

   e) Yes; $0 ✈ 0 = 0$, $1 ✈ 3 = 0$, $2 ✈ 2 = 0$, $3 ✈ 1 = 0$

   f) $(1 ✈ 2) ✈ 3 = 3 ✈ 3 = 2$ and $1 ✈ (2 ✈ 3) = 1 ✈ 1 = 2$

   g) Yes; $3 ✈ 2 = 1 = 2 ✈ 3$

   h) Yes, system satisfies five properties needed.

69. a) {r, s, t, u}   b) ⬭
    c) Yes. All elements in the table are in the
       original set.
    d) Yes, the identity element is t.
    e) Yes; r ⬭ r = t, s ⬭ u = t,
       t ⬭ t = t, u ⬭ s = t
    f) (r ⬭ s) ⬭ u = u ⬭ u = r
       and r ⬭ (s ⬭ u) = r ⬭ t = r
    g) Yes; s ⬭ r = u and r ⬭ s = u
    h) Yes, system satisfies five properties needed.

71. a) {f, r, o, m}   b) ⬭
    c) The system is closed. All elements in the table
       are elements of the set.
    d) (r ⬭ o) ⬭ f = m ⬭ f = m
    e) (f ⬭ r) ⬭ m) = r ⬭ m = f
    f) Identity element is f.
    g) Inverse of r is m since m ⬭ r = f = r ⬭ m.
    h) Inverse of m is r since r ⬭ m = f = m ⬭ r.

73. a) Is closed; all elements in the table are in the
       original set.   b) Identity = ⊡
    c) Inverse: of ⊡ is ⊡ ; of M is M; of ⌂ is ⌂
    d) (M ⊗ ⌂) ⊗ M = ⌂ ⊗ M = M
       M ⊗ (⌂ ⊗ M) = M ⊗ M = ⊡
       Not associative since M ≠ ⊡
    e) ⌂ ⊗ M = M   M ⊗ ⌂ = ⌂
       Not commutative since M ≠ ⌂

75. No inverses for ⊙ and *
    (* ⊡ *) ⊡ T = ⊙ ⊡ T = *
    * ⊡ (* ⊡ T) = * ⊡ T = P
    Not associative since * ≠ P

77. No identity element and therefore no inverses.
    (d ⇔ e) ⇔ d = d ⇔ d = e
    d ⇔ (e ⇔ d) = d ⇔ e = d
    Not associative since e ≠ d
    e ⇔ d = e   d ⇔ e = d
    Not commutative since e ≠ d

79. a)

+	E	O
E	E	O
O	O	E

    b) The system is closed, the identity element is E,
       each element is its own inverse, and the system
       is commutative since the table is symmetric
       about the main diagonal. Since the system has
       fewer than 6 elements and satisfies the above
       properties, it is a commutative group.

81. Student activity - Answers will vary.

83. a) All elements in the table are in the set
       {1, 2, 3, 4, 5, 6} so the system is closed.
       The identity is 6. 5 and 1 are inverses of each
       other, and 2, 3, 4, and 6 are their own inverses.
       Thus, if the associative property is assumed,
       the system is a group.
    b) 4 ∞ 5 = 2, but 5 ∞ 4 = 3

83. Examples of associativity
    (2 ∞ 3) ∞ 4 = 5 ∞ 4 = 3 and
    2 ∞ (3 ∞ 4) = 2 ∞ 5 = 3
    (1 ∞ 3) ∞ 5 = 4 ∞ 5 = 2 and
    1 ∞ (3 ∞ 5) = 1 ∞ 4 = 2

85. a)

*	R	S	T	U	V	I
R	V	T	U	S	I	R
S	U	I	V	R	T	S
T	S	R	I	V	U	T
U	T	V	R	I	S	U
V	I	U	S	T	R	V
I	R	S	T	U	V	I

    b) Is a group: Is closed; identity = I; each element
       has a unique inverse;
       an example of associativity:
       R * (T * V) = R * U = S
       (R * T) * V = U * V = S
    c) R * S = T   S * R = U
       Not Commutative since T ≠ U

87.

+	0	1	2	3	4
0	0	1	2	3	4
1	1	2	3	4	0
2	2	3	4	0	1
3	3	4	0	1	2
4	4	0	1	2	3

88.

+	0	1	2	3	4	5
0	0	1	2	3	4	5
1	1	2	3	4	5	0
2	2	3	4	5	0	1
3	3	4	5	0	1	2
4	4	5	0	1	2	3
5	5	0	1	2	3	4

89. 1) Add number in left column to number in top row   2) Divide this sum by 4

3) Replace remainder in the table.

Exercise Set 10.3

1. A modulo m system consists of m elements, 0 through m − 1, and a binary operation.

3. In a modulo 5 system there will be 5 modulo classes. When a number is divided by 5 the remainder
will be a number from 0 − 4.

0	1	2	3	4
0	1	2	3	4
5	6	7	8	9
10	11	12	13	14

5. In a modulo 12 system there will be 12 modulo classes. When a number is divided by 12 the remainder
will be a number 0 − 11.

7. $27 \equiv 2 \pmod 5$: 2, 12, and 107 have
the same remainder, 2, when divided by 5.

9. Today is Thursday = Day 4, so $30 \equiv 2 \pmod 7$ is
Saturday

11. $365 \div 7 = 52$ remainder 1. So 365 days is 1 day
later than today (Thursday), or Friday.

13. 3 years, 34 days = $(3)(365 + 34)$ days = 1129 days
$1129 \div 7 = 161$ remainder 2. So 3 years, 34 days is
2 days later than today (Thursday), or Saturday.

15. 728 days / 7 = 104 remainder 0. So 728 days is the
same as today, or Thursday.

17. $9 \equiv 9 \pmod{12}$, so 9 months after October is
July.

19. Since 3 years = 36 months and $36 \equiv 0 \pmod{12}$,
3 years 5 months is the same as 5 months; 5
months after October is March.

21. $83 \div 12 = 6$ remainder 11. So 83 months is 11
months after October, which is September.

23. $105 \div 12 = 8$ remainder 9. So 105 months is 9
months after October, which is July.

25. $8 + 6 = 14$ $14 \equiv 4 \pmod 5$

27. $1 + 9 + 12 = 22$ $22 \equiv 2 \pmod 5$

29. $5 - 12 \equiv (10 + 5) - 12 \pmod 5$

$\equiv 15 - 12 \pmod 5$

$\equiv 3 \pmod 5$

31. $8 \bullet 9 = 72$ $72 \equiv 2 \pmod 5$

33. $4 - 8 \equiv (5 + 4) - 8 \pmod 5$

$\equiv 9 - 8 \pmod 5$

$\equiv 1 \pmod 5$

35. $(15 \bullet 4) - 8 = 60 - 8 = 52$ $52 \equiv 2 \pmod 5$

37. $15 \pmod 5 \equiv 0$

39. $84 \pmod{12} \equiv 0$

41. $60 \pmod 9 \equiv 6$

43. $30 \pmod 7 \equiv 2$

45. $-5 \pmod 7 \equiv 2$

47. $-13 \pmod{11} \equiv 9$

49. $135 \pmod{10} \equiv 5$

51. $3 + 4 = 7 \equiv 1 \pmod 6$

53. $2 + 2 \equiv 4 \pmod 5$ 55. $4 - 5 \equiv 5 \pmod 6$ 57. $5 \bullet 5 \equiv 7 \pmod 9$ 59. $3 \bullet \{ \} \equiv 1 \pmod 6$
No solution

61. $4 \bullet \{ \} \equiv 4 \pmod{10}$ 63. $4 - 7 \equiv 9 \pmod{12}$ 65. $3 \bullet 0 \equiv 0 \pmod{10}$
$\{1, 6\}$

67. a) 2016, 2020, 2024,
20,28, 2032

    b) 3004

    c) 2552, 2556, 2560,
2564, 2568, 2572

69.   am/pm am/pm am/pm r am am r r
                                    ↑
                                  today

    a) $28 \equiv 4 \pmod 8$; 4 days after today she rests.

    b) $60 \equiv 4 \pmod 8$; 4 days after today she rests.

    c) $127 \equiv 7 \pmod 8$; 7 days after today she has
morning and afternoon practice.

    d) $82 \equiv 2 \pmod 8$; 2 days after today she has a
morning practice, so she does not have an
off day.

71. The manager's schedule is repeated every seven
weeks. If this is week two of her schedule, then
this is her second weekend that she works, or week
1 in a mod 7 system. Her schedule in mod 7 on any
given weekend is shown in the following table:
Weekend (mod 7):

Work/off	0	1	2	3	4	5	6
	w	w	w	w	w	w	o

    a) If this is weekend 1, then in 5 more weeks
$(1 + 5 = 6)$ she will have the weekend off.

    b) $25 \div 7 = 3$, remainder 4. Thus $25 \equiv 4 \pmod 7$
and 4 weeks from weekend 1 will be weekend 5.
She will not have off.

    c) $50 \div 7 = 7$, remainder 1. One week from
weekend 1 will be weekend 2. It will be 4 more
weeks before she has off. Thus, in 54 weeks she
will have the weekend off.

73. The waiter's schedule in a mod 14 system is given
in the following table:

Day: 0 1 2 3 4 5 6 7 8 9 10 11 12 13
shift: d d d d d e e e d d   d   e   e   e
          ↑

       today

    a) $20 \equiv 6 \pmod{14}$; 6 days from today is an
evening shift.

    b) $52 \equiv 10 \pmod{14}$; 10 days from today is
the day shift.

    c) $365 \equiv 1 \pmod{14}$; 1 day from today is the
day shift.

75. a)

+	0	1	2	3
0	0	1	2	3
1	1	2	3	0
2	2	3	0	1
3	3	0	1	2

    b) Yes. All the numbers in the table are from the
set $\{0, 1, 2, 3\}$.

    c) The identity element is 0.

    d) Yes.   element + inverse = identity
$0 + 0 = 0$ $1 + 3 = 0$ $2 + 2 = 0$ $3 + 1 = 0$

    e) $(1 + 3) + 2 \; 0 + 2 = 2$ $1 + (3 + 2) = 1 + 1 = 2$
Associative since 2 = 2.

    f) Yes, the table is symmetric about the main
diagonal. $1 + 3 = 0 = 3 + 1$

    g) Yes. All five properties are satisfied.

    h) Yes. The modulo system behaves the same no
matter how many elements are in the system.

77. a)

×	0	1	2	3
0	0	0	0	0
1	0	1	2	3
2	0	2	0	2
3	0	3	2	1

    b) Yes. All the elements in the table are from
the set $\{0, 1, 2, 3\}$.

    c) Yes. The identity element is 1.

    d) elem. × inverse = identity
$0 \times$ none $= 1$ $1 \times 1 = 1$ $2 \times$ none $= 1$
$3 \times 3 = 1$ Elements 0 and 2 do not have
inverses.

    e) $(1 \times 3) \times 0 = 3 \times 0 = 0$
$1 \times (3 \times 0) = 1 \times 0 = 0$ Yes, Associative

    f) Yes. $2 \times 3 = 2 = 3 \times 2$

    g) No. Not all elements have inverses.

For the operation of division in modular systems, we define n ÷ d = n • i, where i is the multiplicative inverse of d.

79. $5 \div 7 \equiv$? (mod 9)

Since 7 • 4 = 28 ≡ 1 (mod 9), 4 is the inverse of 7.

$5 \div 7$ (mod 9) ≡ 5 · 4 (mod 9) ≡ 2 (mod 9); ? = 2

81. $? \div ? \equiv 1$ (mod 4)   0 ÷ 0 is undefined.

$1 \div 1 \equiv 1$ (mod 4) $2 \div 2 \equiv 1$ (mod 4)

$3 \div 3 \equiv 1$ (mod 4)   ? = {1, 2, 3}

83. $5k \equiv x$ (mod 5)   5(1) ≡ 0 (mod 5)

5(2) = 10 ≡ 0 (mod 5)   x = 0

85. $4k - 2 \equiv x$ (mod 4)   4(0) – 2 = -2 ≡ 2 (mod 4)

4(1) – 2 = 2 ≡ 2 (mod 4)   4(2) – 2 = 6 ≡ 2 (mod 4)

x = 2

87. (365 days)(24 hrs./day)(60 min./hr.) = 525,600 hrs.

(525,600)/(4) = 131,400 rolls   131400 ≅ 0 (mod 4)

88. 1 yr. 21 days = 365 + 21 = 386 days

386/5 = 77 R 1   Halfway up the mountain

89. If 10 is subtracted from each number on the wheel,

23 11   3 18 10 19   2 10 16   4 24   becomes

13   1 20   8   0   9 19   0   6 21 14   which is equivalent to

M   A   T H   I S   F   U   N

Review Exercises

1. A set of elements and at least one binary operation.

2. A binary operation is an operation that can be performed on two and only two elements of a set. The result is a single element.

3. Yes. The sum of any two integers is always an integer.

4. No. Example: $2 - 3 = - 1$, but $- 1$ is not a natural number.

5. $9 + 10 = 19 \equiv 7$ (mod 12)   6. $5 + 12 = 17 \equiv 5$ (mod 12)   7. $8 - 10 = -2 \equiv 10$ (mod 12)

8. $4 + 7 + 9 = 20 \equiv 8$ (mod 12)   9. $7 - 4 + 6 = 9 \equiv 9$ (mod 12)   10. $2 - 8 - 7 = -13 \equiv 11$ (mod 12)

11. a) The system is closed. If the binary operation is ⊡ then for any elements a and b in the set, a ⊡ b is a member of the set.

   b) There exists an identity element in the set. For any element a in the set, if a ⊡ i = i ⊡ a = a, then i is called the identity element.

   c) Every element in the set has a unique inverse. For any element a in the set, there exists an element b such that a ⊡ b = b ⊡ a = i. Then b is the inverse of a, and a is the inverse of b.

   d) The set is associative under the operation   For elements a, b, and c in the set, (a ⊡ b) ⊡ c = a ⊡ (b ⊡ c).

12. An Abelian group is a group in which the operation has the commutative property.

13 No, the set of positive integers has no identity element under addition.

14. The set of integers with the operation of multiplication does not form a group since not all elements have an inverse. Example: $4 \cdot \dfrac{1}{4} = 1,$ but $\dfrac{1}{4}$ is not an integer. Only 1 and –1 have inverses.

15. Yes. Closure: The sum of any two rational #s is a rational number.

The identity element is zero. Ex.: 5 + 0 = 0 + 5 = 5

Each element has a unique inverse. Ex. : 6 + (– 6) = 0

Yes, Associative. Example: (2 + 3) + 4 = 2 + (3 + 4)

16. The set of rational numbers with the operation of multiplication does not form a group since zero does not have an inverse. $0 \bullet \underline{?} = 1$

17. There is no identity element. Therefore the system does not form a group.

18. Not Associative   Example: $(! \square p) \square ? = p \square ? = !$ and $! \square (p \square ?) = ! \square ! = ?;$ $! \neq ?$

19. Not every element has an inverse.
    Not Associative   Example: $(P\ ?\ P)\ ?\ 4 = L\ ?\ 4 = \#$ $P\ ?\ (P\ ?\ 4) = P\ ?\ L = 4;$ $\# \neq 4$

20. a) $\{\ \vdash,\ \odot,\ ?,\ \Delta\ \}$
    b) $\square\!\!\!\!\!\sqcup$
    c) Yes. All the elements in the table are from the set $\{\ \vdash,\ \odot,\ ?,\ \Delta\ \}$.

    d) The identity element is $\vdash$.
    e) Yes.

    $\vdash\ \square\!\!\!\sqcup\ \vdash\ =\ \vdash$ $\odot\ \square\!\!\!\sqcup\ \Delta\ =\ \vdash$
    $?\ \square\!\!\!\sqcup\ ?\ =\ \vdash$ $\Delta\ \square\!\!\!\sqcup\ \odot\ =\ \vdash$

    Every element has an inverse.

    f) Yes, Associative
    $(\vdash\ \square\!\!\!\sqcup\ ?)\ \square\!\!\!\sqcup\ \Delta = ?\ \square\!\!\!\sqcup\ \Delta = \odot$
    $\vdash\ \square\!\!\!\sqcup\ (?\ \square\!\!\!\sqcup\ \Delta) = \vdash\ \square\!\!\!\sqcup\ \odot = \odot$

    g) Yes. The elements in the table are symmetric around the diagonal.
    $\Delta\ \square\!\!\!\sqcup\ ? = \odot = ?\ \square\!\!\!\sqcup\ \Delta$
    h) Yes, all five properties are satisfied.

21. $21 \div 3 = 7$, remainder 0 $21 \cong 0 \pmod 3$

22. $31 \div 8 = 3$, remainder 7 $31 \cong 7 \pmod 8$

23. $31 \div 6 = 5$, remainder 1 $31 \cong 1 \pmod 6$

24. $59 \div 8 = 7$, remainder 3 $59 \cong 3 \pmod 8$

25. $82 \div 13 = 6$, remainder 4 $82 \cong 4 \pmod{13}$

26. $54 \div 4 = 13$, remainder 2 $54 \cong 2 \pmod 4$

27. $52 \div 12 = 4$, remainder 4 $52 \cong 4 \pmod{12}$

28. $54 \div 14 = 3$, remainder 12 $54 \cong 12 \pmod{14}$

29. $97 \div 11 = 8$, remainder 9 $97 \cong 9 \pmod{11}$

30. $42 \div 11 = 3$, remainder 9 $42 \cong 9 \pmod{11}$

31. $5 + 8 = 13 \cong 4 \pmod 9; ? = 4$

32. $? - 3 \equiv 0 \pmod 5$
    $0 - 3 \equiv 2 \pmod 5$ $1 - 3 \equiv 3 \pmod 5$
    $2 - 3 \equiv 4 \pmod 5$ $3 - 3 \equiv 0 \pmod 5$
    $? = 3$

33. $4 \bullet ? \equiv 3 \pmod 6$
    $4 \bullet 0 \equiv 0 \pmod 6$ $4 \bullet 1 \equiv 4 \pmod 6$
    $4 \bullet 2 = 8 \equiv 2 \pmod 6$ $4 \bullet 3 = 12 \equiv 0 \pmod 6$
    $4 \bullet 4 = 16 \equiv 4 \pmod 6$ $4 \bullet 5 = 20 \equiv 2 \pmod 6$
    There is no solution. $? = \{\ \}$

34. $6 - ? \equiv 5 \pmod 7$
    $6 - 0 \equiv 6 \pmod 7$ $6 - 1 \equiv 5 \pmod 7$
    $6 - 2 \equiv 4 \pmod 7$ $6 - 3 \equiv 3 \pmod 7$
    $6 - 4 \equiv 2 \pmod 7$ $6 - 5 \equiv 1 \pmod 7$
    Replace ? with 1.

35. $? \bullet 4 \equiv 0 \pmod 8$
    $0 \bullet 4 \equiv 0 \pmod 8$ $1 \bullet 4 \equiv 4 \pmod 8$
    $2 \bullet 4 = 8 \equiv 0 \pmod 8$ $3 \bullet 4 = 12 \equiv 4 \pmod 8$
    $4 \bullet 4 = 16 \equiv 0 \pmod 8$ $5 \bullet 4 = 20 \equiv 4 \pmod 8$
    $6 \bullet 4 = 24 \equiv 0 \pmod 8$ $7 \bullet 4 = 28 \equiv 4 \pmod 8$
    Replace ? with $\{0, 2, 4, 6\}$.

36. $10 \bullet 7 \equiv ? \ (\text{mod } 12)$

   $10 \bullet 7 = 70; \quad 70 \quad 12 \equiv 5, \text{remainder } 10$

   Thus, $10 \bullet 7 \equiv 10 \ (\text{mod } 12)$.

   Replace ? with 10.

37. $3 - 5 \equiv ? \ (\text{mod } 7)$

   $3 - 5 = (3+7) - 5 = 5 \equiv 5 \ (\text{mod } 7)$

   Replace ? with 5.

38. $? \bullet 7 \equiv 3 \ (\text{mod } 10)$

$0 \bullet 7 \equiv 0 \ (\text{mod } 10)$	$1 \bullet 7 \equiv 7 \ (\text{mod } 10)$
$2 \bullet 7 = 14 \equiv 4 \ (\text{mod } 10)$	$3 \bullet 7 = 21 \equiv 1 \ (\text{mod } 10)$
$4 \bullet 7 = 28 \equiv 8 \ (\text{mod } 10)$	$5 \bullet 7 = 35 \equiv 5 \ (\text{mod } 10)$
$6 \bullet 7 = 42 \equiv 2 \ (\text{mod } 10)$	$7 \bullet 7 = 49 \equiv 9 \ (\text{mod } 10)$
$8 \bullet 7 = 56 \equiv 6 \ (\text{mod } 10)$	$9 \bullet 7 = 63 \equiv 3 \ (\text{mod } 10)$

   $10 \bullet 7 = 70 \equiv 0 \ (\text{mod } 10)$

   Replace ? with 9.

39. $5 \bullet ? \equiv 3 \ (\text{mod } 8)$

$5 \bullet 0 \equiv 0 \ (\text{mod } 8)$	$5 \bullet 1 \cong 5 \ (\text{mod } 8)$
$5 \bullet 2 = 10 \equiv 2 \ (\text{mod } 8)$	$5 \bullet 3 = 15 \equiv 7 \ (\text{mod } 8)$
$5 \bullet 4 = 20 \equiv 4 \ (\text{mod } 8)$	$5 \bullet 5 = 25 \equiv 1 \ (\text{mod } 8)$
$5 \bullet 6 = 30 \equiv 6 \ (\text{mod } 8)$	$5 \bullet 7 = 35 \equiv 3 \ (\text{mod } 8)$

   Replace ? with 7.

40. $7 \bullet ? \equiv 2 \ (\text{mod } 9)$

$7 \bullet 0 \equiv 0 \ (\text{mod } 9)$	$7 \bullet 1 \equiv 7 \ (\text{mod } 9)$
$7 \bullet 2 = 14 \equiv 5 \ (\text{mod } 9)$	$7 \bullet 3 = 21 \equiv 3 \ (\text{mod } 9)$
$7 \bullet 4 = 28 \equiv 1 \ (\text{mod } 9)$	$7 \bullet 5 = 35 \equiv 7 \ (\text{mod } 9)$
$7 \bullet 6 = 42 \equiv 6 \ (\text{mod } 9)$	$7 \bullet 7 = 49 \equiv 4 \ (\text{mod } 9)$

   $7 \bullet 8 = 56 \equiv 2 \ (\text{mod } 9)$

   Replace ? with 8.

41. a)

+	0	1	2	3	4	5
0	0	1	2	3	4	5
1	1	2	3	4	5	0
2	2	3	4	5	0	1
3	3	4	5	0	1	2
4	4	5	0	1	2	3
5	5	0	1	2	3	4

41. b) Since all the numbers in the table are elements of $\{0, 1, 2, 3, 4, 5\}$, the system has the closure property.

   c) The identity element is 0 and the inverses of each element are $0 - 0$, $1 - 5$, $2 - 4$, $3 - 3$, $4 - 2$, $5 - 1$

   d) The associative property holds as illustrated by the example:

   $(2 + 3) + 5 = 4 = 2 + (3 + 5)$

   e) The commutative property holds since the elements are symmetric about the main diagonal

   f) The modulo 6 system forms a commutative group under addition.

42. a)

×	0	1	2	3
0	0	0	0	0
1	0	1	2	3
2	0	2	0	2
3	0	3	2	1

   b) The identity element is 1, but because 0 and 2 have no inverses, the modulo 4 system does not form a group under multiplication.

43. Day (mod 10): 0 1 2 3 4 5 6 7 8 9

   Work/off : w w w o o w w o o o

   ↑

   today

   a) Since $18 \equiv 8 \ (\text{mod } 10)$, Toni will not be working 18 days after today.

   b) Since $38 \equiv 8 \ (\text{mod } 10)$, Toni will have the evening off in 38 days.

Chapter Test

1. A mathematical system consists of a set of elements and at least one binary operation.

2. Closure, identity element, inverses, associative property, and commutative property.

3. No, the numbers greater than 0 do not have inverses.

4.

+	1	2	3	4	5
1	2	3	4	5	1
2	3	4	5	1	2
3	4	5	1	2	3
4	5	1	2	3	4
5	1	2	3	4	5

5. Yes. It is closed since the only elements in the table are from the set $\{1, 2, 3, 4, 5\}$. The identity element is 5. The inverses are $1 \leftrightarrow 4, 2 \leftrightarrow 3, 3 \leftrightarrow 2$, $4 \leftrightarrow 1$, and $5 \leftrightarrow 5$. The system is associative. The system is commutative since the table is symmetric about the main diagonal. Thus, all five properties are satisfied.

6. $4 + 3 + 2 = 2 + 2 = 4$

7. On a clock with numbers 1 to 5, starting at 2 and moving counterclockwise 5 places ends back at 2. So $2 - 5 = 2$.

8. a) The binary operation is $\square$.

   b) Yes. All elements in the table are from the set $\{W, S, T, R\}$.

   c) The identity element is T, since $T \square x = x = x \square T$, where x is any member of the set $\{W, S, T, R\}$.

   d) The inverse of R is S, since $R \square S = T$ and $S \square R = T$.

   e) $(T \square R) \square W = R \square W = S$

9. The system is not a group. It does not have the closure property since $c * c = d$, and d is not a member of $\{a, b, c\}$.

10. Since all the numbers in the table are elements of $\{1, 2, 3\}$, the system is closed. The commutative property holds since the elements are symmetric about the main diagonal. The identity element is 2 and the inverses are $1 - 3, 2 - 2, 3 - 1$. If it is assumed the associative property holds as illustrated by the example: $(1 ? 2) ? 3 = 2 = 1 ? (2 ? 3)$, then the system is a commutative group.

11. Since all the numbers in the table are elements of $\{ @, \$, \&, \% \}$, the system is closed. The commutative property holds since the elements are symmetric about the main diagonal. The identity element is \$ and the inverses are $@ - \&, \$ - \$, \& - @, \% - \%$. It is assumed the associative property holds as illustrated by the example: $(@ O \$) O \% = \& = @ O (\$ O \%)$, then the system is a commutative group.

12. $64 \div 9 = 7$, remainder 1 $64 \equiv 1 \pmod 9$

13. $58 \div 11 = 5$, remainder 3 $58 \equiv 3 \pmod{11}$

14. $7 + 7 \equiv 6 \pmod 8$

15. $2 - 3 = (5 + 2) - 3 = 4 \equiv 4 \pmod 5$

16. $3 - 5 \equiv 7 \pmod 9$
    $3 - 5 = (3 + 9) - 5 = 12 - 5 \equiv 7 \pmod 9$
    $12 - 5 \equiv 7 \pmod 9$
    Replace ? with 5.

17. $4 \bullet 2 = 8$ and $8 \equiv 2 \pmod 6$
    $4 \bullet 2 \equiv 2 \pmod 6$
    Replace ? with 2.

18. $3 \bullet ? \bullet \equiv 2 \,(\mathrm{mod}\ 6)$

   $3 \bullet 0 \equiv 0 \,(\mathrm{mod}\ 6)$ $3 \bullet 1 \equiv 3 \,(\mathrm{mod}\ 6)$

   $3 \bullet 2 \equiv 0 \,(\mathrm{mod}\ 6)$ $3 \bullet 3 \equiv 3 \,(\mathrm{mod}\ 6)$

   $3 \bullet 4 \equiv 0 \,(\mathrm{mod}\ 6)$ $3 \bullet 5 \equiv 3 \,(\mathrm{mod}\ 6)$

   There is no solution for ?   The answer is { }.

19. $103 \div 7 = 14$, remainder 5

   $103 \equiv 5 \,(\mathrm{mod}\ 7)$

   Replace ? with 5.

20. a)

×	0	1	2	3	4
0	0	0	0	0	0
1	0	1	2	3	4
2	0	2	4	1	3
3	0	3	1	4	2
4	0	4	3	2	1

   b) The system is closed. The identity is 1.
   However, 0 does not have an inverse, so
   the system is <u>not</u> a group.

CHAPTER ELEVEN

CONSUMER MATHEMATICS

Exercise Set 11.1

1. A percent is a ratio of some number to 100.

3. (i) Divide the number by the denominator.
   (ii) Multiply the quotient by 100 (which has the effect of moving the decimal point two places to the right).
   (iii) Add a percent sign.

5. Percent change = $\dfrac{(\text{Amount in latest period}) - (\text{Amount in previous period})}{\text{Amount in previous period}}$ x 100

7. $\dfrac{1}{2} = 0.5 = (0.5)(100)\% = 50\%$

9. $\dfrac{2}{5} = 0.4 = (0.4)(100)\% = 40\%$

11. $0.007654 = (0.007654)(100)\% = 0.8\%$

13. $3.78 = (3.78)(100)\% = 378\%$

15. $4\% = \dfrac{4}{100} = 0.04$

17. $1.34\% = \dfrac{1.34}{100} = 0.0134$

19. $\dfrac{1}{4}\% = 0.25\% = \dfrac{0.25}{100} = 0.0025$

21. $\dfrac{1}{5}\% = 0.2\% = \dfrac{0.2}{100} = 0.002$

23. $1\% = \dfrac{1}{100} = 0.01$

25. $\dfrac{95}{3500} \approx .0271428571 = (0.0271428571 \times 100)\% =$
    2.714%

27. $8(.4125) = 3.3 \qquad 8.0 - 3.3 = 4.7$ g

29. $(591,000)(.08) = \$\,47,280,000 = \$\,47.28$ million

31. $(591 \text{ million})(.27) = \$\,159.57$ million

33. $(32.3 \text{ billion})(.176) = \$\,5.6848$ billion

35. $(32.3 \text{ billion})(.306) = \$\,9.8838$ billion

37. $\dfrac{1,553\,\text{million}}{8,105\,\text{million}} = .1916 \qquad (.1916)(100) = 19.2\%$

39. $\dfrac{1,592\,\text{million}}{8,105\,\text{million}} = .1964 \qquad (.1964)(100) = 19.6\%$

41. $\dfrac{45,793 - 48,622}{48,622} = -0.058 \approx 5.8\%$ decrease

43. a) $\dfrac{9457-8059}{8059} = 0.17347 \approx 17.3\%$

   b) $\dfrac{13,577-9457}{9457} = 0.4357 \approx 43.6\%$

   c) $\dfrac{20,947-13,577}{13,577} = 0.5428 \approx 54.3\%$

   d) $\dfrac{32,240-20,947}{20,947} = 0.5391 \approx 53.9\%$

45. a) $\dfrac{10,403.94-9920}{9920} = 0.0488 \approx 4.9\%$

   b) $\dfrac{7591.93-10,403.94}{10,403.94} = -0.2702 \approx 27\%$ decrease

   c) $\dfrac{7591.93-9920}{9920} = -0.2347 \approx 23.5\%$ decrease

   d) $\dfrac{8397.03-7591.93}{7591.93} = 0.10604 \approx 10.6\%$

47. $(.15)(45) = \$ 6.75$

49. $24/96 = .25;\ (.25)(100) = 25\%$

51. $.05x = 75;\quad x = 75/.05 = 300$

53. a) tax = 6% of $43.50 = (0.06)(43.50) = \$2.61$
   b) total bill before tip = $43.50 + $2.61 = $46.11
   c) tip = 15% of $46.11 = 0.15(46.11) = $6.92
   d) total cost = $46.11 + 6.92 = $53.03

55. $1.50(x) = 18 \qquad x = \dfrac{18}{1.50} = 12$

   12 students got an A on the 2nd test.

57. Mr. Browns' increase was $0.07(36,500) = \$2,555$
   His new salary = $36,500 + $2,555 = $39,055

59. Percent change $= \left(\dfrac{407-430}{430}\right)(100) =$

   $\left(\dfrac{-23}{430}\right)(100) = -5.3\%$

   There was a 5.3% decrease in the # of units sold.

61. $\dfrac{31.1-39.3}{39.3} = -0.2087 \approx 20.9\%$ decrease

63. Percent decrease from regular price =

   $\left(\dfrac{\$439-539.62}{539.62}\right)(100) = \left(\dfrac{-100.62}{539.62}\right)(100) =$

   - 18.6 %
   The sale price is 18.6% lower than the regular price.

65. $(0.18)(\text{sale price}) = \$675$

   sale price $= \dfrac{675}{0.18} = \$3,750$

67. $1000 increased by 10% is $1000 + 0.10($1000) = $1000 + $100 = $1,100.
   $1,100 decreased by 10% is $1,100 − 0.10($1,100) = $1,100 − $110 = $990.
   Therefore if he sells the car at the reduced price he will lose $10.

69. The profit must be $0.40(\$5901.79) = \$2,360.72$. With a 40% profit, the total revenue must be
   $5901.79 + $2360.72 = $8,262.51; the revenue from the first sale would be $100 \times \$9.00 = \$900$, and the
   revenue from the second sale would be $150 \times \$12.50 = \$1,875.00$. The revenue needed for the 250 ties
   would be $8,262.51 − $900.00 − $1,875.00 = $5,487.51. Thus, the selling price should be

   $\dfrac{\$5,487.51}{250} = \$21.95.$

70. a) $200 decreased by 25%   b) $100 increased by 50%   c) $100 increased by 100%

Exercise Set 11.2

1. Interest is the money the borrower pays for the use of the lender's money.

3. Security is anything of value pledged by the borrower that the lender may sell or keep if the borrower does not repay the loan.

5. i = interest, p = principal, r = interest rate, t = time
   The rate and time must be expressed for the same period of time, i.e. days, months or years.

7. The United States Rule states that if a partial payment is made on a loan, interest is computed on the principal from the first day of the loan (or previous partial payment) up to the date of the partial payment. For each partial payment, the partial payment is used to pay the interest first, then the remainder of the payment is applied to the principle. On the due date of the loan the interest is calculated from the date of the last partial payment.

9. $i = prt = (300)(.04)(5) = \$60.00$

11. $(900)(.0375)(30/360) = \$2.81$

13. $i = prt = (587)(0.00045)(60) = \$15.85$

15. $i = (2{,}756.78)(0.1015)\left(\dfrac{103}{360}\right) = \$80.06$

17. $i = (1372.11)(.01375)(6) = \$113.20$

19. $(1500)(r)(3 = 450 \qquad r = \left(\dfrac{450}{4500}\right)(100) = 10\%$

21. $12.00 = p(0.08)\left(\dfrac{3}{12}\right) = p(0.02)$

   $p = \dfrac{12.00}{0.02} = \$600$

23. $124.49 = (957.62)(0.065)(t) = 62.2453t$

   $\dfrac{124.49}{62.2453} = t \qquad t = 2$ years

25. $i = (1000)(.03)\left(\dfrac{6}{12}\right) = \$15.00$

   $\$15 + \$1000 = \$1015$

27. a) $i = prt \qquad i = (3500)(0.075)(6/12) = \$131.25$
    b) $A = p + i \qquad A = 3500 + 131.25 = \$3{,}631.25$

29. a) $i = prt \qquad I = (3650)(0.075)(8/12) = \$182.50$
    b) $3650.00 - 182.50 = \$3467.50$, which is the amount Julie received.
    c) $i = prt \qquad 182.50 = (3467.50)(r)(8/12)$
       $\qquad\qquad\qquad\qquad = 2311.67r$

    $\dfrac{182.50}{2311.67} = r = 0.0789$ or 7.9%

31. Amt. collected $= (470)(4500/2) = \$1{,}057{,}500$
    $i = prt = (1{,}057{,}500)(0.054)(5/12) = \$23{,}793.75$

33. [Jan 17 – July 4]: $185 - 17 = 168$ days

35. [12/08 – 03/17]: $342 - 76 = 266$ days

37. [08/24 – 05/15]: $(365 - 236) + 135 = 129 + 135 = 264$ days

39. [04/15] for 60 days: $105 + 60 = 165$, which is June 14

41. [11/25] for 120 days: $329 + 120 = 449;$
    $449 - 365 = 84 \qquad 84 - 1$ leap year day $=$ day 83, which is March 24

43. [03/01 to 04/01]: $91 - 60 = 31$ days
    $(2000)(.05)(31/360) = 8.61 \qquad 400.00 - 8.61 = 391.39$
    $2000.00 - 391.39 = \$1608.61$

    [04/01 to 05/01]: $121 - 91 = 30$ days
    $(1608.61)(.05)(30/360) = 6.70$
    $1608.61 + 6.70 = \$1615.31$

45. [08/01 to 11/15]: $= 319 - 213 = 106$ days
    $(7000)(.0575)(106/360) = \$118.51$
    $3500.00 - 118.51 = \$3381.49$
    $7000.00 - 3381.49 = \$3618.51$

    [11/15 to 12/15]: $= 349 - 319 = 30$ days
    $(3618.51)(.0575)(30/360) = \$17.34$
    $3618.51 + 17.34 = \$3635.85$

47. [07/15 to 12/27]:   361 – 196 = 165 days
(9000)(.06)(165/360) = $247.50
4000.00 – 247.50 = $3752.50
9000.00 – 3752.50 = $5247.50

[12/27 to 02/01]:   (365 – 361) + 32 = 36 days
(5247.50)(.06)(36/360) = $31.49
5247.50 + 31.49 = $5278.99

49. [08/01 to 09/01]:   244 – 213 = 31 days
(1800)(.15)(31/360) =$ 23.25
500.00 – 23.25 = $476.75
1800.00 – 476.75 = $1323.25

[09/01 to 10/01]:   274 – 244 = 30 days
(1323.25)(.15)(30/360) = $16.54
500.00 – 16.54 = $483.46
1323.25 – 483.46 = $839.79

[10/01 to 11/01]:   305 – 274 = 31 days
(839.79)(.15)(31/360) = $10.85
839.79 + 10.85 = $850.64

51. [03/01 to 08/01]:   213 – 60 = 153 days
(11600)(.06)(153/360) = $95.80
2000.00 – 295.80 = $1704.20
11600.00 – 1704.20 = $9895.80

[08/01 to 11/15]:   319 – 213 = 106 days
(9895.80)(.06)(106/360) = $174.83
4000.00 – 174.83 = $3825.17
9895.8 – 3825.17 = $6070.63

[11/15 to 12/01]:   335 – 319 = 16 days
(6070.63)(.06)(16/360) = $16.19
6070.63 + 16.19 = $6086.82

53. [03/01 to 05/01]:   121 – 60 = 61 days
(6500)(.105)(61/360) = $115.65
1750.00 – 115.65 = $1634.35
6500.00 – 1634.35 = $4865.65

[05/01 to 07/01]:   182 – 121 = 61 days
(4865.65)(.105)(61/360) = $86.57
2350.00 – 86.57 = $2263.43
4865.65 – 2263.43 = $2602.22

Since the loan was for 180 days, the maturity
date occurs in 180 – 61 – 61 = 58 days
(2602.22)(.105)(58/360) = $44.02
2602.22 + 44.02 = $2646.24

55. a) May 5 is day 125   125 + 182 = 307
day 307 is Nov. 3
b) i = (1000)(0.0434)(182/360) = $21.94
Amt. paid = 1000 – 21.94 = $978.06
c) interest = $21.94
d) $r = \dfrac{i}{pt} = \dfrac{21.94}{978.06\left(182/360\right)} = 0.0444$ or 4.44%

57. a) Amt. received = 743.21 – 39.95 = $703.26
i = prt
39.95 = (703.26)(r)(5/360)
39.95 = (9.7675)(r)
r = 39.95/9.7675 = 4.09 or 409%
b) 39.95 = (703.26)(r)10/360)
39.95 = (19.535)(r)
r = 39.95/19.535 = 2.045 or 204.5%
c) 39.95 = (703.26)(r)(20/360)
39.95 = (39.07)r
r = 39.95/39.07 = 1.023 or 102.3%

59. a) $\dfrac{93337}{100000} = 0.93337$
1.00000 – 0.93337 = .06663 or 6.663 %
b) $\dfrac{100000}{93337} = 1.071386$
1.071386 – 1.000000 = .071386 or 7.139 %
c) 100000 – 93337 = $6663.00
d) (6663)(.05)(1) = 33.15
6663.00 + 33.15 = $6696.15

60. a) [08/03/1492 to 12/01/1620]
1492 to 1620 = 127 years = 45720 days
08/03 to 12/31 = 365 – 215 = 150 days
01/01 to 12/01 = 335 days
45720 + 150 + 335 = 46205 days
(1)(.05)(46205/360) = 6.417361 = $6.42
b) [07/04/1776 to 08/03/1492]
284 yrs. minus 30 days = 102,210
(1)(.05)(102,210/360) = 14.1958 = $14.20
c) [08/03/1492 to 12/07/1941]
449 yrs plus 126 days = 161,766 days
(1)(.05)(161,766/360) = $22.47
d) Answers will vary.

Exercise Set 11.3

1. An investment is the use of money or capital for income or profit.

3. For a variable investment neither the principal nor the interest is guaranteed.

5. The effective annual yield on an investment is the simple interest rate that gives the same amount of interest as a compound rate over the same period of time.

7. a) $n = 1, r = 2.0\%, t = 3, p = \$2000$

$$A = 2000\left(1+\frac{0.02}{1}\right)^{1\cdot 3} = \$2122.42$$

b) $i = \$2122.42 - \$2000 = \$122.42$

9. a) $n = 2, r = 3.0\%, t = 4, p = \$3500$

$$A = 3500\left(1+\frac{0.03}{2}\right)^{2\cdot 4} = \$3942.72$$

b) $i = \$3942.72 - \$3500 = \$442.72$

11. a) $n = 4, r = 4.75\%, t = 3, p = \$1500$

$$A = 1500\left(1+\frac{0.0475}{4}\right)^{4\cdot 3} = \$1728.28$$

b) $i = \$1728.28 - \$1500 = \$228.28$

13. a) $n = 12, r = 6.25\%, t = 2, p = \$2500$

$$A = 2500\left(1+\frac{0.0625}{12}\right)^{12\cdot 2} = \$2831.95$$

b) $i = \$2831.95 - \$2500 = \$331.95$

15. a) $n = 360, r = 4.59\%, t = 4$ yr., $p = \$4000$

$$A = 4000\left(1+\frac{0.0459}{360}\right)^{360\cdot 4} = \$4806.08$$

b) $i = \$4806.08 - \$4000 = \$806.08$

17. $A = 7500\left(1+\frac{0.0266}{2}\right)^{2\cdot 4} = \$8336.15$

19. $A = 1500\left(1+\frac{0.039}{12}\right)^{12\cdot 2.5} = \$1653.36$

21. $p = 800 + 150 + 300 + 1000 = \$2250$

$$A = 2250\left(1+\frac{0.02}{360}\right)^{360\cdot 2} = \$2,341.82$$

23. a) $A = 2000\left(1+\frac{0.05}{2}\right)^{2\cdot 15} = \$4,195.14$

b) $A = 2000\left(1+\frac{0.05}{4}\right)^{2\cdot 15} = \$4,214.36$

25. $A = 3000\left(1+\frac{0.0175}{4}\right)^{4\cdot 2} = \$3106.62$

27. $A = 6000\left(1+\frac{0.08}{4}\right)^{12} = \$7,609.45$

29. a) $A = 1000\left(1+\frac{0.02}{2}\right)^{4} = \$1,040.60$

$\quad i = \$1040.60 - \$1000 = \$40.60$

b) $A = 1000\left(1+\frac{0.04}{2}\right)^{4} = \$1,082.43$

$\quad i = \$1082.43 - \$1000 = \$82.43$

c) $A = 1000\left(1+\frac{0.08}{2}\right)^{4} = \$1,169.86$

$\quad i = \$1169.86 - \$1000 = \$169.86$

d) No predictable outcome.

31. a) $A = 1000\left(1+\frac{0.06}{2}\right)^{4} = \$1,125.51$

$\quad i = \$1125.51 - \$1000 = \$125.51$

b) $A = 1000\left(1+\frac{0.06}{2}\right)^{8} = \$1,266.77$

$\quad i = \$1266.77 - \$1000 = \$266.77$

c) $A = 1000\left(1+\frac{0.06}{2}\right)^{16} = \$1,604.71$

$\quad i = \$1604.71 - \$1000 = \$604.71$

d) New amount $= \dfrac{(\text{old amount})^2}{1000}$

33. $A = 1\left(1+\frac{0.035}{12}\right)^{12\cdot 1} \approx 1.0354$

$i = A - 1 = 1.0354 - 1 = 0.0354$

APY $= 3.54\%$

35. $A = 1\left(1+\frac{0.024}{12}\right)^{12\cdot 1} \approx 1.0243$

$i = A - 1 = 1.0243 - 1 = 0.0243$

Yes, APY $= 2.43\%$, not 2.6%

37. The effective rate of the 4.75% account is:

$$A = 1\left(1 + \frac{0.0475}{12}\right)^{12} \approx 1.0485$$

APY = 1.0485 − 1.00 = 0.0485 or 4.85%
Therefore the 5% simple interest account pays
more interest.

39. a) $\dfrac{A}{\left(1 + r/n\right)^{n \cdot t}} = \dfrac{290,000}{\left(1 + 0.0825/2\right)^{20}} = \$129,210.47$

b) surcharge = $\dfrac{129,210.47}{958} = \$134.88$

41. $p = \dfrac{A}{\left(1 + r/n\right)^{n \cdot t}} = \dfrac{30,000}{\left(1 + 0.0515/12\right)^{60}} = \$23,202.23$

43. Present value = $\dfrac{50,000}{\left(1 + \dfrac{0.08}{4}\right)^{72}} = \$12,015.94$

45. p = 1.35, r = 0.025, t = 5, n = 1
$A = 1.35(1 + 0.025)^5 = \$1.53$

47. a) 72/3 = 24 years   b) 72/6 = 12 years
c) 72/8 = 9 years   d) 72/12 = 6 years
e) 72/r = 22
72 = 22r
r = 72/22 = 0.0327
r = 3.27%

49. R = \$500, r = 5.5%, n = 2, t = 17

$$S = 500 \dfrac{\left[\left(1 + \dfrac{0.055}{2}\right)^{34} - 1\right]}{\dfrac{0.055}{2}} =$$

$$500\,[1.51526]\left(\dfrac{2}{0.055}\right) = \$27,550.11$$

51. Use the formula given in exercise 45.
a) R = 150, r = 0.056, n = 12, t = 18
ans. S = \$55,726.01
b) R = 900, r = 0.058, n = 2, t = 18
ans. S = \$55,821.15

52. $i = prt = (100,000)(0.05)(4) = \$20,000$

$$A = 100,000\left(1 + \frac{0.05}{360}\right)^{360 \cdot 4} = \$122,138.58$$

interest earned: \$122,138.58 − 100,000 = \$22,138.58

Select investing at the compounded daily interest because the compound interest is greater by \$2,138.58.

Exercise Set 11.4

1. An open-end installment loan is a loan on which you can make different payment amounts each month. A fixed installment loan is one in which you pay a fixed amount each month for a set number of months.

3. The APR is the true rate of interest charged on a loan.

5. The total installment price is the sum of all the monthly payments and the down payment, if any.

7. The unpaid balance method and the average daily balance method.

9. a) Amount financed = 43,000 − 0.15(43000) = \$36,550.00
From Table 11.2 the finance charge per \$100 at 5.5 % for 60 payments is 14.61.

Total finance charge = $(14.61)\left(\dfrac{36550}{100}\right) = \$5339.96$

b) Total amount due after down payment = 36550.00 + 5339.96 = \$41889.96

Monthly payment = $\dfrac{41889.96}{60}$ = \$698.17

11. a) From Table 11.2, the finance charge per $100 financed at 7.5% for 60 months is $20.23.

Total finance charge is $(20.23)\left(\dfrac{4000}{100}\right) = \$809.20$

  b) Total amount due = 4000 + 809.20 = $4,809.20   Monthly payment = $\dfrac{4809.20}{60} = \$80.15$

13. a) Down payment = 0.20(3200) = $640; Cheryl borrowed 3200 − 640 = $2560
    Total installment price = (60 • 53.14) = $3188.40
    Finance charge = 3188.40 − 2560 = $628.40

  b) $\left(\dfrac{\text{finance charge}}{\text{amt. financed}}\right)(100) = \left(\dfrac{628.40}{2560}\right)(100) = 24.55$

    From Table 11.2 for 60 payments, the value of 24.55 corresponds with an APR of 9.0 %.

15. a) Total installment price = (224)(48) = $10752.00
    Finance charge = 10752.00 − 9000.00 = $1752.00

  b) $\left(\dfrac{\text{finance charge}}{\text{amt. financed}}\right)(100) = \left(\dfrac{1752.00}{9000}\right)(100) = 19.47$

    From Table 11.2, for 48 payments, the value of $19.47 is closest to $19.45 which corresponds with an APR of 9.0 %.

17. Down payment = $0.00   Amount financed = $12,000.00   Monthly payments = $232.00
  a) Installment price = (60)(232) = 13,920
    Finance charge = $13,920.00 − $12,000.00 = $1920.00

    $\left(\dfrac{\text{finance charge}}{\text{amt. financed}}\right)(100) = \left(\dfrac{1920.00}{12000}\right)(100) = 16.00$

    From Table 11.2, for 6 payments, the value of $16.00 corresponds with an APR of 6.0 %.

  b) $u = \dfrac{nPV}{100+V} = \dfrac{(36)(232)(9.52)}{(100+9.52)} = \dfrac{79511.04}{109.52} = 725.9956 = \$726.00$

  c) (232)(23) = 5336;   5336 + 726 = 6062;   13920 − 6062 = $7858.00
19. a) Amount financed = 32000 − 10000 = $22000
    From Table 11.2, the finance charge per 100 financed at 8 % for 36 payments is 12.81.

    Total finance charge = $(12.81)\left(\dfrac{22000}{100}\right) = 2818.20$

  b) Total amt. due = 22000 + 2818.20 = $24,818.20   Monthly payment = $\dfrac{24818.20}{36} = \$689.39$

  c) $u = \dfrac{nPV}{100+V} = \dfrac{(12)(689.89)(4.39)}{100+4.39} = \dfrac{36317.07}{104.39} = \$347.90$

  d) (23)(689.39) = 15855.97;   15855.97 − 347.90 = 16203.87;   24818.20 − 16203.87 = $8614.17
21. a) Amount financed = $7345.00 with no down payment.
    From Table 11.2, the finance charge per 100 financed at 8.5 % for 48 payments is 18.31.

    Total finance charge = $(18.31)\left(\dfrac{7345}{100}\right) = \$1344.87$

  b) Total amt. due = 7345.00 + 1344.87 = $8,689.87   Monthly payment = $\dfrac{8689.87}{48} = \$181.04$

  c) $f = 1344.87, k = 36, n = 48$

    $u = \dfrac{f \cdot k(k+1)}{n(n+1)} = \dfrac{(1344.87)(36)(36+1)}{48(48+1)} = \dfrac{1,791,366.84}{2352} = \$761.64$

  d) (181.04)(48) = 8689.92;   (11)(181.04) = 1991.44;   1991.44 + 761.64 = 2753.08
    8689.92 − 2753.08 = $5936.84

23. a) Interest = 500 + (151.39)(18) - 3000 = $225.02   k = 6, n = 18, and f = 225.02

$$u = \frac{(225.02)(6)(6+1)}{18(18+1)} = \frac{9450.84}{342} = \$27.63$$

b) $908.34 Total of remaining payments   908.34 – 27.63 = 880.71
   880.71 + 151.39 = $1032.10   Total amount due

25. a) Balance due = 365 + 180 + 195 + 84 = $824   min. payment $= \frac{\text{bal. due}}{48} = \frac{824}{48} \approx 17.17 \approx \$18$

b) Bal. due after Dec. 1 payment = 824 – 200 = $624   interest for Dec. = (0.011)(624) = $6.86
   Bal. due Jan. 1 = 624 + 6.86 = $630.86

27. a) Bal. due = 423 + 36 + 145 + 491 = $1095   min. payment $= \frac{\text{bal. due}}{36} = \frac{1095}{36} \approx 30.42 \approx \$31$

b) Bal. due after Mar. 1 payment = 1095 – 548 = $547   interest for March = (0.011)(547) = $6.02
   Bal. due Apr. 1 = 547 + 6.02 = $553.02

29. a) Finance charge = (1097.86)(0.018)(1) = $19.76
    b) Bal. due May 5 = (1097.86 + 19.76 + 425.79) – 800 = $743.41

31. a) Finance charge = (124.78)(0.0125)(1) = $1.56
    b) old balance + finance charge – payment + art supplies + flowers + music CD = new balance
       124.78 + 1.56 – 100.00 + 25.64 + 67.23 + 13.90 = $133.11

33. a)

Date	Balance Due	Number of Days	(Balance)(Days)
May 12	$378.50	1	(378.50)(1) = $ 378.50
May 13	$508.29	2	(508.29)(2) = 1,016.58
May 15	$458.29	17	(458.29)(17) = 7,790.93
June 01	$594.14	7	(594.14)(7) = 4,158.98
June 08	$631.77	4	(631.77)(4) = 2,527.08
		31	sum = $15,872.07

Average daily balance $= \frac{15872.07}{31} =$ $512

b) Finance charge = prt =
   (512.00)(0.013)(1) = $6.66
c) Balance due = 631.77 + 6.66 =
   $638.43

35. a)

Date	Balance Due	Number of Days	(Balance)(Days)
Feb. 03	$124.78	5	(124.78)(5) = $623.90
Feb. 08	$150.42	4	(150.42)(4) = 601.68
Feb. 12	$ 50.42	2	(50.42)(2) = 100.84
Feb. 14	$117.65	11	(117.65)(11) = 1294.15
Feb. 25	$131.55	6	(131.55)(6) = 789.30
Mar 3		28	sum = $3,409.87

Average daily balance $= \frac{3409.87}{28} =$ $121.78

b) Finance charge = prt =
   (121.78)(0.0125)(1) = $1.52
c) Balance due = 131.55 + 1.52 =
   $133.07
d) The finance charge using the avg.
   daily balance method is $0.04 less
   than the finance charge using the
   unpaid balance method.

37. 0.05477 % per day = 0.0005477   a) (600)(0.0005477)(27) = $8.87   b) 600.00 + 8.87 = $608.87

39. $1000.00   5 %   6 payments
   a) State National Bank (SNB):   (1000)(.05)(.5) = $25.00
   b) Consumers Credit Union (CCU):   (1000)(x)(1) = 35.60   (86.30)(12) = 1035.60
      1035.60 – 1000.00 = $35.60

   c) $\left(\frac{25}{1000}\right)(100) = 2.50$ In Table 11.2, $2.49 is the closest value to $2.50, which corresponds to an
      APR of 8.5 %.

   d) $\left(\frac{35.60}{1000}\right)(100) = 3.56$ In Table 11.2, $3.56 corresponds to an APR of 6.5 %.

41. a) Amount financed = 3450 − 862.50 = $2587.50

Month	Finance charge	Payment	Balance
1	None	$432.00	$2155.50
2	(2155.50)(0.013) = $28.02	460.02	1695.48
3	(1695.48)(0.013) = $22.04	454.04	1241.44
4	(1241.44)(0.013) = $16.14	448.14	793.30
5	(793.30)(0.013) = $10.31	442.31	350.99
6	(350.99)(0.013) = $ 4.56	355.55	0.00

a) It will take 6 months to repay the loan.

b) The total amount of interest paid is $81.07

c) The installment loan saves them
81.07 − 34.50 = $46.57

43. $35,000   15 % down payment   60 month fixed loan   APR = 8.5 %
(35000)(.15) = 5250   35000 − 5250 = 29750

a) From Table 11.2, 60 payments at an APR of 8.5 % yields a finance charge of $23.10 per $100.

$$\left(\frac{29750}{100}\right)(23.10) = \$6872.25$$

b) 29750.00 + 6872.25 = 36622.25 $\dfrac{36622.25}{60} = \$610.37$

c) In Table 11.2, 36 payments at an APR of 8.5 % yields a finance charge of $13.64 per $100.

$$u = \frac{(36)(610.37)(13.64)}{100+13.64} = \frac{299716.08}{113.64} = \$2637.42$$

d) $u = \dfrac{f \bullet k(k+1)}{n(n+1)} = \dfrac{(6872.25)(36)(37)}{60(61)} = \dfrac{9153837}{3660} = \$2501.05$

45. With her billing date on the 25th of the month she can buy the camera during the period of June 26 - June 29 and the purchase will be on the July 25th bill. Since she has a 20-day grace period, she can pay the bill on August 5 without paying interest.

Exercise Set 11.5

1. A mortgage is a long-term loan in which the property is pledged as security for payment of the difference between the down payment and the sale price.

3. The major difference between these two types of loans is that the interest rate for a conventional loan is fixed for the duration of the loan, whereas the interest rate for a variable-rate loan may change every period, as specified in the loan agreement.

5. A buyer's adjusted monthly income is found by subtracting any fixed monthly payment with more than 10 months remaining from the gross monthly income.

7. An amortization schedule is a list of the payment number, interest, principal, and balance remaining on the loan.

9. Equity is the difference between the appraised value of your home and the loan balance.

11. a) Down payment = 15% of $250,000
(0.15)(250000) = $35,700

   b) amt. of mortgage = 250000 − 35700 = 212500
Table 11.4 yields $7.65 per $1000 of mortgage

   Monthly payment = $\left(\dfrac{212000}{1000}\right)(7.65) = \$1625.63$

13. a) Down payment = 10% of $210,000
(0.10)(210000) = $21,000

   b) amt. of mortgage = 210000 − 21000 = 189000
Table 11.4 yields $6.60 per $1000 of mortgage

   Monthly payment = $\left(\dfrac{189000}{1000}\right)(6.60) =$

$1247.40

15. a) Down payment = 20% of $195,000
    (0.20)(195000) = $39,000
    b) amt. of mortgage = 195000 – 39000 = 156000
    c) (156000)(.02) = $3120.00

17. $3,200 = monthly income
    a) (25)(335) = $8,375.00   3200 – 335 = $2865
    b) (2865)(.28) = $802.20
    c) Table 11.4 yields $7.91 per $1000 of mortgage

    $$\left(\frac{150000}{1000}\right)(7.91) = \$1186.50$$

    1186.50 + 225.00 = $1411.50
    d) No; $1411.50 > $802.20

19. a) (490.24)(30)(12) = $176,486.40
    176486.40 + 11250.00 = $187,736.40
    b) 187736.40 – 75000 = $112,736.40
    c) i = prt = (63750)(.085)(1/12) = 451.56
    490.24 – 451.56 = $38.68

21. a) down payment = (0.28)(113500) = $31,780
    b) amount of mortgage = 113500 – 31780 = $81,720
    cost of three points = (0.03)(81720) = $2,451.60
    c) 4750 – 420 = $4330.00 adjusted monthly income
    d) maximum monthly payment =
    (0.28)(4330) = $1,212.40
    e) At a rate of 10% for 20 years, Table 11.4
    yields 9.66.

    mortgage payment = $\left(\frac{81720}{1000}\right)$(9.66) = $789.42

    f) 789.42 + 126.67 = $916.09 total mo. Payment
    g) Since $1,212.40 is greater than $916.09, the
    Yakomo's qualify.
    h) interest on first payment = i = prt =
    (81720)(0.10)(1/12) = $681.00
    amount applied to principal = 789.42 – 681.00 =
    $108.42

23. <u>Bank A</u>   Down payment = (0.10)(105000) =
    $10,500
    amount of mortgage   105000 – 10500 = $94,500
    At a rate of 10% for 30 years, Table 11.4 yields
    $8.70 .
    monthly mortgage payment =

    $\left(\frac{94500}{1000}\right)$(8.70) = $822.15

    cost of three points = (0.03)(94500) = $2835
    Total cost of the house =
    10500 + 2835 + (822.15)(12)(30) = $309,309
    <u>Bank B</u>   Down payment = (0.20)(105000) =
    $21,000
    amount of mortgage   105000 – 21000 = $84,000
    At a rate of 11.5% for 25 years, Table 11.4 yields
    $10.16.
    monthly mortgage payment =

    $\left(\frac{84000}{1000}\right)$(10.16) = $853.44

    cost of the house = 21000 + (853.44)(12)(25) =
    $277,032
    The Nagrockis should select Bank B.

25. a) Amount of mortgage = 105000 - 5000 = $100000   Initial monthly payment = $\left(\frac{100000}{1000}\right)$(8.05) = $805.00

    b)
Payment #	Interest	Principal	Balance
1	$750.00	$55.00	$99,945.00
2	749.59	55.41	99,889.59
3	749.17	55.83	99,833.76

    c) effective interest rate = 6.13% + 3.25% = 9.38%.   The new rate is 9.38%.

25. d)
Payment #	Interest	Principal	Balance
4	$780.37	$24.63	$99,809.13
5	780.17	24.83	99,784.30
6	779.98	25.02	99,759.28

   e) New rate = 6.21% + 3.25% = 9.46%

27. a) $\left(\dfrac{\text{amount of mortgage}}{1000} \right)(8.4) = 950$

     amount of mortgage = $113,095.24

  b) (0.75)(total price) = 113,095.24

     total price = $150,793.65

Review Exercises

1. 3/5 = 0.60   (0.60)(100) = 60%

2. 2/3 ≈ 0.667   (0.667)(100) = 66.7%

3. 5/8 = 0.625   (0.625)(100) = 62.5%

4. 0.041   (0.041)(100) = 4.1%

5. 0.0098   (0.0098)(100) = 0.98% ≈ 1.0%

6. 3.141   (3.141)(100) = 314.1%

7. 3% $\dfrac{3}{100} = .03$

8. 12.1% $\dfrac{12.1}{100} = 0.121$

9. 123% $\dfrac{123}{100} = 1.23$

10. $\dfrac{1}{4}\% = 0.25\%$ $\dfrac{.25}{100} = .0025$

11. $\dfrac{5}{6} = 0.8\overline{3}\%$ $\dfrac{0.8\overline{3}}{100} = 0.008\overline{3}$

12. 0.00045% $\dfrac{0.00045}{100} = 0.0000045$

13. $\dfrac{71,500 - 60,790}{60,790} \approx 0.176 = 17.6\%$

14. $\dfrac{51,300 - 46,200}{46,200} \approx 0.1104 = 11.0\%$

15. (x%)(80) = 25   x% = 25/80 = .3125
    (.3125)(100) = 31.25%
    Twenty-five is 31.25% of 80.

16. 0.16x = 44   x = 44/0.16 = 275
    Forty-four is 16% of 275.

17. (0.17)(540) = x   91.8 = x
    Seventeen percent of 540 is 91.8.

18. Tip = 15% of $42.79 = (0.15)(42.79) = $6.42

19. 0.20(x) = 8   x = 8/0.20 = 40
    The original number was 40 people.

20. $\dfrac{(95 - 75)}{75} = \dfrac{20}{75} = .2\overline{6}$ (.267)(100) = 26.7
    The increase was 26.7%.

21. i = (2500)(.04)(60/360) = $16.67

22. 41.56 = (1575)(r)(100/360) = 41.56
    $41.56 = \left(\dfrac{157500}{360} \right)(r)$ r = 0.095 or 9.5%

23. 114.75 = (p)(0.085)(3)
    114.75 = (p)(0.255) $450 = p

24. 316.25 = (5500)(0.115)(t)
    316.25 = (632.50)(t)   t = 0.5 yrs. or 6 mos.

25. $i = (5300)(.0575)(3) = 914.25$

   Total amount due at maturity $= 5300 + 914.25 =$
   $6214.25

26. a) $i = (3000)(0.081)(240/360) = \$162$

   She paid $3000 + 162 = \$3,162$

27. a) $i = (6000)(0.115)(24/120 = \$1380.00$

   b) amount received: $6000.00 - 1380.00 = 4,620.00$

   c) $i = prt$ $1380 = (4620)(r)(24/12) = 9240r$
   $r = (1380)(9240) = .1494$ $(.1494)(100) = 14.9\%$

28. a) $5\frac{1}{2}\% + 2\% = 7\frac{1}{2}\%$

   b) $i = (800)(0.75)(6/12) = \$30$
   $A = \$800 + \$30 = \$830.00$

   c) $x =$ amount of money in the account
   85% of $x = 800$ $0.85x = 800$ $x = \$941.18$

29. a) $A = 1000\left(1 + \frac{.10}{1}\right)^5 = (1.10)^5 = 1610.51$ $1610.51 - 1000 = \$610.51$

   b) $A = 1000\left(1 + \frac{.10}{2}\right)^{10} = (1.05)^{10} = 1628.89$ $1628.89 - 1000 = \$628.89$

   c) $A = 1000\left(1 + \frac{.10}{4}\right)^{20} = (1.025)^{20} = 1638.62$ $1638.62 - 1000 = \$638.62$

   d) $A = 1000\left(1 + \frac{.10}{12}\right)^{60} = (1.008\overline{3})^{60} = 1645.31$ $1645.31 - 1000 = \$645.31$

   e) $A = 1000\left(1 + \frac{.10}{360}\right)^{1800} = (1.0002\overline{7})^{1800} = 1648.61$ $1648.61 - 1000 = \$648.61$

30. $A = p\left(1 + \frac{r}{n}\right)^{nt}$

   $A = 2500\left(1 + \frac{0.0475}{4}\right)^{4 \cdot 15} = \$5,076.35$

31. Let $p = 1.00$. Then $A = 1\left(1 + \frac{0.56}{360}\right)^{360} = 1.05759$

   $i = 1.05759 - 1.00 = 0.05759$
   The effective annual yield is 5.76% .

32. $p\left(1 + \frac{0.055}{4}\right)^{80} = 40000$ $p = \frac{40000}{(1.01375)^{80}} = 13415.00$ You need to invest $13,415.00

33. 48 mo. $176.14/mo. $7500   24 payments

   a) $(176.14)(48) = 8454.72$ $8454.72 - 7500 = \$954.72$ $\left(\frac{954.72}{7500}\right)(100) = \$12.73 /\$100$

   From Table 11.2, $12.73 indicates an APR of 6.0%

   b) $n = 24$, $p = 176.14$, $v = 6.37$ $u = \frac{(24)(176.14)(6.37)}{100 + 6.37} = \frac{26928.28}{106.37} = \$253.16$

   c) $(176.14)(48) = 8454.72$ $(176.14)(23) = 4051.22$ $8454.72 - 4051.22 = \$4403.50$
   $4403.50 - 253.16 = \$4150.34$

34. a) Amount financed $= \$3,500$ Finance charge $= (163.33)(24) - 3500 = \$419.92$

   $f = 419.92$, $k = 12$, $n = 24$ $u = \frac{(419.92)(12)(13)}{(24)(25)} = \$109.18$

   b) $1959.96 - 109.18 = \$1850.78$ $1850.78 + 163.33 = \$2014.11$

35. 24 mo. $111.73/mo. Down payment = $860 24 payments

  a) $3420 - 860 = \$2560.00$ $(111.73)(24) = 2681.52$ $2681.52 - 2560.00 = \$121.52$

  $\left(\dfrac{121.52}{2560}\right)(100) = \$4.75\,/\$100$

  From Table 11.2, $4.75 indicates an APR of 4.5%

  b) $n = 12, p = 111.73, v = 2.45$ $u = \dfrac{(12)(111.73)(2.45)}{100 + 2.45} = \dfrac{3284.86}{102.45} = \$32.06$

  c) $(111.73)(11) = 1229.03$ $2681.52 - 1229.03 = 1452.49$ $1452.49 - 32.06 = \$1420.43$

36. Balance = $485.75 as of June 01 $i = 1.3\%$

  June 04: $485.75 - 375.00) = \$110.75$ June 08: $110.75 + 370.00 = \$480.75$
  June 21: $480.75 + 175.80 = \$656.55$ June 28: $656.55 + 184.75 = \$841.30$

  a) $(485.75)(.013)(1) = \$6.31$ b) $841.30 + 6.31 = \$847.61$

  c) $(485.75)(3) + (110.75)(4) + (480.75)(13) + (656.55)(7) + (841.30)(3) = \$15269.75$
    $15269.75/30 = \$508.99$

  d) $(508.99)(.013)(1) = \$6.62$ e) $841.30 + 6.62 = \$847.92$

37. a) $i = (185.72)(0.14)(1) = \$2.60$

  b) Aug. 01: $185.72
    Aug. 05: $185.72 + 2.60 = \$188.32$
    Aug. 08: $188.32 + 85.75 = \$274.07$
    Aug. 10: $274.07 - 75.00 = \$199.07$
    Aug. 15: $199.07 + 72.85 = \$271.92$
    Aug. 21: $271.92 + 275.00 = \$546.92$
    As of Sep 5, the new account balance
    is $546.92.

  c)
Date	Balance	# of Days	(Balance)(Days)
Aug. 05	185.72	3	$(185.72)(3) =$ 557.16
Aug. 08	271.47	2	$(271.47)(2) =$ 542.94
Aug. 10	196.47	5	$(196.47)(5) =$ 982.35
Aug. 15	269.32	6	$(269.32)(6) =$ 1615.92
Aug. 21	544.32	15	$(544.32)(15) = 8164.80$
Sep 05		31	sum = $11,863.17

  avg. daily balance = $\dfrac{11,863.17}{31} = \$382.68$

  d) $i = (\$382.68)(0.014)(1) = \$5.36$

  e) $544.32 + 5.36 = \$549.68$ is the amount due on Sep 5.

38. a) $(\$52,000)(0.20) = \$10,400$

  b) $52,000 - 10,400 = \$41,600$

  c) $(930.02)(48) = \$44,640.96$;
    $44,640.96 - 41,600 = \$3040.96$

  d) $\left(\dfrac{\text{finance charge}}{\text{amount financed}}\right)(100)$

    $= \left(\dfrac{3040.96}{41,600}\right)(100) = 7.31$;

    Using Table 11.2, where no. of
    payments is 48, APR = 3.5%

39. a) $275 - 50 = \$225$; $(19.62)(12) = \$235.44$;
    $235.44 - 225 = \$10.44$ interest paid

  b) $\left(\dfrac{\text{finance charge}}{\text{amount financed}}\right)(100)$

    $= \left(\dfrac{10.44}{225}\right)(100) = 4.64$;

    Using Table 11.2, where no. of payments is
    12, $4.64 is closest to $4.66, which yields an
    APR of 8.5%

40. a) down payment = (0.25)(135700) = $33,925
    b) gross monthly income = 64000/12 = $5,333.33
       adjusted monthly income:
       5333.33 − 528.00 = $4,805.33
    c) maximum monthly payment:
       (0.28)(4805.33) = $1,345.49
    d) $\left(\dfrac{101,775}{1000}\right)(8.41) = \$855.93$
    e) total monthly payment:
       855.93 + 316.67 = $1172.60
    f) Yes, $1345.49 is greater than $1172.60.

41. a) down payment = (0.15)(89900) = $13,485
    b) amount of mortgage = 89,900 - 13,485 =
       $76,415
       At 11.5% for 30 years, Table 11.4 yields 9.90.
       monthly mortgage payment:
       $\left(\dfrac{76415}{1000}\right)(9.90) = \$756.51$
    c) i = prt = (76415)(0.115)(1/12) = $732.31
       amount applied to principal:
       756.51 − 732.51 = $24.20
    d) total cost of house: 13485 + (756.51)(12)(30) =
       $285,828.60
    e) total interest paid: 285,828.60 − 89900 =
       $195,928.60

42. a) amount of mortgage: 105,000 − 26,250 = $78,750   First payment = $\left(\dfrac{78,750}{1000}\right)(6.99) = \$550.46$

    b) 5.00% + 3.00% = 8.00%   c) 4.75% + 3.00% = 7.75%

Chapter Test

1. i = (2000)(0.04)(1/2) = $40.00

2. 288 = (1200)(0.08)(t)   288 = 96t   t = 3 years

3. i = prt = (5000)(0.085)(18/12) = $637.50

4. Total amount paid to the bank
   5000 + 637.50 = $5,637.50

5. Partial payment on Sept. 15 (45 days)
   i = (5400)(0.125)(45/360) = $84.375
   $3000.00 - 84.375 = $2,915.625
   5400.00 − 2915.625 = $2484.375

   i = (2484.375)(0.125)(45/360) = $38.82
   2484.38 + 38.82 = $2523.20

6. 84.38 + 38.82 = $123.20

7. $A = 7500 \left(1+\dfrac{0.03}{4}\right)^{8} = \$7961.99$

   interest = 7961.99 − 7500.00 = $461.99

8. $A = 2500 \left(1+\dfrac{0.065}{12}\right)^{36} = \$3036.68$

   interest = 3036.68 − 2500.00 = $536.68

9. (2350)(.15) = 352.50
   2350 − 352.50 = $1997.50

10. (90.79)(24) = 2178.96
    2178.96 − 1997.50 = $181.46

11. $\left(\dfrac{181.46}{1997.50}\right)(100) = \$9.08 \ /\$100$

    In Table 11.2, $9.08 is closest to $9.09 which
    yields an APR of 8.5% .

12. $6750 $1550 down payment   12 mo.
    6750 − 1550 = $5200
    5590.20 − 5200.00 = 390.20
    a) $u = \dfrac{f \bullet k(k+1)}{n(n+1)} = \dfrac{(390.20)(6)(7)}{12(13)} = \$105.05$

    b) (465.85)(5) = $2329.25
       5590.20 − 2329.25 = $3260.95
       3260.95 − 105.05 = $3155.90

13. $7500   36 mo. $223.10 per mo. $(223.10)(36) = 8031.60$ $8031.60 - 7500.00 = 531.60$

a) $\left(\dfrac{181.46}{1997.50}\right)(100) = \$9.08$ In Table 11.2, $9.08 yields an APR of 4,5% .

b) $u = \dfrac{n \cdot P \cdot V}{100 + V} = \dfrac{(12)(223.10)(2.45)}{100 + 2.45} = \dfrac{6559.14}{102.45} = \$64.02$

c) $(223.10)(23) = 5131.30$ $8031.60 - 5131.30 = \$2900.30$ $2900.30 - 64.02 = \$2836.28$

14. Mar. 23: $878.25

Mar. 26: $878.25 + 95.89 = \$974.14$

Mar. 30: $974.14 + 68.76 = \$1042.90$

Apr. 03: $1042.90 - 450.00 = \$592.90$

Apr. 15: $592.90 + 90.52 = \$683.42$

Apr. 22: $683.42 + 450.85 = \$1134.27$

a) $i = (878.25)(.014)(1) = \$12.30$

b) $1134.27 + 12.30 = \$1146.57$

c)

Date	Balance	# of Days	Balance-Days	
Mar. 23	878.25	3	$(878.25)(3) =$	2634.75
Mar. 26	974.14	4	$(974.14)(4) =$	3896.56
Mar. 30	1042.90	4	$(1042.90)(4) =$	4171.60
Apr. 03	592.90	12	$(592.90)(12) =$	7114.80
Apr. 15	683.42	7	$(683.42)(7) =$	4783.94
Apr. 22	1134.27	1	$(1134.27)(1) =$	1134.27
Apr 23		31	sum $=$	$23,735.92

avg. daily balance $= \dfrac{23,735.92}{31} = \$765.67$

d) $(765.67)(.014)(1) = \$10.72$

e) $1134.27 + 10.72 = \$1144.99$

15. down payment $= (0.15)(144500) = \$21,675.00$

16. gross monthly income $= 86500 \div 12 = \$7208.33$

$7,208.33 - 605.00 = \$6,603.33$ adj. mo. income

17. maximum monthly payment $= (0.28)(6603.33) =$ $1,848.93

18. At 10.5% interest for 30 years, Table 11.4 yields $9.15.

amount of loan $= 144500 - 21675 = \$122,825$

monthly payments $= \left(\dfrac{122825}{1000}\right)(9.15) = \$1,123.85$

19. $1123.85 + 304.17 = \$1428.02$ total mo. payment

20. Yes, the bank feels he can afford $1,848.93 per month and his payments would be $1,428.02.

21. a) Total cost of the house:

$21675 + (1123.85)(12)(30) = \$426,261$

b) interest $= 426,261 - 144,500 = \$281,761$

CHAPTER TWELVE

PROBABILITY

Exercise Set 12.1

1. An experiment is a controlled operation that yields a set of results.

3. Empirical probability is the relative frequency of occurrence of an event. It is determined by actual observation of an experiment.

$$P(E) = \frac{\text{number of times the event occurred}}{\text{number of times the experiment was performed}}$$

5. Relative frequency over the long run can accurately be predicted, not individual events or totals.

7. Not necessarily, but it does mean that if a coin was flipped many times, about one-half of the tosses would land heads up.

9. Not necessarily, but it does mean that the average person with traits similar to Mr. Duncan's will live another 43.21 years.

11. AWV 13. AWV

15. Of 30 birds: 14 finches 10 cardinals 6 blue jays
   a) P(f) = 14/30 = 7/15   b) P(c) = 10/30 = 1/3   c) P(bj) = 6/30 = 1/5

17. Of 95 animals: 40 are dogs. 35 are cats 15 are birds 5 are iguanas

   a) P(dog) = 40/95 = 8/19   b) P(cat) = 35/95 = 7/19   c) P(iguana) = 5/95 = 1/19

19. Of 900 people: 19% like bananas 32% like apples 22% like oranges 27% like others
   a) Percents = the relative frequencies of the events occurring.

   b) $P(a) = \frac{32}{100} = 0.32$ c) $P(o) = \frac{22}{100} = 0.22$ d) $P(b) = \frac{19}{100} = 0.19$

21. a) $P(\text{increase}) = \frac{\text{freq. of increases}}{\text{no. of observations}} = \frac{12}{12} = \frac{1}{1} = 1$ b) Yes, the answer in part (a) is only an estimate based on observation.

23. Of 80 votes: 22 for Allison 18 for Emily 20 for Kimberly 14 for Johanna 6 for others
   a) P(A) = 22/80 = 11/40   b) P(E) = 18/80 = 9/40   c) P(K) = 20/80 = 1/4   d) P(J) = 14/80 = 7/40
   e) P(others) = 6/80 = 3/40

25. a) $P(\text{bulls-eye}) = \frac{6}{20} = \frac{3}{10}$

   b) $P(\text{not bulls-eye}) = \frac{14}{20} = \frac{7}{10}$

   c) $P(\text{at least 20 pts.}) = \frac{14}{20} = \frac{7}{10}$

   d) $P(\text{does not score}) = \frac{2}{20} = \frac{1}{10}$

27. a) $P(\text{affecting circular}) = \frac{0}{150} = 0$

   b) $P(\text{affecting elliptical}) = \frac{50}{250} = 0.2$

   c) $P(\text{affecting irregular}) = \frac{100}{100} = 1$

29. a) P(white flowers) = $\dfrac{224}{929} = 0.24$

    b) P(purple flowers) = $\dfrac{705}{929} = 0.76$

31. Answers will vary.

32. Answers will vary.

Exercise Set 12.2

1. If each outcome of an experiment has the same chance of occurring as any other outcome, they are said to be equally likely outcomes.

3. P(A) + P(not A) = 1

5. P(event will not occur) = $1 - 0.3 = 0.7$

7. P(event will occur) = $1 - \dfrac{5}{12} = \dfrac{12}{12} - \dfrac{5}{12} = \dfrac{7}{12}$

9. None of the possible outcomes is the event in question.

11. All probabilities are between 0 and 1.

13. a) P(correct) = 1/5   b) P(correct) = 1/4

15. P(you win) = $\dfrac{\text{one choice}}{50 \text{ possible choices}} = \dfrac{1}{50}$

17. P(7) = $\dfrac{4}{52} = \dfrac{1}{13}$

19. P(not 7) = $\dfrac{48}{52} = \dfrac{12}{13}$

21. P(black) = $\dfrac{13+13}{52} = \dfrac{26}{52} = \dfrac{1}{2}$

23. P(red or black) = $\dfrac{26+26}{52} = \dfrac{52}{52} = \dfrac{1}{1} = 1$

25. P(>4 and <9) = P(5,6,7,8) = $\dfrac{16}{52} = \dfrac{4}{13}$

27. a) P(red) = $\dfrac{2}{4} = \dfrac{1}{2}$ b) P(green) = $\dfrac{1}{4}$

    c) P(yellow) = $\dfrac{1}{4}$ d) P(blue) = 0

29. a) P(red) = $\dfrac{2}{6} = \dfrac{1}{3}$ b) P(green) = $\dfrac{1}{6}$

    c) P(yellow) = $\dfrac{2}{6} = \dfrac{1}{3}$ d) P(blue) = $\dfrac{1}{6}$

Of 100 cans: 30 are cola (c)   40 are orange (o)   10 are ginger ale (ga)   20 are root beer (rb)

31. P(o) = $\dfrac{40}{100} = \dfrac{2}{5}$ 33. P(c, rb, o) = $\dfrac{90}{100} = \dfrac{9}{10}$

35. P(600) = $\dfrac{1}{12}$ 37. P(lose/bankrupt) = $\dfrac{2}{12} = \dfrac{1}{6}$

Of 50 tennis balls: 23 are Wilson (w)   17 are Penn (p)   10 are other (o)

39. P(w) = $\dfrac{23}{50}$ 41. P(not p) = $\dfrac{33}{50}$

For a traffic light: 25 seconds on red (r)   5 seconds on yellow (y)   55 seconds on green (g)

43. P(g) = $\dfrac{55}{85} = \dfrac{11}{17}$ 45. P(not r) = $\dfrac{60}{85} = \dfrac{12}{17}$

Of 11 letters: 1 = m   4 = i   4 = s   2 = p

47. P(s) = $\dfrac{4}{11}$ 49. P(vowel) = $\dfrac{4}{11}$ 51. P(not v) = 1   52. P(w) = 0

53. P(= 60) = $\dfrac{1}{11}$ 55. P(> 50 and < 250) = $\dfrac{4}{11}$

57. $P(15) =$

$\dfrac{1}{26}$

59. $P(\geq 22) =$

$\dfrac{5}{26}$

61. $P(\text{male}) =$

$\dfrac{345}{715} = \dfrac{69}{143}$

63. $P(\text{GM, Ford, C-D}) =$

$\dfrac{533}{715}$

65. $P(\text{female} - \text{other}) =$

$\dfrac{97}{715}$

67. $P(\text{Jif}) =$

$\dfrac{50}{159}$

69. $P(\text{chunky}) =$

$\dfrac{66}{159} = \dfrac{22}{53}$

71. $P(\text{Peter Pan - chunky})$

$= \dfrac{23}{159}$

73. $P(\text{red}) = \dfrac{2}{18} + \dfrac{1}{12} + \dfrac{1}{6} = \dfrac{4}{36} + \dfrac{3}{36} + \dfrac{6}{36} = \dfrac{13}{36}$

75. $P(\text{yellow}) = \dfrac{1}{6} + \dfrac{1}{12} + \dfrac{1}{12} = \dfrac{2}{12} + \dfrac{2}{12} = \dfrac{4}{12} = \dfrac{1}{3}$

77. $P(\text{yellow or green}) = \dfrac{1}{3} + \dfrac{11}{36} = \dfrac{23}{36}$

79. a) $P(CC) = 0$ b) $P(CC) = 1$

81. a) $P(R/R) = \dfrac{2}{4} \cdot \dfrac{2}{4} = \dfrac{4}{16} = \dfrac{1}{4}$

   b) $P(G/G) = \dfrac{2}{4} \cdot \dfrac{2}{4} = \dfrac{4}{16} = \dfrac{1}{4}$

   c) $P(R/G) = \dfrac{2}{4} \cdot \dfrac{2}{4} = \dfrac{4}{16} = \dfrac{1}{4}$

83. $4 \cdot 7 + 1 = 29$

Exercise Set 12.3

1. The odds against an event are found by dividing the probability that the event does not occur by the probability that the event does occur. The probabilities used should be expressed in fractional form.

3. Odds against are more commonly used.

5. $9 : 5$ or 9 to 5

7. a) $P(\text{event occurs}) = \dfrac{1}{1+1} = \dfrac{1}{2}$

   b) $P(\text{event fails to occur}) = \dfrac{1}{1+1} = \dfrac{1}{2}$

9. a) $P(\text{tie goes well}) = \dfrac{8}{27}$

   b) $P(\text{tie does not go well}) = \dfrac{19}{27}$

   c) odds against tie going well =

   $\dfrac{P(\text{tie does not go well})}{P(\text{tie goes well})} = \dfrac{19/27}{8/27} = \dfrac{19}{27} \cdot \dfrac{27}{8} = \dfrac{19}{8}$

   d) odds in favor of it going well are $8:19$.

11. $5 : 1$

13. odds against rolling less than 3 $= \dfrac{P(3 \text{ or greater})}{P(\text{less than 3})} =$

$$\dfrac{4/6}{2/6} = \dfrac{4}{6} \cdot \dfrac{6}{2} = \dfrac{4}{2} = \dfrac{2}{1} \quad \text{or} \quad 2:1$$

15. odds against a queen $=$
$$\dfrac{P(\text{failure to pick a queen})}{P(\text{pick a queen})} =$$

$$\dfrac{48/52}{4/52} = \dfrac{48}{52} \cdot \dfrac{52}{4} = \dfrac{48}{4} = \dfrac{12}{1} \quad \text{or} \quad 12:1$$

Therefore, odds in favor of picking a queen are 1:12.

17. odds against a picture card $=$
$$\dfrac{P(\text{failure to pick a picture})}{P(\text{pick a picture})} = \dfrac{40/52}{12/52} = \dfrac{40}{12} = \dfrac{10}{3}$$

or 10:3

Therefore, odds in favor of picking a picture card are 3:10.

19. odds against red $=$
$$\dfrac{P(\text{not red})}{P(\text{red})} = \dfrac{1/2}{1/2} = \dfrac{1}{2} \cdot \dfrac{2}{1} = \dfrac{2}{2} = \dfrac{1}{1} \quad \text{or} \quad 1:1$$

21. odds against red $= \dfrac{P(\text{not red})}{P(\text{red})} = \dfrac{5/8}{3/8} = \dfrac{5}{8} \cdot \dfrac{8}{3} = \dfrac{5}{3}$

or 5:3

23. a) odds against selecting female $=$
$$\dfrac{P(\text{failure to select female})}{P(\text{select female})} = \dfrac{16/30}{14/30} = \dfrac{16}{14} = \dfrac{8}{7}$$

or 8 : 7 .

b) odds against selecting male $=$
$$\dfrac{P(\text{failure to select male})}{P(\text{select male})} = \dfrac{14/30}{16/30} = \dfrac{14}{16} = \dfrac{7}{8}$$

or 7 : 8 .

25. odds against a stripe $= \dfrac{P(\text{not a stripe})}{P(\text{stripe})} =$

$$\dfrac{8/15}{7/15} = \dfrac{8}{15} \cdot \dfrac{15}{7} = \dfrac{8}{7} \quad \text{or} \quad 8:7$$

27. odds in favor of not the 8 ball are
$$\dfrac{P(\text{not the 8 ball})}{P(\text{the 8 ball})} = \dfrac{14/15}{1/15} = \dfrac{14}{15} \cdot \dfrac{15}{1} = \dfrac{14}{1} \quad \text{or} \quad 14:1$$

29. odds against a ball with 9 or greater are
$$\dfrac{P(\text{less than 9})}{P(\text{9 or greater})} = \dfrac{8/15}{7/15} = \dfrac{8}{15} \cdot \dfrac{15}{7} = \dfrac{8}{7} \quad \text{or} \quad 8:7$$

31. a) $P(> \$5 \text{ M}) = \dfrac{5}{9}$

b) Odds against payout $> \$5$ M $\qquad$ 4 : 5

33. The odds against testing negative $=$
$$\dfrac{P(\text{test positive})}{P(\text{test negative})} = \dfrac{4/76}{72/76} = \dfrac{4}{72} = \dfrac{1}{18} \quad \text{or} \quad 1 : 18$$

35. a) $P(\text{June wins}) = \dfrac{8}{8+5} = \dfrac{8}{13}$

b) $P(\text{June loses}) = \dfrac{5}{8+5} = \dfrac{5}{13}$

37. Odds against $\quad 4 : 11 \quad P(\text{promoted}) = \dfrac{11}{4+11} = \dfrac{11}{15}$

39. $P(G) = \dfrac{15}{75} = \dfrac{1}{5}$

41. Odds in favor of G $= \dfrac{P(G)}{P(\text{not G})} = \dfrac{1/5}{4/5} = \dfrac{1}{4} \quad \text{or} \quad 1:4$

43. Odds against B9 $=$
$$\dfrac{P(\text{not B9})}{P(\text{B9})} = \dfrac{74/75}{1/75} = \left(\dfrac{74}{75}\right)\left(\dfrac{75}{1}\right) = \dfrac{74}{1} \quad \text{or} \quad 74:1$$

45. $P(A+) = \dfrac{34}{100} = 0.34$

47. $\dfrac{66}{34} = \dfrac{33}{17} \quad \text{or} \quad 33 : 17$

49. $P(O \text{ or } O\text{-}) = \dfrac{43}{100} = \dfrac{43}{43+57}$ or $43:57$

51. If $P(\text{selling out}) = 0.9 = \dfrac{9}{10}$, then

$P(\text{do not sell your car this week}) = 1 - \dfrac{9}{10} = \dfrac{1}{10}$.

The odds against selling out $= \dfrac{1/10}{9/10} = \dfrac{1}{9}$ or 1:9.

53. If $P(\text{all parts are present}) = \dfrac{7}{8}$, then the odds in favor of all parts being present are $7:1$.

55. a) $P(\text{Douglas is a male}) = \dfrac{20}{21}$

    b) Odds against being a female are $20:1$.

57. $P(\#1 \text{ wins}) = \dfrac{2}{9}$ $P(\#2 \text{ wins}) = \dfrac{1}{3}$

     $P(\#3 \text{ wins}) = \dfrac{1}{16}$ $P(\#4 \text{ wins}) = \dfrac{5}{12}$

     $P(\#5 \text{ wins}) = \dfrac{1}{2}$

59. $119648 + 6742 + 506 + 77 = 126{,}973$ multiple births

     $.03x = 126973$ $x = 4{,}232{,}433$ total births

     Odds against a multiple birth

     $\dfrac{(4232433 - 126973)}{126973} = \dfrac{4105460}{126973} = \dfrac{97}{3}$ or $97:3$.

Exercise Set 12.4

1. Expected value is used to determine the average gain or loss of an experiment over the long run.

3. The fair price is the amount charged for the game to be fair and result in an expected value of 0.

5. To obtain fair price, add the cost to play to the expected value.

7. $0.50. Since you would lose $1.00 on average for each game you played, the fair price of the game should be $1.00 less. Then the expected value would be 0, and the game would be fair.

9. $3(-\$0.40) = -\$1.20$

11. $E = P_1A_1 + P_2A_2 = 0.70(200) + 0.30(120) = 140 + 36 = 176$ people

13. $E = P_1A_1 + P_2A_2 = 0.50(78) + 0.50(62) = 39 + 31 = 70$ points

15. $E = P_1A_1 + P_2A_2 = 0.40(1.2\,M) + 0.60(1.6\,M) = .48\,M + .96\,M = 1.44\,M$ viewers

17. a) $E = P_1A_1 + P_2A_2 = (.60)(10000) + (.10)(0) + (.30)(-7200) = 6000 + 0 + -2160 = \$3840$

19. $E = P_1A_1 + P_2A_2 + P_3A_3 = P(\$1 \text{ off})(\$1) + P(\$2 \text{ off})(\$2) + P(\$5 \text{ off})(\$5)$
     $E = (1/10)(1) + (2/10)(2) + (1/10)(5) = 7/10 + 4/10 + 5/10 = 16/10 = \$1.60$

21. a) $(2/6)(8) + (4/6)(-5) = 8/3 - 20/6 = 16/6 - 20/6 =$
        $-4/6 = -\$.67$

    b) $(2/6)(-8) + (4/6)(5) = -8/3 + 20/6 = 16/6 + 20/6$
        $= \$.67$

23. a) $(1/5)(5) + (0)(0) + (4/5)(-1) = 1 - 4/5 = 1/5$
        Yes, positive expectations $= 1/5$

    b) $(1/4)(5) + (0)(0) + (3/4)(-1) = 5/4 - 3/4 = 1/2$
        Yes, positive expectations $= 1/2$

25. a) $\left(\dfrac{1}{500}\right)(398) + \left(\dfrac{499}{500}\right)(-2) = \dfrac{398 - 998}{500} =$

     $\dfrac{-600}{500} = \dfrac{-300}{250} = -\$1.20$

    b) Fair price $= -1.20 + 2.00 = \$.80$

27. a) $\left(\dfrac{1}{2000}\right)(997) + \left(\dfrac{2}{2000}\right)(497) + \left(\dfrac{1997}{2000}\right)(-3) = \dfrac{997 + 994 - 5991}{2000} = \dfrac{-4000}{2000} = -\$2.00$

    b) Fair price $= -2.00 + 3.00 = \$1.00$

29. $\frac{1}{2}(1) + \frac{1}{2}(10) = \frac{1}{2} + 5 = 5.5 = \$5.50$

31. $\frac{1}{2}(10) + \frac{1}{4}(-5) + \frac{1}{4}(-20) = 5 - 1.25 - 5 = -\$1.25$

33. a) $\frac{1}{2}(-1) + \frac{1}{2}(3) = -.50 + 1.50 = \$1.00$

   b) Fair price = 1.00 + 2.00 = \$3.00

35. a) $\frac{1}{2}(-1) + \frac{1}{4}(3) + \frac{1}{4}(8) = -.50 + .75 + 2.00 = \$2.25$

   b) Fair price = 2.25 + 2.00 = \$4.25

37. $E = P_1A_1 + P_2A_2 + P_3A_3 + P_4A_4 + P_5A_5 =$
   $0.17(1) + 0.10(2) + 0.02(3) + 0.08(4) + 0.63(0) =$
   0.75 base

39. $E = P_1A_1 + P_2A_2 + P_3A_3$

   $= \frac{3}{10}(4) + \frac{5}{10}(3) + \frac{2}{10}(1) = 1.2 + 1.5 + 0.2$

   $= 2.9$ points

41. $(0.34)(850) + (0.66)(140) = 289 + 92.4 =$
   381.4 employees

43. $(.11)(10) + (.65)(15) + (.24)(20) = 1.1 + 9.75 + 4.8$
   $= 15.65$ minutes

45. $E = P(1)(1) + P(2)(2) + P(3)(3) + P(4)(4) + P(5)(5)$
   $+ P(6)(6)$

   $= \frac{1}{6}(1) + \frac{1}{6}(2) + \frac{1}{6}(3) + \frac{1}{6}(4) + \frac{1}{6}(5) + \frac{1}{6}(6)$

   $= \frac{21}{6} = 3.5$ points

47. $E = P_1A_1 + P_2A_2 + P_3A_3$

   $= \frac{200}{365}(110) + \frac{100}{365}(160) + \frac{65}{365}(210)$

   $= 60.27 + 43.84 + 37.40 = 141.51$ calls/day

49. a) $P(1) = \frac{1}{2} + \frac{1}{16} = \frac{8}{16} + \frac{1}{16} = \frac{9}{16}$, $P(10) = \frac{1}{4} = \frac{4}{16}$,

   $P(\$20) = \frac{1}{8} = \frac{2}{16}$, $P(\$100) = \frac{1}{16}$

   b) $E = P_1A_1 + P_2A_2 + P_3A_3 + P_4A_4$

   $= \frac{9}{16}(\$1) + \frac{4}{16}(\$10) + \frac{2}{16}(\$20) + \frac{1}{16}(\$100)$

   $= \frac{9}{16} + \frac{40}{16} + \frac{40}{16} + \frac{100}{16} = \frac{189}{16} = \$11.81$

   c) fair price = expected value – cost to play =
   \$11.81 – 0 = \$11.81

51. $E = P(\text{insured lives})(\text{cost}) + P(\text{insured dies})(\text{cost} - \$40,000)$
   $= 0.97(\text{cost}) + 0.03(\text{cost} - 40,000)$
   $= 0.97(\text{cost}) + 0.03(\text{cost}) - 1200$
   $= 1.00(\text{cost}) - 1200$

   Thus, in order for the company to make a profit,
   the cost must exceed \$1,200

53. $E = P(\text{win})(\text{amount won}) + P(\text{lose})(\text{amount lost})$

   $= \left(\frac{1}{38}\right)(35) + \left(\frac{37}{38}\right)(-1) = \frac{35}{38} - \frac{37}{38} = -\frac{2}{38}$

   $= -\$0.053$

55. a) $E = \frac{1}{12}(100) + \frac{1}{12}(200) + \frac{1}{12}(300) + \frac{1}{12}(400) + \frac{1}{12}(500) + \frac{1}{12}(600) + \frac{1}{12}(700) + \frac{1}{12}(800) + \frac{1}{12}(900)$

   $\frac{1}{12}(1000) = \left(\frac{5500}{12}\right) = \$458.3\overline{3}$

   b) $E = \frac{1}{12}(5500) + \frac{1}{12}(-1800) = \frac{3700}{12} = \$308.33$

Exercise Set 12.5

1. If a first experiment can be performed in M distinct way and a second experiment can be performed in N distinct ways, then the two experiments in that specific order can be performed in M · N distinct ways.

3. (2)(7) = 14 ways. Using the counting principle.

5. The first selection is made. Then the second selection is made before the first selection is returned to the group of items being selected from.

7. a) (50)(50) = 2500   b) (50)(49) = 2450

9. a) (6)(6)(6) = 216   b) (6)(5)(4) = 120

11. a) (2)(2) = 4 points

b)

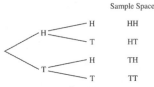

c) P(no heads) = 1/4

d) P(exactly one head) = 2/4 = 1/2

e) P(two heads) = 1/4

13. a) (3)(3) = 9 points

b)

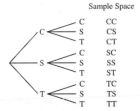

c) P(two circles) = 1/9

d) P(square and then triangle) = 1/9

e) P(at least one circle) = 5/9

15. a) (4)(3) = 12 points

b)

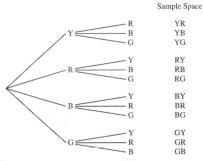

c) P(exactly one red) = 6/12 = ½

d) P(at least one is not red) = 12/12 = 1

e) P(no green) = 6/12 = 1/2

17. a) (2)(2)(2) = 8 points

b

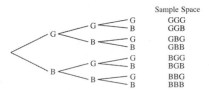

c) P(no boys) = 1/8

d) P(at least one girl) = 7/8

e) P(either exactly 2 boys or 2 girls) = 6/8 = ¾

f) P(boy 1st and boy 2nd and girl 3rd) = 1/8

19. a) (6)(6) = 36 points

b)

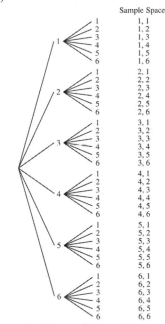

c) P(double) = 6/36 = 1/6

d) P(sum of 7) = 6/36 = 1/6

e) P(sum of 2) = 1/36

; the P(sum of 2) < P(sum of 7)

21. a) $(3)(2)(1) = 6$ points

b)

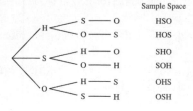

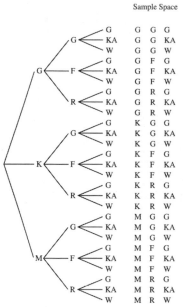

c) P(Sears – 1st) = 2/6 = 1/3

d) P(Home Depot – 1st / Outback - last) = 1/6

e) P(Sears,Outback,Home Depot) = 1/6

23. a) $(4)(4) = 16$ points

b)

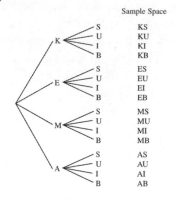

c) P(M.K. or E.C.) = 8/16 = ½

d) P(MGM or Univ.) = 8/16 = ½

e) P(M.K. and (S.W. or B.G.)) = 2/16 = 1/8

25. a) $(3)(3)(3) = 27$

b)

c) P(GE,GE,GE) = 1/27

d) P(not GE) = 8/27

e) P(at least 1 GE) = 19/27

27. a) $(2)(4)(3) = 24$ sample points

b)

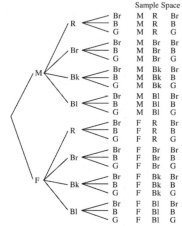

c) P(M, black, blue) = 1/24

d) P(F, blonde) = 3/24 = 1/8

29. a) P(white) = 1/3   b) P(red) = 2/3

c) No; P(white) < P(red)

d)

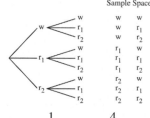

$$P(w, w) = \frac{1}{9}; P(r, r) = \frac{4}{9}$$

31. 1 red, 1 blue, and 1 brown   32. $60/(2 \cdot 6) = 5$; 5 faces

Exercise Set 12.6

1. a) A occurs, B occurs, or both occur   b) "and" means both events, A and B, must occur.

3. a) Two events are mutually exclusive if it is impossible for both events to occur simultaneously.

   b) $P(A \text{ or } B) = P(A) + P(B)$

5. We assume that event A has already occurred.

7. Two events are dependent if the occurrence of either event affects the probability of occurrence of the other event. Ex. Select two cards from a deck (without replacement); find P (King and King).

9. a) No.   b) Yes.

11. If the events are mutually exclusive, the events cannot happen simultaneously and thus $P(A \text{ and } B)=0$.

13. $P(A \text{ and } B) = 0.3$
$$P(A \text{ or } B) = P(A) + P(B) - P(A \text{ and } B)$$
$$= 0.6 + 0.4 - (0.6)(0.4)$$
$$= 1.0 - 0.3 = 0.7$$

15. $P(B) = P(A \text{ or } B) + P(A \text{ and } B) - P(A)$
$$= 0.8 + 0.1 - 0.4 = 0.5$$

17. $P(2 \text{ or } 5) = 1/6 + 1/6 = 2/6 = 1/3$

19. $P(\text{greater than 4 or less than 2}) = P(5, 6. \text{ or } 1) = 2/6 + 1/6 = 3/6 = 1/2$

21. Since these events are mutually exclusive,
$$P(\text{ace or } 2) = P(\text{ace}) + P(2) =$$
$$= \frac{4}{52} + \frac{4}{52} = \frac{8}{52} = \frac{2}{13}$$

23. Since it is possible to obtain a card that is a picture card and a red card, these events are not mutually exclusive.
$P(\text{picture or red})$
$$= P(\text{pict.}) + P(\text{red}) - P(\text{pict. \& red})$$
$$= \frac{12}{52} + \frac{26}{52} - \frac{6}{52} = \frac{32}{52} = \frac{8}{13}$$

25. Since it is possible to obtain a card less than 7 that is a club, these events are not mutually exclusive.
$$P(< 7 \text{ or club}) = \frac{24}{52} + \frac{13}{52} - \frac{6}{52} = \frac{31}{52}$$

27. a) P(frog and frog)= $\dfrac{5}{20}\cdot\dfrac{5}{20}=\dfrac{1}{4}\cdot\dfrac{1}{4}=\dfrac{1}{16}$

    b) P(frog and frog)= $\dfrac{5}{20}\cdot\dfrac{4}{19}=\dfrac{1}{4}\cdot\dfrac{4}{19}=\dfrac{1}{19}$

29. a) P(lion and bird) = $\dfrac{5}{20}\cdot\dfrac{5}{20}=\dfrac{1}{4}\cdot\dfrac{1}{4}=\dfrac{1}{16}$

    b) P(lion and bird) = $\dfrac{5}{20}\cdot\dfrac{5}{19}=\dfrac{1}{4}\cdot\dfrac{5}{19}=\dfrac{5}{76}$

31. a) P(red bird and monkey) =
$\dfrac{3}{20}\cdot\dfrac{5}{20}=\dfrac{3}{20}\cdot\dfrac{1}{4}=\dfrac{3}{80}$

    b) P(red bird and monkey) =
$\dfrac{3}{20}\cdot\dfrac{5}{19}=\dfrac{15}{380}=\dfrac{3}{76}$

33. a) P(odd and odd) = $\dfrac{12}{20}\cdot\dfrac{12}{20}=\dfrac{3}{5}\cdot\dfrac{3}{5}=\dfrac{9}{25}$

    b) P(odd and odd) = $\dfrac{12}{20}\cdot\dfrac{11}{19}=\dfrac{3}{5}\cdot\dfrac{11}{19}=\dfrac{33}{95}$

35. P(monkey or even) = $\dfrac{5}{20}+\dfrac{8}{20}-\dfrac{2}{20}=\dfrac{11}{20}$

37. P(lion or a 2) = $\dfrac{5}{20}+\dfrac{4}{20}-\dfrac{1}{20}=\dfrac{8}{20}=\dfrac{2}{5}$

39. P(2 reds) = $\dfrac{1}{2}\bullet\dfrac{1}{2}=\dfrac{1}{4}$

41. P(red and green) = $\dfrac{1}{4}\bullet\dfrac{1}{2}=\dfrac{1}{8}$

43. P(2 yellows) = P(red and red) = $\dfrac{3}{8}\bullet\dfrac{3}{8}=\dfrac{9}{64}$

45. P(2 reds) = $\dfrac{1}{2}\bullet\dfrac{1}{4}=\dfrac{1}{8}$

47. P(both not yellow) = $\dfrac{1}{2}\bullet\dfrac{3}{4}=\dfrac{3}{8}$

49. P(3 girls) = P(1st girl) • P(2nd girl) • P(3rd girl)
$=\dfrac{1}{2}\bullet\dfrac{1}{2}\bullet\dfrac{1}{2}=\dfrac{1}{8}$

51. P(G,G,B) = P(1st girl) • P(2nd girl) • P(3rd boy)
$=\dfrac{1}{2}\bullet\dfrac{1}{2}\bullet\dfrac{1}{2}=\dfrac{1}{8}$

53. a) P(5 boys) = P(b) • P(b) • P(b) • P(b) • P(b)
$=\dfrac{1}{2}\bullet\dfrac{1}{2}\bullet\dfrac{1}{2}\bullet\dfrac{1}{2}\bullet\dfrac{1}{2}=\dfrac{1}{32}$

    b) P(next child is a boy) = $\dfrac{1}{2}$

55. a) P(Titleist/Pinnacle) = $\dfrac{4}{7}\bullet\dfrac{1}{7}=\dfrac{4}{49}$

    b) P(Titleist/Pinnacle) = $\dfrac{4}{7}\bullet\dfrac{1}{6}=\dfrac{4}{42}=\dfrac{2}{21}$

57. a) P(at least 1 Top Flite) =
$\dfrac{2}{7}\bullet\dfrac{5}{7}+\dfrac{5}{7}\bullet\dfrac{2}{7}+\dfrac{2}{7}\bullet\dfrac{2}{7}=\dfrac{24}{49}$

    b) P(at least 1 Top Flite)=
$\dfrac{2}{7}\bullet\dfrac{5}{6}+\dfrac{5}{7}\bullet\dfrac{2}{6}+\dfrac{2}{7}\bullet\dfrac{1}{6}=\dfrac{11}{21}$

59. P(neither had trad. ins.) = $\dfrac{26}{40}\bullet\dfrac{25}{39}=\dfrac{10}{24}=\dfrac{5}{12}$

61. P(at least one trad.) = 1 – P(neither trad.) =
$1-\dfrac{5}{12}=\dfrac{7}{12}$ (see Exercise 59)

63. P(all recommended) = $\dfrac{19}{30}\bullet\dfrac{18}{29}\bullet\dfrac{17}{28}=\dfrac{969}{4060}$

65. P(no/no/not sure) = $\dfrac{6}{30}\bullet\dfrac{5}{29}\bullet\dfrac{5}{28}=\dfrac{5}{812}$

67. The probability that any individual reacts favorably is 70/100 or 0.7.
P(Mrs. Rivera reacts favorably) = 0.7

69. P(all 3 react favorably) = $0.7 \bullet 0.7 \bullet 0.7 = 0.343$

71. Since each question has four possible answers of which only one is correct, the probability of guessing correctly on any given question is 1/4.
P(correct answer on any one question) = ¼

73. P(only the 3rd and 4th questions correct) =
$$\left(\frac{3}{4}\right)\left(\frac{3}{4}\right)\left(\frac{1}{4}\right)\left(\frac{1}{4}\right)\left(\frac{3}{4}\right) = \frac{27}{1024}$$

75. P(none of the 5 questions correct) =
$$\left(\frac{3}{4}\right)\left(\frac{3}{4}\right)\left(\frac{3}{4}\right)\left(\frac{3}{4}\right)\left(\frac{3}{4}\right) = \frac{243}{1024}$$

77. P(bell on 1st reel) = 3/22

79. P(no bar/no bar/no bar)
$$= \left(\frac{20}{22}\right)\left(\frac{20}{22}\right)\left(\frac{21}{22}\right) = \frac{1050}{1331}$$

81. P(yellow/yellow) = $\left(\frac{1}{8}\right)\left(\frac{2}{12}\right) = \frac{2}{96} = \frac{1}{48}$

83. P(not red on outer and not red on inner) =
$$\frac{8}{12} \cdot \frac{5}{8} = \frac{5}{12}$$

85. P(no hit/no hit) = $(0.6)(0.6) = 0.36$

87. P(both hit) = $(0.4)(0.9) = 0.36$

89. a) No; The probability of the 2nd depends on the outcome of the first.
b) P(one afflicted) = .001
c) P(both afflicted) = $(.001)(.04) = .00004$
d) P(not afflicted/afflicted) = $(.999)(.001) = .000999$
e) P(afflicted/not afflicted) = $(0.001)(.96) = (.00096)$
f) P(not affl/not affl) = $(.999)(.999) = .99801$

91. P(audit this year) = .032

93. P(audit/no audit) = $(.032)(.968) = .030976$

95. P(2 - same color) = P(2 r) + P(2 b) + P(2 y)
$$= \left(\frac{5}{10}\right)\left(\frac{4}{9}\right) + \left(\frac{3}{10}\right)\left(\frac{2}{9}\right) + \left(\frac{2}{10}\right)\left(\frac{1}{9}\right)$$
$$= \left(\frac{20}{90}\right) + \left(\frac{6}{90}\right) + \left(\frac{2}{90}\right) = \frac{28}{90} = \frac{14}{45}$$

96. P(at least 1 yen) =
$$= \left(\frac{3}{10}\right)\left(\frac{7}{9}\right) + \left(\frac{7}{10}\right)\left(\frac{3}{9}\right) + \left(\frac{3}{10}\right)\left(\frac{2}{9}\right)$$
$$= \left(\frac{21}{90}\right) + \left(\frac{21}{90}\right) + \left(\frac{6}{90}\right) = \frac{48}{90} = \frac{8}{15}$$

97. P(no diamonds) = $\left(\frac{39}{52}\right)\left(\frac{38}{51}\right) = \frac{1482}{2652} = .56$

The game favors the dealer since the probability of no diamonds is greater than 1/2.

98. The other card could be the ace or the queen and it is equally likely that it is either one. Thus, the probability the card is the queen is 1/2.

99. P(2/2) = (2/6)(2/6) = 4/36 = 1/9

101. P(even or < 3) = 2/6 + 3/6 − 2/6 = 3/6 = 1/2

Exercise Set 12.7

1. The probability of E_2 given that E_1 has occurred.

3. $P(E_2 \mid E_1) = \dfrac{n(E_1 \text{ and } E_2)}{n(E_1)} = \dfrac{4}{12} = \dfrac{1}{3}$

5. $P(5 \mid \text{orange}) = 1/3$

7. $P(\text{even} \mid \text{not orange}) = 2/3$

9. $P(\text{red} \mid \text{orange}) = 2/3$

11. $P(\text{circle} \mid \text{odd}) = 3/4$

13. $P(\text{red} \mid \text{even}) = 2/3$

15. $P(\text{circle or square} \mid < 4) = 2/3$

17. $P(5 \mid \text{red}) = 1/3$

18. $P(\text{even} \mid \text{red}) = 1/3$

19. $P(\text{purple} \mid \text{odd}) = 2/6 = 1/3$

21. $P(> 4 \mid \text{purple}) = 3/5$

23. $P(\text{gold} \mid > 5) = 1/7$

25. $P(1 \text{ and } 1) = (1/4)(1/4) = 1/16$

27. $P(5 \mid \text{at least a } 5) = 1/7$

29. $P(\text{sum} = 6) = 5/36$

31. $P(6 \mid 3) = 1/6$

33. $P(> 7 \mid 2^{\text{nd}} \text{ die} = 5) = 4/6 = 2/3$

35. $P(\text{Pepsi}) = 107/217$

37. $P(\text{Coke} \mid \text{woman}) = 50/112 = 25/56$

39. $P(\text{man} \mid \text{prefers Coke}) = 60/110 = 12/22 = 6/11$

41. $P(\text{girl}) = 160/360 = 4/9$

43. $P(\text{elephant} \mid \text{boy}) = 110/200 = 11/20$

45. $P(\text{boy} \mid \text{elephant}) = 110/195 = 22/39$

47. $P(\text{only tapes}) = 133/300$

49. $P(\text{DVD} \mid < 30) = 60/120 = 1/2$

51. $P(< 30 \mid \text{both VTs and DVDs}) = 21/43$

53. $P(\text{Air Force}) = 8833/27630 = 0.3197$

55. $P(\text{acquitted} \mid \text{Army}) = 434/5458 = 0.0795$

57. $P(\text{Army} \mid \text{convicted}) = 5024/26056 = 0.1928$

59. $P(\text{good}) = \dfrac{300}{330} = \dfrac{10}{11}$

61. $P(\text{defective} \mid 20 \text{ watts}) = \dfrac{15}{95} = \dfrac{3}{19}$

63. $P(\text{good} \mid 50 \text{ or } 100 \text{ watts}) = \dfrac{220}{235} = \dfrac{44}{47}$

65. $P(\text{ABC or NBC}) = \dfrac{110}{270} = \dfrac{11}{27}$

67. $P(\text{ABC or NBC} \mid \text{man}) = \dfrac{50}{145} = \dfrac{10}{29}$

69. $P(\text{ABC,NBC,or CBS} \mid \text{man}) = \dfrac{55}{145} = \dfrac{11}{29}$

71. $P(\text{large company stock}) = 93/200$

73. $P(\text{blend} \mid \text{medium co. stock}) = 15/52$

75. a) $n(A) = 140$ b) $n(B) = 120$
   c) $P(A) = 140/200 = 7/10$
   d) $P(B) = 120/200 = 6/10 = 3/5$
   e) $P(A \mid B) = \dfrac{n(B \text{ and } A)}{n(B)} = \dfrac{80}{120} = \dfrac{2}{3}$
   f) $P(B \mid A) = \dfrac{n(A \text{ and } B)}{n(B)} = \dfrac{80}{140} = \dfrac{4}{7}$
   g) $P(A) \bullet P(B) = \left(\dfrac{7}{10}\right)\left(\dfrac{3}{5}\right) = \dfrac{21}{50}$

   $P(A \mid B) \quad P(A) \bullet P(B) \qquad \dfrac{2}{3} \neq \dfrac{21}{50}$

   A and B are not independent events.

77. a) $P(A \mid B) = \dfrac{n(B \text{ and } A)}{n(B)} = \dfrac{0.12}{0.4} = 0.3$

   b) $P(B \mid A) = \dfrac{n(A \text{ and } B)}{n(A)} = \dfrac{0.12}{0.3} = 0.4$

   c) Yes, $P(A) = P(A \mid B)$ and $P(B) = P(B \mid A)$.

78. $P(\text{green circle} \mid +) = 1/3$

79. $P(+ \mid \text{orange circle}) = 1/2$

80. $P(\text{yellow circle} \mid -) = 1/3$

81. $P(\text{green} + \mid +) = 1/3$

82. $P(\text{green or orange circle} \mid \text{green} +) = 1$

83. $P(\text{orange circle w/green} + \mid +) = 0$

Exercise Set 12.8

1. Answers will vary.

3. $n! = n(n-1)(n-2) \cdots 3 \cdot 2 \cdot 1$

5. The number of permutations of n items taken r at a time.

7. $_nP_r = \dfrac{n!}{(n-r)!}$

9. $6! = 720$

11. $_6P_2 = \dfrac{6!}{4!} = 6 \cdot 5 = 30$

13. $0! = 1$

15. $_8P_0 = \dfrac{8!}{8!} = 1$

17. $_9P_4 = \dfrac{9!}{5!} = 9 \cdot 8 \cdot 7 \cdot 6 = 3024$

19. $_8P_5 = \dfrac{8!}{3!} = 8 \cdot 7 \cdot 6 \cdot 5 \cdot 4 = 6720$

21. $(10)(10)(10)(10) = 10000$

23. a) $(26)(25)(24)(10)(9)(8) = 11,232,000$
    b) $(26)(26)(26)(10)(10)(10) = 17,576,000$

25. a) $5^5 = 3125$

    b) $\dfrac{1}{3125} = 0.00032$

27. $(34)(36)(36)(36)(36) = 57,106,944$

29. $8 \bullet 10 \bullet 9 = 720$ systems

31. a) $6! = 720$ b) $5! = 120$
    c) $4! = 24$ d) $5! \bullet 5 = 600$

33. $_{10}P_3 = \dfrac{10!}{(10-3)!} = \dfrac{10!}{7!} = \dfrac{10 \cdot 9 \cdot 8 \cdot 7!}{7!} = 720$

35. a) There are 12 individuals and they can be arranged in $12! = 479,001,600$ ways
    b) $10! = 3,628,800$ different ways
    c) $5! \cdot 5! = 14,400$ different ways

37. $(26)(25)(10)(9)(8)(7) = 650 \bullet 5040 = 3,276,000$

39. $(26)(10)(9)(8)(7) = 131,040$

41. $(10)(10)(10)(26)(26) = 676,000$

43. $(5)(4)(8)(26)(25) = 104,000$

45. a) $(8)(10)(10)(10)(10)(10)(10) = 8,000,000$
    b) $(8)(10)(10)(8,000,000) = 6,400,000,000$
    c) $(8)(10)(10)(8)\left(10^{10}\right) = (64)\left(10^{12}\right)$
    $= 64,000,000,000,000$

47. $_{15}P_6 = \dfrac{15!}{9!} = \dfrac{(15)(14)(13) \cdot (12) \cdot (11)(10)(9!)}{9!}$
    $= 3,603,800$

49. $_7P_7 = \dfrac{7!}{0!} = \dfrac{7!}{1} = 7! = 5,040$

51. $(5)(4)(7)(2) = 280$ systems

53. $_9P_9 = \dfrac{9!}{0!} = 9! = 362,880$

55. $\dfrac{12!}{4!3!2!} = \dfrac{479001600}{(24)(6)(2)} = 1,663,200$

57. $\dfrac{7!}{2!2!2!} = \dfrac{(7)(6)(5)(4)(3)(2)(1)}{(2)(1)(2)(1)(2)(1)} = 630$

59. The order of the flags is important. Thus, it is a permutation problem.
    $$_8P_5 = \dfrac{8!}{(8-5)!} = \dfrac{8!}{3!} = \dfrac{40320}{6} = 6,720$$

61. a) Since the pitcher must bat last, there is only one possibility for the last position. $- - - - - - - - \underline{1}$
    There are 8 possible batters left for the 1st position. Once the 1st batter has been selected, there are
    7 batters left for the 2nd position, 6 for the third, etc. $\underline{(8)}\ \underline{(7)}\ \underline{(6)}\ \underline{(5)}\ \underline{(4)}\ \underline{(3)}\ \underline{(2)}\ \underline{(1)}\ \underline{(1)} = 40{,}320$

    b) $9! = (9)(8)(7)(6)(5)(4)(3)(2)(1) = 362{,}880$

63. a) $5^5 = 3125$ different keys

    b) $400{,}000 \div 3{,}125 = 128$ cars

    c) $\dfrac{1}{3125} = 0.00032$

65. $_7P_5 = \dfrac{7!}{2!} = \dfrac{7 \cdot 6 \cdot 5 \cdot 4 \cdot 3 \cdot 2!}{2!} = 2{,}520$ different

    letter permutations; $2500 \times \dfrac{1}{12} = 210$ minutes

    or $3\dfrac{1}{2}$ hours

67. No, Ex. $_3P_2 \neq {_3P_{(3-2)}}$

    $\dfrac{3!}{1!} \neq \dfrac{3!}{2!}$ because $6 \neq 3$

68. A ○ ○ ○ ○ ○ ○ B

    $(8)(7) = 56$ tickets

69. $(25)(24) = 600$ tickets

Exercise Set 12.9

1. The selection of a certain number of items without regard to their order.

3. $_nC_r = \dfrac{n!}{(n-r)!\,r!}$

5. If the order of the items is important then it is a permutation problem. If order is not important then it is a combination problem.

7. $_5C_3 = \dfrac{5!}{(5-3)!\,3!} = \dfrac{(5)(4)(3)(2)(1)}{(2)(1)(3)(2)(1)} = 10$

9. a) $_6C_4 = \dfrac{6!}{4!\,2!} = \dfrac{(6)(5)}{(2)(1)} = 15$

   b) $_6P_4 = \dfrac{6!}{(6-4)!} = \dfrac{6!}{2!} = (6)(5)(4)(3) = 360$

11. a) $_8C_0 = \dfrac{8!}{8!\,0!} = 1$

    b) $_8P_0 = \dfrac{8!}{(8-0)!} = \dfrac{8!}{8!} = 1$

13. a) $_{10}C_3 = \dfrac{10!}{7!\,3!} = \dfrac{(10)(9)(8)(7!)}{(7!)(3)(2)(1)} = 120$

    b) $_{10}P_3 = \dfrac{10!}{(10-3)!} = \dfrac{(10)(9)(8)(7!)}{7!} = 720$

15. $\dfrac{_5C_3}{_5P_3} = \dfrac{\frac{5!}{2!3!}}{\frac{5!}{2!}} = \left(\dfrac{5!}{2!3!}\right)\left(\dfrac{2!}{5!}\right) = \dfrac{1}{3!} = \dfrac{1}{6}$

17. $\dfrac{_8C_5}{_8C_2} = \dfrac{\frac{8!}{3!5!}}{\frac{8!}{6!2!}} = \left(\dfrac{8!}{3!5!}\right)\left(\dfrac{6!2!}{8!}\right) = \dfrac{6}{3} = 2$

19. $\dfrac{_9P_5}{_{10}C_4} = \dfrac{\frac{9!}{4!}}{\frac{10!}{6!4!}} = \dfrac{(9)(8)(7)(6)(5)}{\frac{(10)(9)(8)(7)}{(4)(3)(2)(1)}} = \dfrac{144}{2} = 72$

21. $_9C_6 = \dfrac{9!}{3!6!} = \dfrac{(9)(8)(7)(6!)}{(3)(2)(1)(6!)} = \dfrac{504}{6} = 84$ ways

23. $_5C_4 = \dfrac{5!}{1!4!} = 5$

25. $_{10}C_4 = \dfrac{10!}{6!4!} = \dfrac{(10)(9)(8)(7)}{(4)(3)(2)(1)} = 210$

27. $_9C_5 = \dfrac{9!}{4!5!} = \dfrac{(9)(8)(7)(6)}{(4)(3)(2)(1)} = 126$

29. $_{12}C_8 = \dfrac{12!}{4!8!} = \dfrac{(12)(11)(10)(9)}{(4)(3)(2)(1)} = 495$

29. $_{12}C_8 = \dfrac{12!}{4!8!} = \dfrac{(12)(11)(10)(9)}{(4)(3)(2)(1)} = 495$

31. $_{10}C_8 = \dfrac{10!}{2!8!} = \dfrac{(10)(9)}{(2)(1)} = 45$

33. $_8C_2 = \dfrac{8!}{6!2!} = \dfrac{(8)(7)}{(2)(1)} = 28$ tickets

35. $_{12}C_3 \bullet {}_8C_2 =$

$\left(\dfrac{12!}{9!3!}\right)\left(\dfrac{8!}{6!2!}\right) = \left(\dfrac{(12)(11)(10)}{(3)(2)(1)}\right)\left(\dfrac{(8)(7)}{(2)(1)}\right) = 6160$

37. Mathematics: $\quad _8C_5 = \dfrac{8!}{3!5!} = \dfrac{(8)(7)(6)}{(3)(2)(1)} = 56$

Computer Sci. $\quad _5C_3 = \dfrac{5!}{2!3!} = \dfrac{(5)(4)}{(2)(1)} = 10$

$(56)(10) = 560$ different choices

39. Teachers: $\quad _6C_2 = \dfrac{6!}{4!2!} = \dfrac{(6)(5)}{(2)(1)} = 15$

Students: $\quad _{50}C_3 =$

$\dfrac{50!}{47!3!} = \dfrac{(50)(49)(48)}{(3)(2)(1)} = 19600$

$(1)(19,600) = 294,000$ ways to select the comm.

41. $_8C_3 \bullet {}_5C_2 =$

$\left(\dfrac{8!}{5!3!}\right)\left(\dfrac{5!}{3!2!}\right) = \left(\dfrac{(8)(7)(6)}{(3)(2)(1)}\right)\left(\dfrac{(5)(4)}{(2)(1)}\right) = 560$

43. $_6C_3 \bullet {}_5C_2 \bullet {}_4C_2 =$

$\left(\dfrac{6!}{3!3!}\right)\left(\dfrac{5!}{3!2!}\right)\left(\dfrac{4!}{2!2!}\right) =$

$\left(\dfrac{(6)(5)(4)}{(3)(2)(1)}\right)\left(\dfrac{(5)(4)}{(2)(1)}\right)\left(\dfrac{(4)(3)}{(2)(1)}\right) = 1200$

45. a) $_{10}C_8 = \dfrac{10!}{2!8!} = \dfrac{(10)(9)}{(2)(1)} = 45$

b) $_{10}C_9 = \dfrac{10!}{1!9!} = \dfrac{(10)(9!)}{(1)(9!)} = 10$

$_{10}C_{10} = \dfrac{10!}{10!} = 1$

$_{10}C_8 + {}_{10}C_9 + {}_{10}C_{10} = 45 + 10 + 1 = 56$

49. a) $4! = 24$ 　　 b) $4!/4 = 6$

47. a)

```
            1
         1     1
      1     2     1
   1     3     3     1
1     4     6     4     1
```

b) 　 1 　 5 　 10 　 10 　 5 　 1

51. a) The order of the numbers is important. For example: if the combination is 12 - 4 - 23, the lock will not open if 4 - 12 - 23 is used. Since repetition is permitted, it is not a true permutation problem.

b) $(40)(40)(40) = 64,000$

c) $(40)(39)(38) = 59,280$

Exercise Set 12.10

1. P(4 red balls) = $\dfrac{\text{no. of 4 red ball comb.}}{\text{no. of 4 ball comb.}} = \dfrac{_6C_4}{_{10}C_4}$

3. P(3 vowels) = $\dfrac{\text{no. of 3 vowel comb.}}{\text{no. of 3 letter comb.}} = \dfrac{_5C_3}{_{26}C_3}$

5. P(all 7 are Palaminos) =

$\dfrac{\text{no. of 5 Palamino comb.}}{\text{no. of 5 horse comb.}} = \dfrac{_{10}C_5}{_{18}C_5}$

7. P(none of the 9 are oak) =

$\dfrac{\text{no. of 9 non-oak comb.}}{\text{no. of 9 tree comb.}} = \dfrac{_{14}C_9}{_{30}C_9}$

9. $_5C_3 = \dfrac{5!}{2!3!} = \dfrac{(5)(4)}{(2)(1)} = 10$

$_9C_3 = \dfrac{9!}{6!3!} = \dfrac{(9)(8)(7)}{(3)(2)(1)} = 84$

P(3 reds) = $\dfrac{10}{84} = \dfrac{5}{42}$

11. $_8C_5 = \dfrac{8!}{3!5!} = \dfrac{(8)(7)(6)}{(3)(2)(1)} = 56$

$_{14}C_5 = \dfrac{14!}{5!9!} = \dfrac{(14)(13)(12)(11)(10)}{(5)(4)(3)(2)(1)} = 2002$

P(5 men's names) = $\dfrac{56}{2002} = \dfrac{4}{143}$

13. $_5C_3 = \dfrac{5!}{2!3!} = \dfrac{(5)(4)}{(2)(1)} = 10$

$_{10}C_3 = \dfrac{10!}{7!3!} = \dfrac{(10)(9)(8)}{(3)(2)(1)} = 120$

P(3 greater than 4) = $\dfrac{10}{120} = \dfrac{1}{12}$

15. $_6C_3 = \dfrac{6!}{3!3!} = \dfrac{(6)(5)(4)}{(3)(2)(1)} = 20$

$_{11}C_3 = \dfrac{11!}{8!3!} = \dfrac{(11)(10)(9)}{(3)(2)(1)} = 165$

P(all from manufacturing) = $\dfrac{20}{165} = \dfrac{4}{33}$

17. $_{46}C_6 = \dfrac{46!}{40!6!} = 9,366,819 \quad _6C_6 = 1$

P(win grand prize) = $\dfrac{1}{9,366,819}$

19. $_3C_2 = \dfrac{3!}{1!2!} = 3 \quad _5C_2 = \dfrac{5!}{3!2!} = \dfrac{(5)(4)}{(2)(1)} = 10$

P(no cars) = $\dfrac{3}{10}$

21. P(at least 1 car) = 1 – P(no cars) =

$1 - 1 - \dfrac{3}{10} = \dfrac{7}{10}$

23. $_6C_3 = \dfrac{6!}{3!3!} = \dfrac{(6)(5)(4)}{(3)(2)(1)} = 20$

$_{25}C_3 = \dfrac{25!}{3!22!} = \dfrac{(25)(24)(23)}{(3)(2)(1)} = 2300$

P(3 infielders) = $\dfrac{20}{2300} = \dfrac{1}{115}$

25. $_{10}C_2 = \dfrac{10!}{8!2!} = 45 \quad _6C_1 = \dfrac{6!}{5!1!} = 6$

P(2 pitchers and 1 infielder) = $\dfrac{(45)(6)}{2300} = \dfrac{27}{230}$

For problems 27 – 29, use the fact that $_{39}C_{12} = \dfrac{39!}{27!12!} = 3,910,797,436$

27. $_{22}C_{12} = \dfrac{22!}{10!12!} = 646,646$

   P(all women) $= \dfrac{646646}{3910797436} = 0.0001653$

29. $_{17}C_6 = \dfrac{17!}{11!6!} = 12,376$

   $_{22}C_6 = \dfrac{22!}{16!6!} = 74,613$

   P(6 men/6 women) $= \dfrac{(12376)(74613)}{3910797436} = 0.236$

For problems 31 – 33, use the fact that $_{15}C_5 = \dfrac{15!}{10!5!} = \dfrac{(15)(14)(13)(12)(11)}{(5)(4)(3)(2)(1)} = 3003$

31. $_4C_3 = \dfrac{4!}{3!1!} = 4 \qquad _6C_2 = \dfrac{6!}{4!2!} = \dfrac{(6)(5)}{(2)(1)} = 15$

   P(3 in FL/2 in VA) $= \dfrac{(4)(15)}{3003} = \dfrac{60}{3003} = \dfrac{20}{1001}$

33. $_5C_2 = \dfrac{5!}{2!3!} = \dfrac{(5)(4)}{(2)(1)} = 10$

   $_4C_1 = \dfrac{4!}{3!1!} = 4$

   $_6C_2 = \dfrac{6!}{2!4!} = \dfrac{(6)(5)}{(2)(1)} = 15$

   P(1 in FL/2 in KY/2 in VA) $=$
   $\dfrac{(10)(4)(15)}{3003} = \dfrac{200}{1001}$

For problems 35 – 37, use the fact that $_{11}C_5 = \dfrac{11!}{6!5!} = \dfrac{(11)(10)(9)(8)(7)}{(5)(4)(3)(2)(1)} = 462$

35. $_6C_5 = \dfrac{6!}{1!5!} = 6$

   P(5 women first) $= \dfrac{6}{462} = \dfrac{1}{77}$

37. Any one of the 6 women can sit in any one of the five seats - 30 possibilities.

   P(exactly 1 woman) $= \dfrac{30}{462} = \dfrac{5}{77}$

39. $_{24}C_6 = \dfrac{24!}{18!6!} = 134,596; \quad _3C_3 = 1$

   $_{21}C_3 = \dfrac{21!}{18!3!} = 1,330$

   P(3 brothers) $=$
   $\dfrac{_3C_3 \cdot _{21}C_3}{_{24}C_6} = \dfrac{(1)(1330)}{134596} = 0.00988$

41. a) P(royal spade flush) $= \dfrac{_{47}C_2}{_{52}C_7} = \dfrac{1}{123760}$

   b) P(any royal flush) $= \dfrac{4}{123760} = \dfrac{1}{30,940}$

43. a) $\left(\dfrac{(_4C_2)(_4C_2)(_{44}C_1)}{_{52}C_5}\right) = \dfrac{1584}{2598960} = \dfrac{33}{54,145}$

P(2 aces/2 8's/other card ace or 8) = $\dfrac{33}{54,145}$

b) P(aces of spades and clubs/8's of spades and clubs/9 of diamonds) =

$\dfrac{1}{_{52}C_5} = \dfrac{1}{2,598,960}$

45. a) $\left(\dfrac{1}{15}\right)\left(\dfrac{1}{14}\right)\left(\dfrac{1}{13}\right)\left(\dfrac{5}{12}\right)\left(\dfrac{4}{11}\right)\left(\dfrac{3}{10}\right)\left(\dfrac{2}{9}\right)\left(\dfrac{1}{8}\right)$

$= \dfrac{120}{259459200} = \dfrac{1}{2,162,160}$

b) P(any 3 of 8 for officers) =

$\dfrac{(8)(7)(6)}{2162160} = \dfrac{1}{6435}$

47. Since there are more people than hairs, 2 or more people must have the same number of hairs.

Exercise Set 12.11

1. A probability distribution shows the probability associated with each specific outcome of an experiment. In a probability distribution every possible outcome must be listed and the sum of all the probabilities must be 1.

3. $P(x) = {}_nC_x p^x q^{n-x}$

5. $P(2) = {}_4C_2 (0.3)^2 (0.7)^{4-2}$

$= \dfrac{4!}{2!2!}(.09)(.49) = 0.2646$

7. $P(2) = {}_5C_2 (0.4)^2 (0.6)^{5-2}$

$= \dfrac{5!}{2!3!}(.16)(.216) = 0.3456$

9. $P(0) = {}_6C_0 (0.5)^0 (0.5)^{6-0}$

$= \dfrac{6!}{0!6!}(1)(.0156252) = 0.015625$

11. $p = 0.14$, $q = 1 - p = 1 - 0.14 = 0.86$

a) $P(x) = {}_nC_x (0.14)^x (0.86)^{n-x}$

b) $n = 12$, $x = 2$, $p = 0.14$, $q = 0.86$

$P(2) = {}_{12}C_2 (0.14)^2 (0.86)^{12-2}$

13. $P(4) = {}_6C_4 (0.3)^4 (0.7)^{6-4}$

$= \dfrac{6!}{4!2!}(.0081)(.49) = 0.05954$

15. $P(2) = {}_3C_2 (0.96)^2 (0.04)^{3-2}$

$= \dfrac{3!}{2!1!}(.9216)(.04) = 0.1106$

17. $P(4) = {}_6C_4 (0.92)^4 (0.08)^{6-4}$

$= \dfrac{6!}{4!2!}(.7164)(.0064) = 0.06877$

19. $P(4) = {}_5C_4 (.8)^4 (.2)^{5-4}$

$= \dfrac{5!}{1!4!}(.4096)(.2) = 0.4096$

21. a) $P(0) = {}_5C_0 (0.6)^0 (0.4)^{5-0}$

$= \dfrac{5!}{5!}(1)(.01024) = 0.01024$

b) P(at least 1) = $1 - P(0) = 0.98976$

23. a) $P(3) = {}_6C_3 \left(\dfrac{12}{52}\right)^3 \left(\dfrac{40}{52}\right)^3$

$\dfrac{6!}{3!3!}(.01229)(.45517) = 0.11188$

b) $P(2) = {}_6C_2 \left(\dfrac{13}{52}\right)^2 \left(\dfrac{39}{52}\right)^4$

$= \dfrac{6!}{2!4!}(.0625)(.3164) = 0.29663$

25. The probability that the sun would be shining would equal 0 because 72 hours later would occur at midnight.

Review Exercises

1. Relative frequency over the long run can accurately be predicted, not individual events or totals.

2. Roll the die many times then compute the relative frequency of each outcome and compare with the expected probability of 1/6.

3. P(mountain bike) = $\dfrac{8}{40} = \dfrac{1}{5}$

4. Answers will vary.

5. P(watches ABC) = $\dfrac{80}{200} = \dfrac{2}{5}$

6. P(even) = $\dfrac{5}{10} = \dfrac{1}{2}$

7. P(odd or > 5) = $\dfrac{5}{10} + \dfrac{4}{10} - \dfrac{2}{10} = \dfrac{7}{10}$

8. P(> 2 or < 5) = $\dfrac{7}{10} + \dfrac{5}{10} - \dfrac{2}{10} = \dfrac{10}{10} = 1$

9. P(even and > 4) = $\dfrac{2}{10} = \dfrac{1}{5}$

10. P(Grand Canyon) = $\dfrac{50}{240} = \dfrac{5}{24}$

11. P(Yosemite) = $\dfrac{40}{240} = \dfrac{1}{6}$

12. P(Rocky Mtn. or Smoky Mtn.) =
$\dfrac{35}{240} + \dfrac{45}{240} = \dfrac{80}{240} = \dfrac{1}{3}$

13. P(not Grand Canyon) = $\dfrac{190}{240} = \dfrac{19}{24}$

14. a) 9:1   b) 1:9

15. 5:3

16. P(wins Triple Crown) = $\dfrac{3}{85}$

17. 7:3

18. a) E = P(win $200)•$198 + P(win $100)•$98
$+ P(lose)•(-\$2)$
$= (.003)(198) + (.002)(98) - (.995)(2)$
$= .594 + .196 - 1.990 = -1.200 \rightarrow -\$1.20$

b) The expectation of a person who purchases three tickets would be 3(–1.20) = –$3.60.

c) Expected value = Fair price – Cost
–1.20 = Fair price – 2.00 $.80 = Fair price

19. a) $E_{Cameron}$ = P(pic. card)($9) +
P(not pic. card)(–$3)
$= \left(\dfrac{12}{52}\right)(9) - \left(\dfrac{40}{52}\right)(3) = \approx -\$0.23$

b) $E_{Lindsey}$ = P(pic. card)(–$9) + P(not pic. card)($3)
$= \dfrac{-27}{13} + \dfrac{30}{13} = \dfrac{3}{13} \approx \$0.23$

c) Cameron can expect to lose $(100)\left(\dfrac{3}{13}\right) \approx \$23.08$

20. E = P(sunny)(1000) + P(cloudy)(500) + P(rain)(100) = 0.4(1000) + 0.5(500) + 0.1(100) =
400 + 250 + 10 = 660 people

21. a)

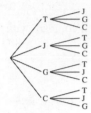

   b) Sample space:

   {TJ,TG,TC,JT,JG,JC,GT,GJ,GC,CT,CJ,CG}

   c) P(Gina is Pres. and Jake V.P.) = 1/12

23. P(even and even) = (4/8)(4/8) = 16/64 = 1/4

25. P(outer odd and inner < 6)

$$= \text{P(outer odd) P(inner < 6)} = \frac{4}{8} \cdot \frac{5}{8} = \frac{1}{2} \cdot \frac{5}{8} = \frac{5}{16}$$

27. P(inner even and not green) =

$$\frac{1}{2} + \frac{6}{8} - \frac{2}{8} = \frac{1}{2} + \frac{4}{8} = 1$$

29. P(all 3 are Hersheys) = $\dfrac{5}{12} \cdot \dfrac{4}{11} \cdot \dfrac{3}{10} = \dfrac{60}{1320} = \dfrac{1}{22}$

31. P(at least one is Nestle) = 1 – P(none are Nestle)

$$= 1 - \frac{14}{55} = \frac{55}{55} - \frac{14}{55} = \frac{41}{55}$$

33. P(yellow) = 1/4

35. $5 for red; $10 for yellow; $20 for green

   P(green) = ½; P(yellow) = ¼; P(red) = ¼

   EV = (1/4)(5) + (1/4)(10) + (1/2)(20) = $13.75

37. P(not green) = 1/4 + 1/4 + 1/8 = 5/8

39. E = P(green)($10) + P(red)($5) +

   P(yellow)(–$20)

   = (3/8)(10) + (1/2)(5) – (1/8)(20)

   = (15/4) + (10/4) – (10/4) = 15/4 → $3.75

22. a)

   b) Sample space:

   {H1,H2,H3,H4,T1,T2,T3,T4}

   c) P(heads and odd) = (1/2)(2/4) = 2/8 = ¼

   d) P(heads or odd) = (1/2)(2/4) + (1/2)(2/4)

   = 4/8 + 2/8 = 6/8 = 3/4

24. P(outer is greater than 5 and inner is greater than 5)

$$= \text{P(outer is > 5)} \cdot \text{P(inner is > 5)} = \frac{3}{8} \cdot \frac{3}{8} = \frac{9}{64}$$

26. P(outer is even or less than 6)

   = P(even) + P(< 6) – P(even and < 6)

$$= \frac{4}{8} + \frac{5}{8} - \frac{2}{8} = \frac{7}{8}$$

28. P(outer gold and inner not gold)

$$= \left(\frac{2}{8}\right)\left(\frac{6}{8}\right) = \left(\frac{1}{4}\right)\left(\frac{3}{4}\right) = \frac{3}{16}$$

30. P(none are Nestle) = $\dfrac{8}{12} \cdot \dfrac{7}{11} \cdot \dfrac{6}{10} = \dfrac{336}{1320} = \dfrac{14}{55}$

32. P(Hershey and Hershey and Reese)

$$= \frac{5}{12} \cdot \frac{4}{11} \cdot \frac{3}{10} = \frac{60}{1320} = \frac{1}{22}$$

34. Odds against yellow   3:1

   Odds for yellow   1:3

36. P(red, then green) = P(red)P(green)

   = (1/4)(1/2) = 1/8

38. Odds in favor of green   3:5

   Odds against green   5:3

40. P(at least one red) = 1 – P(none are red)

   = 1 – (1/2)(1/2)(1/2) = 1-1/8 = 7/8

41. P($<$ 6 defects $|$ American built) = 89/106 = 0.84

42. P($<$ 6 defects $|$ foreign built) = 55/74 = 0.74

43. P($\geq$ 6 defects $|$ foreign built) = 19/74 = 0.26

44. P($\geq$ 6 defects $|$ American built) = 17/106 = 0.16

45. P(right handed) = $\dfrac{230}{400} = \dfrac{23}{40}$

46. P(left brained $|$ left handed) = $\dfrac{30}{170} = \dfrac{3}{17}$

47. P(right handed $|$ no predominance) = $\dfrac{60}{80} = \dfrac{3}{4}$

48. P(right brained $|$ left handed) = $\dfrac{120}{170} = \dfrac{12}{17}$

49. a) $4! = (4)(3)(2)(1) = 24$

   b) $E = (1/4)(10K) + (1/4)(5K) + (1/4)(2K) + (1/4)(1K) = (1/4)(18K) = \$4,500.00$

50. # of possible arrangements = $(_5C_2)(_3C_2)(_1C_1)$

   $= \left(\dfrac{5!}{3!2!}\right)\left(\dfrac{3!}{1!2!}\right)\left(\dfrac{1!}{1!}\right) = \dfrac{(5)(4)(3)}{(2)(1)} = 30$

51. $_{10}P_3 = \dfrac{10!}{7!} = (10)(9)(8) = 720$

52. $_9P_3 = \dfrac{9!}{6!} = \dfrac{(9)(8)(7)(6!)}{6!} = (9)(8)(7) = 504$

53. $_6C_3 = \dfrac{6!}{3!3!} = \dfrac{(6)(5)(4)}{(3)(2)(1)} = 20$

54. a) $_{15}C_{10} = \dfrac{15!}{5!10!} = \dfrac{(15)(14)(13)(12)(11)}{(5)(4)(3)(2)(1)} = 3003$

   b) number of arrangements = $10! = 3,628,800$

55. a) P(match 5 numbers) = $\dfrac{1}{_{52}C_5}$

   $= \dfrac{1}{\dfrac{52!}{47!5!}} = \dfrac{47!5!}{52!} = \dfrac{1}{2,598,960}$

   b) P(Big game win) = P(match 5 #s and Big #)
   = P(match 5 #s) • P(match Big #)

   $= \left(\dfrac{1}{2,598,960}\right)\left(\dfrac{1}{52}\right) = \dfrac{1}{135,145,920}$

56. $(_8C_2)(_{10}C_4) =$

   $\left(\dfrac{8!}{6!2!}\right)\left(\dfrac{10!}{6!4!}\right) = \dfrac{(8)(7)(10)(9)(8)(7)}{(2)(1)(4)(3)(2)(1)} = 5880$ combos.

57. $(_8C_3)(_5C_2) =$

   $\left(\dfrac{8!}{5!3!}\right)\left(\dfrac{5!}{2!3!}\right) = \dfrac{(8)(7)(6)(5)(4)}{(3)(2)(1)(2)(1)} = 560$

58. P(two aces) = $\dfrac{_4C_2}{_{52}C_2} = \dfrac{\dfrac{4!}{2!2!}}{\dfrac{52!}{50!2!}}$

   $= \left(\dfrac{4!}{2!2!}\right)\left(\dfrac{50!2!}{52!}\right) = \dfrac{1}{221}$

59. P(all three are red) = $\left(\dfrac{5}{10}\right)\left(\dfrac{4}{9}\right)\left(\dfrac{3}{8}\right) = \dfrac{1}{12}$

60. P(1st 2 are red/3rd is blue) = $\left(\dfrac{5}{10}\right)\left(\dfrac{4}{9}\right)\left(\dfrac{2}{8}\right) = \dfrac{1}{18}$

61. P(1st red, 2nd white, 3rd blue)

   $= \left(\dfrac{5}{10}\right)\left(\dfrac{3}{9}\right)\left(\dfrac{2}{8}\right) = \dfrac{1}{24}$

62. P(at least one red) = 1 − P(none are red)

   $= 1 - \left(\dfrac{5}{10}\right)\left(\dfrac{4}{9}\right)\left(\dfrac{3}{8}\right) = 1 - \dfrac{1}{12} = \dfrac{11}{12}$

63. P(3 N&WRs) =

$$\frac{_5C_3}{_{14}C_3} = \frac{\dfrac{5!}{3!2!}}{\dfrac{14!}{3!11!}} = \frac{5!3!11!}{3!2!14!} = \frac{(5)(4)(3)}{(14)(13)(12)} = \frac{5}{182}$$

64. P(2 NWs & 1 Time) =

$$\frac{(_6C_2)(_3C_1)}{_{14}C_3} = \frac{\left(\dfrac{6!}{2!4!}\right)\left(\dfrac{3!}{1!2!}\right)}{\dfrac{14!}{3!11!}}$$

$$= \frac{(6)(5)(3)(3)(2)(1)}{(2)(1)(14)(13)(12)} = \frac{45}{364}$$

65. $$\frac{_8C_3}{_{14}C_3} = \frac{\dfrac{8!}{3!5!}}{\dfrac{14!}{3!11!}} = \frac{8!3!11!}{3!5!14!}$$

$$= \frac{(8)(7)(6)}{(14)(13)(12)} = \frac{336}{2184} = \frac{2}{13}$$

66. $1 - \dfrac{2}{13} = \dfrac{11}{13}$

67. a) $P(x) = {_nC_x}\,(0.6)^x\,(0.4)^{n-x}$

   b) $P(75) = {_{100}C_{75}}(0.6)^{75}(0.4)^{25}$

68. n = 5, x = 3, p = 1/5, q = 4/5

$$P(3) = {_5C_3}\left(\frac{1}{5}\right)^3\left(\frac{4}{5}\right)^2 = 10\cdot\left(\frac{1}{5}\right)^3\left(\frac{4}{5}\right)^2 =$$
0.0512

69. a) n = 4, p = 0.6, q = 0.4

$$P(0) = {_4C_0}(0.6)^0(0.4)^4$$

$$= (1)(1)(0.4)^4 = 0.0256$$

   b) P(at least 1) = 1 – P(0) = 1 – 0.0256 = 0.9744

Chapter Test

1. P(fishing for bass) = $\dfrac{22}{30} = \dfrac{11}{15}$

2. $(P > 7) = \dfrac{2}{9} \approx 0.22$

3. P(odd) = $\dfrac{5}{9} \approx 0.55$

4. $P(\geq 4) = \dfrac{7}{9} \approx 0.78$

5. P(odd and > 4) = $\dfrac{3}{9} = \dfrac{1}{3} \approx 0.33$

6. P(both > 5) = $\dfrac{4}{9}\cdot\dfrac{3}{8} = \dfrac{12}{72} = \dfrac{1}{6}$

7. P(both even) = $\dfrac{4}{9}\cdot\dfrac{3}{8} = \dfrac{1\cdot1}{3\cdot2} = \dfrac{1}{6}$

8. P(1st odd, 2nd even) = $\dfrac{5}{9}\cdot\dfrac{4}{8} = \dfrac{5}{9}\cdot\dfrac{1}{2} = \dfrac{5}{18}$

9. P(neither > 6) = $\dfrac{6}{9}\cdot\dfrac{5}{8} = \dfrac{1\cdot5}{3\cdot4} = \dfrac{5}{12}$

10. P(red or picture)
    = P(red) + P(picture) – P(red and picture)
    $$= \frac{26}{52} + \frac{12}{52} - \frac{6}{52} = \frac{32}{52} = \frac{8}{13}$$

11. 1 die   (6)(3) = 18

12.

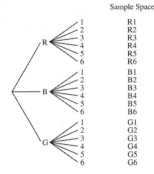

Sample Space

R	1 R1
	2 R2
	3 R3
	4 R4
	5 R5
	6 R6
B	1 B1
	2 B2
	3 B3
	4 B4
	5 B5
	6 B6
G	1 G1
	2 G2
	3 G3
	4 G4
	5 G5
	6 G6

13. P(blue and 1) = $\dfrac{1}{18}$

14. P(blue or 1) = $\dfrac{6}{18} + \dfrac{3}{18} - \dfrac{1}{18} = \dfrac{8}{18} = \dfrac{4}{9}$

15. P(not red or odd) = $\dfrac{12}{18} + \dfrac{9}{18} - \dfrac{6}{18} = \dfrac{15}{18} = \dfrac{5}{6}$

16. Number of codes = (9)(26)(26)(10)(10)
    = 608,400

17. a) 5:4   b) 5:4

18. odds against Aimee winning are 5:2 or

$\dfrac{5}{2} = \dfrac{5/7}{2/7} = \dfrac{P(\text{not winning})}{P(\text{winning})}$

Therefore, P(Aimee wins) = 2/7

19. E = P(club) ($8) + P(heart) ($4)
        + P(spade or diamond) (–$6)

$= \left(\dfrac{1}{4}\right)(8) + \left(\dfrac{1}{4}\right)(4) + \left(\dfrac{2}{4}\right)(-6)$

$= \dfrac{8}{4} + \dfrac{4}{4} - \dfrac{12}{4} = \$0.00$

20. d) P(GW Bridge | car) = $\dfrac{120}{214} = \dfrac{60}{107}$

20. a) P(car) = $\dfrac{214}{456} = \dfrac{107}{228}$

b) P(Golden Gate) = $\dfrac{230}{456} = \dfrac{115}{228}$

c) P(SUV | Golden Gate) = $\dfrac{136}{230} = \dfrac{68}{115}$

21. $_6P_3 = \dfrac{6!}{(6-3)!} = \dfrac{6!}{3!} = 6 \cdot 5 \cdot 4 = 120$

22. P(neither is good) = $\dfrac{8}{20} \cdot \dfrac{7}{19} = \dfrac{2}{5} \cdot \dfrac{7}{19} = \dfrac{14}{95}$

23. P($\geq$ 1 good) = 1 – P(neither -good) =
    $1 - \dfrac{14}{95} = \dfrac{81}{95}$

24. $_7C_3 = \dfrac{7!}{4!3!} = \dfrac{(7)(6)(5)}{(3)(2)(1)} = 35$

$_5C_2 = \dfrac{5!}{3!2!} = \dfrac{(5)(4)}{(2)(1)} = 10$

$_{12}C_5 = \dfrac{12!}{7!5!} = \dfrac{(12)(11)(10)(9)(8)}{(5)(4)(3)(2)(1)} = 792$

P(3 red and 2 green) = $\dfrac{(35)(10)}{792} = \dfrac{350}{792} = \dfrac{175}{396}$

25. (0.1)(0.1)(0.1) = 0.001
    (0.1)(0.1)(0.1)(0.9)(0.9) = 0.00081

$_5C_3 = \dfrac{5!}{3!2!} = \dfrac{(5)(4)}{(2)(1)} = 10$

(10)(.00081) = 0.0081

CHAPTER THIRTEEN

STATISTICS

Exercise Set 13.1

1. **Statistics** is the art and science of gathering, analyzing, and making inferences (predictions) from numerical information obtained in an experiment.

3. Answers will vary.

5. Insurance companies, sports, airlines, stock market, medical profession

7. a) A **population** consists of all items or people of interest.
   b) A **sample** is a subset of the population.

9. a) A **random sample** is a sample drawn in such a way that each item in the population has an equal chance of being selected.
   b) Number each item in the population. Write each number on a piece of paper and put each numbered piece of paper in a hat. Select pieces of paper from the hat and use the numbered items selected as your sample.

11. a) A **stratified sample** is one that includes items from each part (or strata) of the population.
    b) First identify the strata you are interested in. Then select a random sample from each strata.

13. a) An **unbiased sample** is one that is a small replica of the entire population with regard to income, education, gender, race, religion, political affiliation, age, etc.

15. Stratified sample

17. Cluster sample

19. Systematic sample

21. Convenience sample

23. Random sample

25. a) – c) Answers will vary.

27. President; four out of 42 U.S. presidents have been assassinated (Lincoln, Garfield, McKinley, Kennedy).

28. Answers will vary.

Exercise Set 13.2

1. Answers will vary.

3. There may have been more car thefts in Baltimore, Maryland than Reno, Nevada because many more people live in Baltimore than in Reno. But, Reno may have more car thefts per capita than Baltimore.

5. Although the cookies are fat free, they still contain calories. Eating many of them may still cause you to gain weight.

7. More people drive on Saturday evening. Thus, one might expect more accidents.

9. People with asthma may move to Arizona because of its climate. Therefore, more people with asthma may live in Arizona.

11. Although milk is less expensive at Star Food Markets than at Price Chopper Food Markets, other items may be more expensive at Star Food Markets.

13. There may be deep sections in the pond, so it may not be safe to go wading.

15. Half the students in a population are expected to be below average.

17. a)

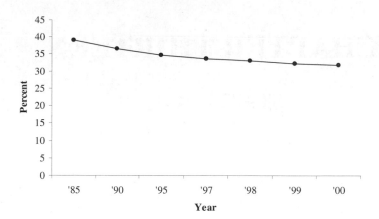

b)

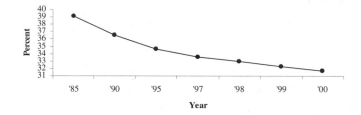

19. a)

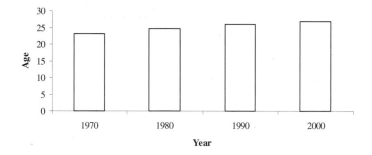

b)

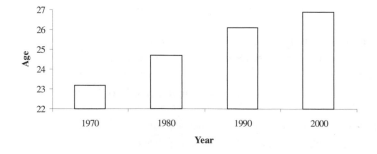

21. a)

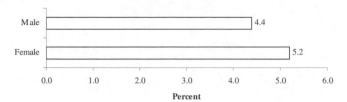

Percent of Survey Respondents That Purchased Clothing Accessories Online, Nov. 2000 - Jan. 2001

b) Yes. The new graph gives the impression that the percents are closer together.

23. A decimal point

Exercise Set 13.3

1. A **frequency distribution** is a listing of observed values and the corresponding frequency of occurrence of each value.

3. a) 7   b) 16-22   c) 16   d) 22

5. The **modal class** is the class with the greatest frequency.

7. a) Number of observations = sum of frequencies = 18
   b) Width = $16 - 9 = 7$
   c) $\dfrac{16 + 22}{2} = \dfrac{38}{2} = 19$
   d) The modal class is the class with the greatest frequency. Thus, the modal class is 16 - 22.
   e) Since the class widths are 7, the next class would be 51 - 57.

9.

Number Sold	Number of Days
0	3
1	8
2	3
3	5
4	2
5	7
6	2
7	3
8	4
9	1
10	2

11.

I.Q.	Number of Students
78 - 86	2
87 - 95	15
96 - 104	18
105 - 113	7
114 - 122	6
123 - 131	1
132 - 140	1

13.

I.Q.	Number of Students
80 - 90	8
91 - 101	22
102 - 112	11
113 - 123	7
124 - 134	1
135 - 145	1

15.

Placement test scores	Number of Students
472 - 492	9
493 - 513	9
514 - 534	5
535 - 555	2
556 - 576	3
577 - 597	2

17.

Placement test scores	Number of Students
472 - 487	4
488 - 503	9
504 - 519	7
520 - 535	3
536 - 551	2
552 - 567	2
568 - 583	2
584 - 599	1

19.

Circulation (thousands)	Number of Newspapers
209 - 458	36
459 - 708	8
709 - 958	3
959 - 1208	1
1209 - 1458	0
1459 - 1708	0
1709 - 1958	1
1959 - 2208	1

21.

Circulation (thousands)	Number of Newspapers
209 - 408	34
409 - 608	9
609 - 808	3
809 - 1008	1
1009 - 1208	1
1209 - 1408	0
1409 - 1608	0
1609 - 1808	1
1809 - 2008	0
2009 - 2208	1

23.

Population (millions)	Number of Counties
1.4 - 2.1	15
2.2 - 2.9	6
3.0 - 3.7	2
3.8 - 4.5	0
4.6 - 5.3	0
5.4 - 6.1	1
6.2 - 6.9	0
7.0 - 7.7	0
7.8 - 8.5	0
8.6 - 9.3	0
9.4 - 10.1	1

25.

Population (millions)	Number of Counties
1.0 - 2.5	19
2.6 - 4.1	4
4.2 - 5.7	1
5.8 - 7.3	0
7.4 - 8.9	0
9.0 - 10.5	1

27.

Price ($)	Number of States
0.35 - 0.44	6
0.45 - 0.54	10
0.55 - 0.64	11
0.65 - 0.74	3
0.75 - 0.84	2
0.85 - 0.94	4
0.95 - 1.04	1
1.05 - 1.14	2
1.15 - 1.24	2
1.25 - 1.34	1
1.35 - 1.44	0
1.45 - 1.54	1

29.

Price ($)	Number of States
0.35 - 0.54	16
0.55 - 0.74	14
0.75 - 0.94	6
0.95 - 1.14	3
1.15 - 1.34	3
1.35 - 1.54	1

31. February, since it has the fewest number of days

32. a) **Did You Know?, page 762:** There are 6 F's.
     b) Answers will vary.

Exercise Set 13.4

1. Answers will vary.

3. Answers will vary.

5. a) Answers will vary.
    b)

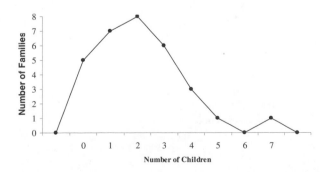

Children in Selected Families

7. a) Answers will vary.

    b) | Observed Values | Frequency |
        |:---:|:---:|
        | 45 | 3 |
        | 46 | 0 |
        | 47 | 1 |
        | 48 | 0 |
        | 49 | 1 |
        | 50 | 1 |
        | 51 | 2 |

9. Occasionally: $0.59(500) = 295$

    Most Times: $0.25(500) = 125$

    Every Time: $0.07(500) = 35$

    Never: $0.09(500) = 45$

11. Travelocity: $\dfrac{175}{500} = 0.35 = 35\%$ Priceline: $\dfrac{85}{500} = 0.17 = 17\%$

    Expedia: $\dfrac{125}{500} = 0.25 = 25\%$ Other: $\dfrac{115}{500} = 0.23 = 23\%$

13.   a) and b)

Height of Male High School Seniors

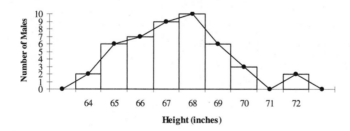

15.   a) and b)

DVDs Owned

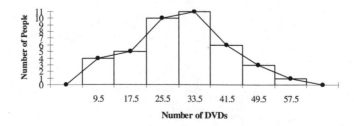

17. a) The total number of people surveyed:
    2 + 7 + 8 + 5 + 4 + 3 + 1 = 30
    b) Four people purchased four soft drinks.
    c) The modal class is 2 because more people
    purchased 2 soft drinks than any other
    number of soft drinks.
    d) Two people bought 0 soft drinks   0
    Seven people bought 1 soft drink   7
    Eight people bought 2 soft drinks 16
    Five people bought 3 soft drinks   15
    Four people bought 4 soft drinks   16
    Three people bought 5 soft drinks   15
    One person bought 6 soft drinks   6
    Total number of soft drinks purchased: 75

e)

Number of Soft Drinks Purchased	Number of People
0	2
1	7
2	8
3	5
4	4
5	3
6	1

19. a) 7 calls
    b) Adding the number of calls responded to in 6, 5, 4, or 3 minutes gives: 4 + 7 + 3 + 2 = 16 calls
    c) The total number of calls surveyed: 2 + 3 + 7 + 4 + 3 + 8 + 6 + 3 = 36

    d)
Response Time (min.)	Number of Calls
3	2
4	3
5	7
6	4
7	3
8	8
9	6
10	3

e)

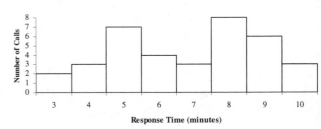

21.

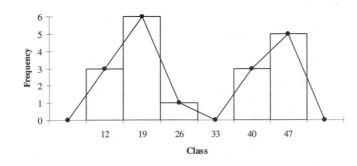

23. 1|5 represents 15

1| 0   5   7
2| 4   4
3| 6   0   3
4| 8   5   2   5   8
5| 3   4
6| 0   2   0

25. a)

Salaries (in $1000)	Number of Companies
27	1
28	7
29	4
30	3
31	2
32	3
33	3
34	2

b) and c)

Starting Salaries for 25 Different Social Workers

d) 2|3 represents 23

2| 7 8 8 8 8 8 8 8 9 9 9 9
3| 0 0 0 1 1 2 2 2 3 3 3 4 4

27. a)

Advertising Spending (millions of dollars)	Number of Companies
597 - 905	19
906 - 1214	14
1215 - 1523	7
1524 - 1832	3
1833 - 2141	2
2142 - 2450	3
2451 - 2759	1
2760 - 3068	0
3069 - 3377	1

b) and c)

Advertising Spending

29. a) - e) Answers will vary.

Exercise Set 13. 5

1. **Ranked data** are data listed from the lowest value to the highest value or from the highest value to the lowest value.

3. The **median** is the value in the middle of a set of ranked data. To find the median, rank the data and select the value in the middle.

5. The **mode** is the most common piece of data. The piece of data that occurs most frequently is the mode.

7. The median should be used when there are some values that differ greatly from the rest of the values in the set, for example, salaries.

9. The mean is used when each piece of data is to be considered and "weighed" equally, for example, weights of adult males.

	mean	median	mode	midrange
11.	$\dfrac{99}{9} = 11$	10	10	$\dfrac{5+23}{2} = 14$
13.	$\dfrac{485}{7} \approx 69.3$	72	none	$\dfrac{42+90}{2} = 66$
15.	$\dfrac{64}{8} = 8$	$\dfrac{7+9}{2} = 8$	none	$\dfrac{1+15}{2} = 8$
17.	$\dfrac{118}{9} \approx 13.1$	11	1	$\dfrac{1+36}{2} = 18.5$
19.	$\dfrac{95}{8} \approx 11.9$	$\dfrac{12+13}{2} = 12.5$	13	$\dfrac{6+17}{2} = 11.5$
21.	$\dfrac{65}{10} = 6.5$ weeks	$\dfrac{5+5}{2} = 5$ weeks	3 and 5 weeks	$\dfrac{2+19}{2} = 10.5$ weeks
23. a)	$\dfrac{34}{7} \approx 4.9$	5	5	$\dfrac{1+11}{2} = 6$
b)	$\dfrac{37}{7} \approx 5.3$	5	5	$\dfrac{1+11}{2} = 6$
c)	Only the mean			
d)	$\dfrac{33}{7} \approx 4.7$	5	5	$\dfrac{1+10}{2} = 5.5$

The mean and the midrange

25. A 79 mean average on 10 quizzes gives a total of 790 points. An 80 mean average on 10 quizzes requires a total of 800 points. Thus, Jim missed a B by 10 points not 1 point.

27. a) Mean: $\dfrac{87.7}{10} \approx 8.8$ million   b) Median: $\dfrac{7.8+8.2}{2} = 8.0$ million

c) Mode: none

d) Midrange: $\dfrac{4.6+19.7}{2} \approx 12.2$ million

29. a) Mean: $\dfrac{\$55.9}{11} \approx \$5.1$ billion   b) Median: $\$2.3$ billion

c) Mode: $\$2.3$ billion and $\$1.5$ billion

d) Midrange: $\dfrac{\$1.5+\$26.5}{2} = \$14$ billion

e) Answers will vary.

31. Let x = the sum of his scores

$\dfrac{x}{6} = 85$

$x = 85(6) = 510$

33. One example is 72, 73, 74, 76, 77, 78.

Mean: $\dfrac{450}{6} = 75$, Median: $\dfrac{74+76}{2} = 75$, Midrange: $\dfrac{72+78}{2} = 75$

35. a) **Yes**
    b) **Cannot** be found since we do not know the middle two numbers in the ranked list
    c) **Cannot** be found without knowing all of the numbers
    d) **Yes**
    e) **Mean:** $\dfrac{24,000}{120} = 200$; Midrange: $\dfrac{50 + 500}{2} = 275$

37. a) **For** a mean average of 60 on 7 exams, she must have a total of $60 \times 7 = 420$ points. Sheryl presently has $49 + 72 + 80 + 60 + 57 + 69 = 387$ points. Thus, to pass the course, her last exam must be $420 - 387 = 33$ or greater.
    b) A C average requires a total of $70 \times 7 = 490$ points. Sheryl has 387. Therefore, she would need $490 - 387 = 103$ on her last exam. If the maximum score she can receive is 100, she cannot obtain a C.
    c) **For** a mean average of 60 on 6 exams, she must have a total of $60 \times 6 = 360$ points. If the lowest score on an exam she has already taken is dropped, she will have a total of $72 + 80 + 60 + 57 + 69 = 338$ points. Thus, to pass the course, her last exam must be $360 - 338 = 22$ or greater.
    d) **For** a mean average of 70 on 6 exams, she must have a total of $70 \times 6 = 420$ points. If the lowest score on an exam she has already taken is dropped, she will have a total of 338 points. Thus, to obtain a C, her last exam must be $420 - 338 = 82$ or greater.

39. One example is 1, 2, 3, 3, 4, 5 changed to 1, 2, 3, 4, 4, 5.

    First set of data: Mean: $\dfrac{18}{6} = 3$, Median: $\dfrac{3+3}{2} = 3$, Mode: 3

    Second set of data: Mean: $\dfrac{19}{6} = 3.1\overline{6}$, Median: $\dfrac{3+4}{2} = 3.5$, Mode: 4

41. No, by changing only one piece of the six pieces of data you cannot alter both the median and the midrange.

43. The data must be arranged in either ascending or descending order.

45. He is taller than approximately 35% of all kindergarten children.

47. a) $Q_2 = \text{Median} = \$430$

    b) \$290, \$300, \$300, \$330, \$350, \$350, \$350, \$350, \$350, \$400

    $Q_1 = \text{Median of the data listed below} = \dfrac{\$350 + \$350}{2} = \$350$

    c) \$450, \$450, \$500, \$600, \$650, \$650, \$700, \$700, \$750, \$800

    $Q_3 = \text{Median of the data listed above} = \dfrac{\$650 + \$650}{2} = \$650$

49. Second quartile, median

51. a) **\$490**   b) \$500   c) 25%   d) 25%   e) 17%   f) $100 \times \$510 = \$51,000$

53. a) **Ruth:** $\approx 0.290, 0.359, 0.301, 0.272, 0.315$
       Mantle: $\approx 0.300, 0.365, 0.304, 0.275, 0.321$
    b) Mantle's is greater in every case.
    c) **Ruth:** $\dfrac{593}{1878} \approx 0.316$; Mantle: $\dfrac{760}{2440} \approx 0.311$; Ruth's is greater.
    d) Answers will vary.
    e) **Ruth:** $\dfrac{1.537}{5} \approx 0.307$; Mantle: $\dfrac{1.565}{5} = 0.313$; Mantle's is greater.

    f) and g) Answers will vary.

55. $\Sigma xw = 84(0.40) + 94(0.60) = 33.6 + 56.4 = 90$

    $\Sigma w = 0.40 + 0.60 = 1.00$ weighted average $= \dfrac{\Sigma xw}{\Sigma w} = \dfrac{90}{1.00} = 90$

57. a) – c) Answers will vary.

58. a) Answers will vary. One example is 2, 3, 5, 7, 7.

   b) Answers will vary. The answers for the example given in part a) above are as follows:

   Mean: $\dfrac{24}{5} = 4.8$, Median = 5, Mode = 7

Exercise Set 13.6

1. To find the **range**, subtract the lowest value in the set of data from the highest value.

3. Answers will vary.

5. It may be important to determine the consistency of the data.

7. σ

9. In manufacturing or anywhere else where a minimum variability is desired

11. They would be the same since the spread of data about each mean is the same.

13. a) The grades will be centered about the same number since the mean, 75.2, is the same for both classes.

   b) The spread of the data about the mean is greater for the evening class since the standard deviation is greater for the evening class.

15. Range = $13 - 2 = 11$

   $\bar{x} = \dfrac{35}{5} = 7$

x	$x - \bar{x}$	$(x - \bar{x})^2$
7	0	0
5	-2	4
2	-5	25
8	1	1
13	6	36
	0	66

   $\dfrac{66}{4} = 16.5, s = \sqrt{16.5} \approx 4.06$

17. Range = $126 - 120 = 6$

   $\bar{x} = \dfrac{861}{7} = 123$

x	$x - \bar{x}$	$(x - \bar{x})^2$
120	-3	9
121	-2	4
122	-1	1
123	0	0
124	1	1
125	2	4
126	3	9
	0	28

   $\dfrac{28}{6} \approx 4.67, s = \sqrt{4.67} \approx 2.16$

19. Range = $15 - 4 = 11$

   $\bar{x} = \dfrac{60}{6} = 10$

x	$x - \bar{x}$	$(x - \bar{x})^2$
4	-6	36
8	-2	4
9	-1	1
11	1	1
13	3	9
15	5	25
	0	76

   $\dfrac{76}{5} = 15.2, \ s = \sqrt{15.2} \approx 3.90$

21. Range = 12 − 7 = 5

$$\bar{x} = \frac{63}{7} = 9$$

x	$x - \bar{x}$	$(x - \bar{x})^2$
7	-2	4
9	0	0
7	-2	4
9	0	0
9	0	0
10	1	1
12	3	9
	0	18

$$\frac{18}{6} = 3, s = \sqrt{3} \approx 1.73$$

23. Range = 50 − 18 = $32

$$\bar{x} = \frac{360}{10} = \$36$$

x	$x - \bar{x}$	$(x - \bar{x})^2$
28	-8	64
28	-8	64
50	14	196
45	9	81
30	-6	36
45	9	81
48	12	144
18	-18	324
45	9	81
23	-13	169
	0	1240

$$\frac{1240}{9} \approx 137.78, s = \sqrt{137.78} \approx \$11.74$$

25. Range = 200 − 50 = $150

$$\bar{x} = \frac{1100}{10} = \$110$$

x	$x - \bar{x}$	$(x - \bar{x})^2$
50	-60	3600
130	20	400
60	-50	2500
55	-55	3025
75	-35	1225
200	90	8100
110	0	0
125	15	225
175	65	4225
	0	23,400

$$\frac{23,400}{9} = 2600, \; s = \sqrt{2600} \approx \$50.99$$

27. a) Range = 68 - 5 = $63

$$\bar{x} = \frac{204}{6} = \$34$$

x	$x - \bar{x}$	$(x - \bar{x})^2$
32	-2	4
60	26	676
14	-20	400
25	-9	81
5	-29	841
68	34	1156
	0	3158

$$\frac{3158}{5} = 631.6, s = \sqrt{631.6} \approx \$25.13$$

b) New data: 42, 70, 24, 35, 15, 78

The range and standard deviation will be the same. If each piece of data is increased by the same number, the range and standard deviation will remain the same.

c) Range = 78 - 15 = $63

$$\bar{x} = \frac{264}{6} = \$44$$

x	$x - \bar{x}$	$(x - \bar{x})^2$
42	-2	4
70	26	676
24	-20	400
35	-9	81
15	-29	841
78	34	1156
	0	3158

$$\frac{3158}{5} = 631.6, s = \sqrt{631.6} \approx \$25.13$$

The answers remain the same.

29. a) - c) Answers will vary.

d) If each number in a distribution is multiplied by n, both the mean and standard deviation of the new distribution will be n times that of the original distribution.

e) The mean of the second set is $4 \times 5 = 20$, and the standard deviation of the second set is $2 \times 5 = 10$.

31. a) The standard deviation increases. There is a greater spread from the mean as they get older.

b) ≈ 133 lb

c) $\frac{175 - 90}{4} = 21.25 \approx 21$ lb

d) The mean weight is about 100 pounds and the normal range is about 60 to 140 pounds.

e) The mean height is about 62 inches and the normal range is about 53 to 68 inches.

f) 100% - 95% = 5%

33.   a)   East

Number of oil changes made	Number of days
15-20	2
21-26	2
27-32	5
33-38	4
39-44	7
45-50	1
51-56	1
57-62	2
63-68	1

                                West

Number of oil changes made	Number of days
15-20	0
21-26	0
27-32	6
33-38	9
39-44	4
45-50	6
51-56	0
57-62	0
63-68	0

b)

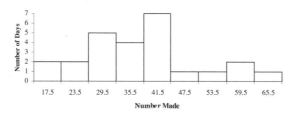

Number of Oil Changes Made Daily at East Store

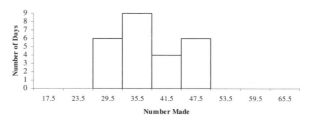

Number of Oil Changes Made Daily at West Store

c) They appear to have about the same mean since they are both centered around 38.

d) The distribution for East is more spread out. Therefore, East has a greater standard deviation.

e) East: $\dfrac{950}{25} = 38$, West: $\dfrac{950}{25} = 38$

33. f)

	East				West	
x	$x-\bar{x}$	$(x-\bar{x})^2$		x	$x-\bar{x}$	$(x-\bar{x})^2$
33	-5	25		38	0	0
30	-8	64		38	0	0
25	-13	169		37	-1	1
27	-11	121		36	-2	4
40	2	4		30	-8	64
44	6	36		45	7	49
49	11	121		28	-10	100
52	14	196		47	9	81
42	4	16		30	-8	64
59	21	441		46	8	64
19	-19	361		38	0	0
22	-16	256		39	1	1
57	19	361		40	2	4
67	29	841		34	-4	16
15	-23	529		31	-7	49
41	3	9		45	7	49
43	5	25		29	-9	81
27	-11	121		38	0	0
42	4	16		38	0	0
43	5	25		39	1	1
37	-1	1		37	-1	1
38	0	0		42	4	16
31	-7	49		46	8	64
32	-6	36		31	-7	49
35	-3	9		48	10	100
	0	3832			0	858

$$\frac{3832}{24} \approx 159.67, \quad s = \sqrt{159.67} \approx 12.64 \qquad \frac{858}{24} = 35.75, \quad s = \sqrt{35.75} \approx 5.98$$

34. Answers will vary.   35. 6, 6, 6, 6, 6

Exercise Set 13.7

1. A **rectangular distribution** is one where all the values have the same frequency.

3. A **bimodal distribution** is one where two nonadjacent values occur more frequently than any other values in a set of data.

5. A **distribution skewed to the left** is one that has "a tail" on its left.

7. a) *B*
   b) *C*
   c) *A*

9. The distribution of outcomes from the roll of a die

11. J shaped right - consumer price index; J shaped left - value of the dollar

13. Normal   15.   Skewed right

17. The mode is the lowest value, the median is greater than the mode, and the mean is greater than the median. The greatest frequency appears on the left side of the curve. Since the mode is the value with the greatest frequency, the mode would appear on the left side of the curve (where the lowest values are). Every value in the set of data is considered in determining the mean. The values on the far right of the curve would increase the value of the mean. Thus, the value of the mean would be farther to the right than the mode. The median would be between the mode and the mean.

19. Answers will vary.

21. In a normal distribution the mean, median, and the mode all have the same value.
23. A z-score will be negative when the piece of data is less than the mean.
25. 0

27. 0.500

29. $0.477 + 0.341 = 0.818$

31. $0.500 - 0.466 = 0.034$

33. $0.500 - 0.463 = 0.037$

35. $0.500 - 0.481 = 0.019$

37. $0.500 - 0.447 = 0.053$

39. $0.261 = 26.1\%$

41. $0.410 + 0.488 = 0.898 = 89.8\%$

43. $0.500 + 0.471 = 0.971 = 97.1\%$

45. $0.500 + 0.475 = 0.975 = 97.5\%$

47. $0.466 - 0.437 = 0.029 = 2.9\%$

49. a) Jake, Sarah, and Carol scored above the mean because their z-scores are positive.
    b) Marie and Kevin scored at the mean because their z-scores are zero.
    c) Omar, Justin, and Kim scored below the mean because their z-scores are negative.

51. $0.500 = 50\%$

53. $z_{23} = \dfrac{23-18}{4} = \dfrac{5}{4} = 1.25$

    $0.500 - 0.394 = 0.106 = 10.6\%$

55. $z_{1650} = \dfrac{1650-1600}{100} = \dfrac{50}{100} = 0.50$

    $0.500 + 0.192 = 0.692 = 69.2\%$

57. $z_{1650} = 0.50$ and $z_{1750} = 1.50$
    (See Exercises 55 and 56.)
    $0.433 - 0.192 = 0.241 = 24.1\%$

59. $z_{1500} = \dfrac{1500-1600}{100} = \dfrac{-100}{100} = -1.00$

    $z_{1625} = \dfrac{1625-1600}{100} = \dfrac{25}{100} = 0.25$

    $0.341 = 0.099 = 0.44 = 44.0\%$

61. $z_{7.4} = \dfrac{7.4-7.6}{0.4} = \dfrac{-0.2}{0.4} = -0.50$

    $z_{7.7} = \dfrac{7.7-7.6}{0.4} = \dfrac{0.1}{0.4} = 0.25$

    $0.192 + 0.099 = 0.291 = 29.1\%$

63. $z_{7.7} = 0.25$ (See Exercise 61.)

    $0.500 + 0.099 = 0.599 = 59.9\%$

65. $0.500 = 50.0\%$

67. $z_{191} = \dfrac{191-206}{12} = \dfrac{-15}{12} = -1.25$

    $0.500 - 0.394 = 0.106 = 10.6\%$

69. 10.6% of females have a cholesterol level less than 191. (See Exercise 67.)
    approximately $0.106(200) = 21.2 \approx 21$ women

71. $z_{30,750} = \dfrac{30,750-35,000}{2500} = \dfrac{-4250}{2500} = -1.70$

    $z_{38,300} = \dfrac{38,300-35,000}{2500} = \dfrac{3300}{2500} = 1.32$

    $0.455 + 0.407 = 0.862 = 86.2\%$

73. The tires that last less than 30,750 miles will fail to live up to the guarantee.
    $z_{30,750} = -1.70$ (See Exercise 71.)
    $0.500 - 0.455 = 0.045 = 4.5\%$

75. $z_{3.1} = \dfrac{3.1-3.7}{1.2} = \dfrac{-0.6}{1.2} = -0.50$

    $0.192 + 0.500 = 0.692 = 69.2\%$

77. $z_{6.7} = \dfrac{6.7 - 3.7}{1.2} = \dfrac{3.0}{1.2} = 2.50$

$0.500 - 0.494 = 0.006 = 0.6\%$

79. 69.2% of the children are older than 3.1 years. (See Exercise 75.)

approximately $0.692(120) = 83.04 \approx 83$ children

81. Customers will be able to claim a refund if they lose less than 5 lb.

$z_5 = \dfrac{5 - 6.7}{0.81} = \dfrac{-1.7}{0.81} = -2.10$

$0.500 - 0.482 = 0.018 = 1.8\%$

83. The standard deviation is too large. There is too much variation.

85. a) Katie: $z_{28,408} = \dfrac{28,408 - 23,200}{2170} = \dfrac{5208}{2170} = 2.4$

Stella: $z_{29,510} = \dfrac{29,510 - 25,600}{2300} = \dfrac{3910}{2300} = 1.7$

b) Katie. Her z-score is higher than Stella's z-score. This means her sales are further above the mean than Stella's sales.

87. Answers will vary.

89. Using Table 13.7, the answer is -1.18.

90. Answers will vary.

91. $\dfrac{0.77}{2} = 0.385$

Using the table in Section 13.7, an area of 0.385 has a z-score of 1.20.

$z = \dfrac{x - \overline{x}}{s}$

$1.20 = \dfrac{14.4 - 12}{s}$

$1.20 = \dfrac{2.4}{s}$

$\dfrac{1.20s}{1.20} = \dfrac{2.4}{1.20}$

$s = 2$

Exercise Set 13.8

1. The **correlation coefficient** measures the strength of the linear relationship between the quantities.

3. 1   5. 0

7. A positive correlation indicates that as one quantity increases, the other quantity increases.

9. The **level of significance** is used to identify the cutoff between results attributed to chance and results attributed to an actual relationship between the two variables.

11. No correlation

13. Strong positive

15. Yes, $\left| 0.76 \right| > 0.684$

17. Yes, $\left| -0.73 \right| > 0.707$

19. No, $\left| -0.23 \right| < 0.254$

21. No, $\left| 0.82 \right| < 0.917$

Note: The answers in the remainder of this section may differ slightly from your answers, depending upon how your answers are rounded and which calculator you used.

23. a)

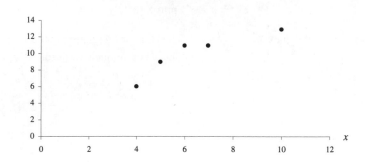

b)

x	y	x^2	y^2	xy
4	6	16	36	24
5	9	25	81	45
6	11	36	121	66
7	11	49	121	77
10	13	100	169	130
32	50	226	528	342

$$r = \frac{5(342) - 32(50)}{\sqrt{5(226) - 1024}\sqrt{5(528) - 2500}} = \frac{110}{\sqrt{106}\sqrt{140}} \approx 0.903$$

c) Yes, $|0.903| > 0.878$ d) No, $|0.903| < 0.959$

25. a)

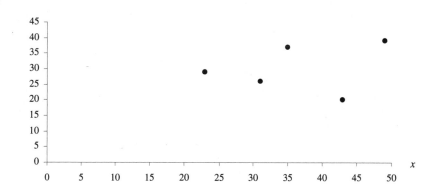

b)

x	y	x^2	y^2	xy
23	29	529	841	667
35	37	1225	1369	1295
31	26	961	676	806
43	20	1849	400	860
49	39	2401	1521	1911
181	151	6965	4807	5539

$$r = \frac{5(5539) - 181(151)}{\sqrt{5(6965) - 32{,}761}\sqrt{5(4807) - 22{,}801}} = \frac{364}{\sqrt{2064}\sqrt{1234}} \approx 0.228$$

c) No, $|0.228| < 0.878$ d) No, $|0.228| < 0.959$

27.   a)

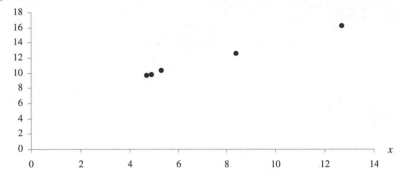

b)

x	y	x^2	y^2	xy
5.3	10.3	28.09	106.09	54.59
4.7	9.6	22.09	92.16	45.12
8.4	12.5	70.56	156.25	105
12.7	16.2	161.29	262.44	205.74
4.9	9.8	24.01	96.04	48.02
36	58.4	306.04	712.98	458.47

$$r = \frac{5(458.47) - 36(58.4)}{\sqrt{5(306.04) - 1296}\sqrt{5(712.98) - 3410.56}} = \frac{189.95}{\sqrt{234.2}\sqrt{154.34}} \approx 0.999$$

c) Yes, $|0.999| > 0.878$

d) Yes, $|0.999| > 0.959$

29.   a)

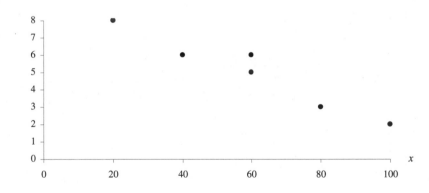

b)

x	y	x^2	y^2	xy
100	2	10,000	4	200
80	3	6400	9	240
60	5	3600	25	300
60	6	3600	36	360
40	6	1600	36	240
20	8	400	64	160
360	30	25,600	174	1500

$$r = \frac{6(1500) - 360(30)}{\sqrt{6(25,600) - 129,600}\sqrt{6(174) - 900}} = \frac{-1800}{\sqrt{24,000}\sqrt{144}} \approx -0.968$$

c) Yes, $|-0.968| > 0.811$

d) Yes, $|-0.968| > 0.917$

31. From # 23: $m = \dfrac{5(342) - 32(50)}{5(226) - 1024} = \dfrac{110}{106} \approx 1.0$

$b = \dfrac{50 - \dfrac{110}{106}(32)}{5} \approx 3.4, \quad y = 1.0x + 3.4$

33. From # 25: $m = \dfrac{5(5539) - 181(151)}{5(6965) - 32{,}761} = \dfrac{364}{2064} \approx 0.2$

$b = \dfrac{151 - \dfrac{364}{2064}(181)}{5} \approx 23.8, \quad y = 0.2x + 23.8$

35. From # 27: $m = \dfrac{5(458.47) - 36(58.4)}{5(306.04) - 1296} = \dfrac{189.95}{234.2} \approx 0.8$

$b = \dfrac{58.4 - \dfrac{189.95}{234.2}(36)}{5} \approx 5.8, \quad y = 0.8x + 5.8$

37. From # 29: $m = \dfrac{6(1500) - 360(30)}{6(25{,}600) - 129{,}600} = \dfrac{-1800}{24{,}000} \approx -0.1$

$b = \dfrac{30 - \dfrac{-1800}{24{,}000}(360)}{6} \approx 9.5, \quad y = -0.1x + 9.5$

39. a)

x	y	x^2	y^2	xy
8	15	64	225	120
20	28	400	784	560
9	20	81	400	180
15	25	225	625	375
16	28	256	784	448
2	5	4	25	10
70	121	1030	2843	1693

$r = \dfrac{6(1693) - 70(121)}{\sqrt{6(1030) - 4900}\sqrt{6(2843) - 14{,}641}} = \dfrac{1688}{\sqrt{1280}\sqrt{2417}} \approx 0.960$

b) Yes, $|\,0.960\,| > 0.811$

c) $m = \dfrac{6(1693) - 70(121)}{6(1030) - 4900} = \dfrac{1688}{1280} \approx 1.3$, $b = \dfrac{121 - \dfrac{1688}{1280}(70)}{6} \approx 4.8, \quad y = 1.3x + 4.8$

41. a)

x	y	x^2	y^2	xy
20	40	400	1600	800
40	45	1600	2025	1800
50	70	2500	4900	3500
60	76	3600	5776	4560
80	92	6400	8464	7360
100	95	10,000	9025	9500
350	418	24,500	31,790	27,520

$$r = \frac{6(27,520)-350(418)}{\sqrt{6(24,500)-122,500}\sqrt{6(31,790)-174,724}} = \frac{18,820}{\sqrt{24,500}\sqrt{16,016}} \approx 0.950$$

b) Yes, $\left|0.950\right| > 0.917$

c) $m = \dfrac{6(27,520)-350(418)}{6(24,500)-122,500} = \dfrac{18,820}{24,500} \approx 0.8$, $\quad b = \dfrac{418 - \dfrac{18,820}{24,500}(350)}{6} \approx 24.9$, $\quad y = 0.8x + 24.9$

43. a)

x	y	x^2	y^2	xy
20	8	400	64	160
12	10	144	100	120
18	12	324	144	216
15	9	225	81	135
22	6	484	36	132
10	15	100	225	150
20	7	400	49	140
12	18	144	324	216
129	85	2221	1023	1269

$$r = \frac{8(1269)-129(85)}{\sqrt{8(2221)-16,641}\sqrt{8(1023)-7225}} = \frac{-813}{\sqrt{1127}\sqrt{959}} \approx -0.782$$

b) Yes, $\left|-0.782\right| > 0.707$

c) $m = \dfrac{8(1269)-129(85)}{8(2221)-16,641} = \dfrac{-813}{1127} \approx -0.7$, $\quad b = \dfrac{85 - \dfrac{-813}{1127}(129)}{8} \approx 22.3$, $\quad y = -0.7x + 22.3$

d) $y = -0.7(14) + 22.3 = 12.5$ muggings

45. a)

x	y	x^2	y^2	xy
89	22	7921	484	1958
110	28	12,100	784	3080
125	30	15,625	900	3750
92	26	8464	676	2392
100	22	10,000	484	2200
95	21	9025	441	1995
108	28	11,664	784	3024
97	25	9409	625	2425
816	202	84,208	5178	20,824

$$r = \frac{8(20{,}824) - 816(202)}{\sqrt{8(84{,}208) - 665{,}856}\sqrt{8(5178) - 40{,}804}} = \frac{1760}{\sqrt{7808}\sqrt{620}} \approx 0.800$$

b) Yes, $|\,0.800\,| > 0.707$

c) $m = \dfrac{8(20{,}824) - 816(202)}{8(84{,}208) - 665{,}856} = \dfrac{1760}{7808} \approx 0.2$, $b = \dfrac{202 - \dfrac{1760}{7808}(816)}{8} \approx 2.3$, $y = 0.2x + 2.3$

d) $y = 0.2(115) + 2.3 = 25.3 \approx 25$ units

47. a)

x	y	x^2	y^2	xy
1	80.0	1	6400.0	80.0
2	76.2	4	5806.4	152.4
3	68.7	9	4719.7	206.1
4	50.1	16	2510.0	200.4
5	30.2	25	912.0	151.0
6	20.8	36	432.6	124.8
21	326	91	20,780.7	914.7

$$r = \frac{6(914.7) - 21(326)}{\sqrt{6(91) - 441}\sqrt{6(20{,}780.7) - 106{,}276}} = \frac{-1357.8}{\sqrt{105}\sqrt{18{,}408.2}} \approx -0.977$$

b) Yes, $|-0.977| > 0.917$

c) $m = \dfrac{6(914.7) - 21(326)}{6(91) - 441} = \dfrac{-1357.8}{105} \approx -12.9$, $b = \dfrac{326 - \dfrac{-1357.8}{105}(21)}{6} \approx 99.6$, $y = -12.9x + 99.6$

d) $y = -12.9(4.5) + 99.6 = 41.55 \approx 41.6\%$

49. a) and b) Answers will vary.

c)

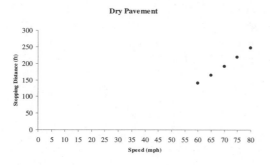

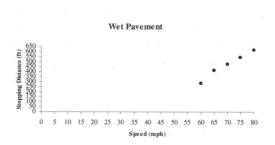

d)

x	y	x^2	y^2	xy
60	140	3600	19,600	8400
65	164	4225	26,896	10,660
70	190	4900	36,100	13,300
75	218	5625	47,524	16,350
80	247	6400	61,009	19,760
350	959	24,750	191,129	68,470

$$r = \frac{5(68,470) - 350(959)}{\sqrt{5(24,750) - 122,500}\sqrt{5(191,129) - 919,681}} = \frac{6700}{\sqrt{1250}\sqrt{35,964}} \approx 0.999$$

e)

x	y	x^2	y^2	xy
60	280	3600	78,400	16,800
65	410	4225	168,100	26,650
70	475	4900	225,625	33,250
75	545	5625	297,025	40,875
80	618	6400	381,924	49,440
350	2328	24,750	1,151,074	167,015

$$r = \frac{5(167,015) - 350(2328)}{\sqrt{5(24,750) - 122,500}\sqrt{5(1,151,074) - 5,419,584}} = \frac{20,275}{\sqrt{1250}\sqrt{335,786}} \approx 0.990$$

f) Answers will vary.

g) $m = \dfrac{5(68,470) - 350(959)}{5(24,750) - 122,500} = \dfrac{6700}{1250} \approx 5.4$, $\quad b = \dfrac{959 - \dfrac{6700}{1250}(350)}{5} = -183.4$, $\quad y = 5.4x - 183.4$

h) $m = \dfrac{5(167,015) - 350(2328)}{5(24,750) - 122,500} = \dfrac{20,275}{1250} \approx 16.2$, $\quad b = \dfrac{2328 - \dfrac{20,275}{1250}(350)}{5} = -669.8$, $\quad y = 16.2x - 669.8$

i) Dry: $y = 5.4(77) - 183.4 = 232.4$ ft

 Wet: $y = 16.2(77) - 669.8 = 577.6$ ft

51. Answers will vary.

53. a)

x	y	x^2	y^2	xy
1996	157	3,984,016	24,649	313,372
1997	161	3,988,009	25,921	321,517
1998	163	3,992,004	26,569	325,674
1999	167	3,996,001	27,889	333,833
2000	172	4,000,000	29,584	344,000
2001	177	4,004,001	31,329	354,177
11,991	997	23,964,031	165,941	1,992,573

$$r = \frac{6(1,992,573) - 11,991(997)}{\sqrt{6(23,964,031) - 143,784,081}\sqrt{6(165,941) - 994,009}} = \frac{411}{\sqrt{105}\sqrt{1637}} \approx 0.991$$

b) Should be the same.

c)

x	y	x^2	y^2	xy
0	157	0	24,649	0
1	161	1	25,921	161
2	163	4	26,569	326
3	167	9	27,889	501
4	172	16	29,584	688
5	177	25	31,329	885
15	997	55	165,941	2561

$$r = \frac{6(2561) - 15(997)}{\sqrt{6(55) - 225}\sqrt{6(165,941) - 994,009}} = \frac{411}{\sqrt{105}\sqrt{1637}} \approx 0.991$$

The values are the same.

Review Exercises

1. a) A **population** consists of all items or people of interest.

   b) A **sample** is a subset of the population.

2. A **random sample** is one where every item in the population has the same chance of being selected.

3. The candy bars may have lots of calories, or fat, or sodium. Therefore, it may not be healthy to eat them.

4. Sales may not necessarily be a good indicator of profit. Expenses must also be considered.

5. a)

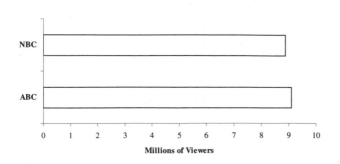

b)

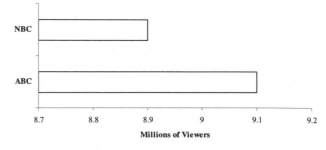

6. a)

Class	Frequency
35	1
36	3
37	6
38	2
39	3
40	0
41	4
42	1
43	3
44	1
45	1

b) and c)

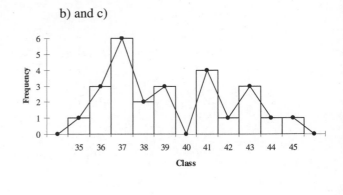

7. a)

High Temperature	Number of Cities
30 - 39	5
40 - 49	8
50 - 59	5
60 - 69	6
70 - 79	6
80 - 89	10

b) and c)

Average Daily High Temperature in January for Selected Cities

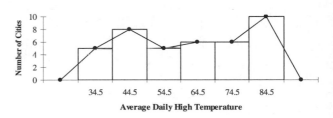

d) 3 | 6 represents 36

```
3 | 0 3 4 5 6
4 | 1 2 2 3 4 7 8 8
5 | 0 4 4 5 6
6 | 5 6 6 7 8 9
7 | 3 5 5 7 7 9
8 | 0 1 3 3 4 6 6 7 8 9
```

8. $\bar{x} = \dfrac{480}{6} = 80$

9. $\dfrac{79+83}{2} = 81$

10. None

11. $\dfrac{63+93}{2} = 78$

12. $93 - 63 = 30$

13.

x	$x - \bar{x}$	$(x - \bar{x})^2$
63	-17	289
76	-4	16
79	-1	1
83	3	9
86	6	36
93	13	169
	0	520

$\dfrac{520}{5} = 104$, $s = \sqrt{104} \approx 10.20$

14. $\bar{x} = \dfrac{156}{12} = 13$

15. $\dfrac{12+14}{2} = 13$

16. 12 and 7

17. $\dfrac{4+23}{2} = 13.5$

18. $23 - 4 = 19$

19.

x	$x - \bar{x}$	$(x - \bar{x})^2$
4	-9	81
5	-8	64
7	-6	36
7	-6	36
12	-1	1
12	-1	1
14	1	1
15	2	4
17	4	16
19	6	36
21	8	64
23	10	100
	0	440

$$\dfrac{440}{11} = 40, \quad s = \sqrt{40} \approx 6.32$$

20. $z_{37} = \dfrac{37-42}{5} = \dfrac{-5}{5} = -1.00$

$z_{47} = \dfrac{47-42}{5} = \dfrac{5}{5} = 1.00$

$0.341 + 0.341 = 0.682 = 68.2\%$

21. $z_{32} = \dfrac{32-42}{5} = \dfrac{-10}{5} = -2.00$

$z_{52} = \dfrac{52-42}{5} = \dfrac{10}{5} = 2.00$

$0.477 + 0.477 = 0.954 = 95.4\%$

22. $z_{50} = \dfrac{50-42}{5} = \dfrac{8}{5} = 1.60$

$0.500 + 0.445 = 0.945 = 94.5\%$

23. $z_{50} = \dfrac{50-42}{5} = \dfrac{8}{5} = 1.60$

$0.500 - 0.445 = 0.055 = 5.5\%$

24. $z_{39} = \dfrac{39-42}{5} = \dfrac{-3}{5} = -.60$

$0.500 + 0.226 = 0.726 = 72.6\%$

25. $z_{20} = \dfrac{20-20}{5} = \dfrac{0}{5} = 0$

$z_{25} = \dfrac{25-20}{5} = \dfrac{5}{5} = 1.00$

$0.341 = 34.1\%$

26. $z_{18} = \dfrac{18-20}{5} = \dfrac{-2}{5} = -0.40$

$0.500 - 0.155 = 0.345 = 34.5\%$

27. $z_{22} = \dfrac{22-20}{5} = \dfrac{2}{5} = 0.40$

$z_{28} = \dfrac{28-20}{5} = \dfrac{8}{5} = 1.60$

$0.445 - 0.155 = 0.29 = 29.0\%$

28. $z_{30} = \dfrac{30-20}{5} = \dfrac{10}{5} = 2.00$

$0.500 - 0.477 = 0.023 = 2.3\%$

29. a)

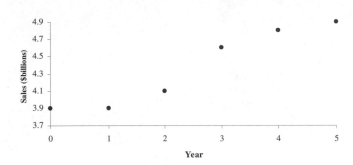

b)   Yes; positive because generally as the year increases, the sales increase.

c)

x	y	x^2	y^2	xy
0	3.9	0	15.21	0
1	3.9	1	15.21	3.9
2	4.1	4	16.81	8.2
3	4.6	9	21.16	13.8
4	4.8	16	23.04	19.2
5	4.9	25	24.01	24.5
15	26.2	55	115.44	69.6

$$r = \frac{6(69.6)-15(26.2)}{\sqrt{6(55)-225}\sqrt{6(115.44)-686.44}} = \frac{24.6}{\sqrt{105}\sqrt{6.2}} \approx 0.964$$

d) Yes, $\mid 0.964 \mid > 0.811$

e) $m = \dfrac{6(69.6)-15(26.2)}{6(55)-225} = \dfrac{24.6}{105} \approx 0.2$

$$b = \frac{26.2 - \dfrac{24.6}{105}(15)}{6} \approx 3.8, \quad y = 0.2x + 3.8$$

30. a)

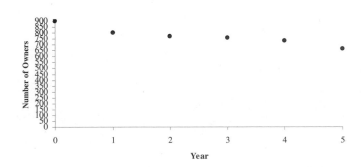

b)   Yes; negative because generally as the year increases, the number of owners decreases.

30. c)

x	y	x^2	y^2	xy
0	897	0	804,609	0
1	800	1	640,000	800
2	770	4	592,900	1540
3	760	9	577,600	2280
4	735	16	540,225	2940
5	663	25	439,569	3315
15	4625	55	3,594,903	10,875

$$r = \frac{6(10,875) - 15(4625)}{\sqrt{6(55) - 225}\sqrt{6(3,594,903) - 21,390,625}} = \frac{-4125}{\sqrt{105}\sqrt{178,793}} \approx -0.952$$

d) Yes, $\left| -0.952 \right| > 0.811$

e) $m = \dfrac{6(10,875) - 15(4625)}{6(55) - 225} = \dfrac{-4125}{105} \approx -39.3$

$b = \dfrac{4625 - \dfrac{-4125}{105}(15)}{6} \approx 869.0, \quad y = -39.3x + 869.0$

31. a)

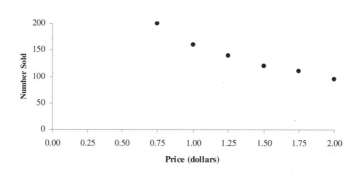

b)   Yes; negative because generally as the price increases, the number sold decreases.

c)

x	y	x^2	y^2	xy
0.75	200	0.5625	40,000	150
1.00	160	1	25,600	160
1.25	140	1.5625	19,600	175
1.50	120	2.25	14,400	180
1.75	110	3.0625	12,100	192.5
2.00	95	4	9025	190
8.25	825	12.4375	120,725	1047.5

$$r = \frac{6(1047.5) - 8.25(825)}{\sqrt{6(12.4375) - 68.0625}\sqrt{6(120,725) - 680,625}} = \frac{-521.25}{\sqrt{6.5625}\sqrt{43,725}} \approx -0.973$$

d) Yes, $\left| -0.973 \right| > 0.811$

31. e) $m = \dfrac{6(1047.5) - 8.25(825)}{6(12.4375) - 68.0625} = \dfrac{-521.25}{6.5625} \approx -79.4$

$b = \dfrac{825 - \dfrac{-521.25}{6.5625}(8.25)}{6} \approx 246.7, \quad y = -79.4x + 246.7$

f) $y = -79.4(1.60) + 246.7 = 119.66 \approx 120$ sold

32. Mode = 175 lb

33. Median = 180 lb

34. 25%

35. 25%

36   100% - 86% = 14%

37. approximately 100(187) = 18,700 lb

38. 187 + 2(23) = 233 lb

39. 187 - 1.8(23) = 145.6 lb

41. 2

40. $\bar{x} = \dfrac{150}{42} \approx 3.57$

42. $\dfrac{3+3}{2} = 3$

43. $\dfrac{0+14}{2} = 7$

44. 14 - 0 = 14

45.

x	$x-\bar{x}$	$(x-\bar{x})^2$	x	$x-\bar{x}$	$(x-\bar{x})^2$	x	$x-\bar{x}$	$(x-\bar{x})^2$
0	-3.6	12.96	2	-1.6	2.56	4	0.4	0.16
0	-3.6	12.96	2	-1.6	2.56	5	1.4	1.96
0	-3.6	12.96	3	-0.6	0.36	5	1.4	1.96
0	-3.6	12.96	3	-0.6	0.36	5	1.4	1.96
0	-3.6	12.96	3	-0.6	0.36	6	2.4	5.76
0	-3.6	12.96	3	-0.6	0.36	6	2.4	5.76
1	-2.6	6.76	3	-0.6	0.36	6	2.4	5.76
1	-2.6	6.76	3	-0.6	0.36	6	2.4	5.76
2	-1.6	2.56	4	0.4	0.16	6	2.4	5.76
2	-1.6	2.56	4	0.4	0.16	7	3.4	11.56
2	-1.6	2.56	4	0.4	0.16	8	4.4	19.36
2	-1.6	2.56	4	0.4	0.16	10	6.4	40.96
2	-1.6	2.56	4	0.4	0.16	14	10.4	108.16
2	-1.6	2.56	4	0.4	0.16			332.32
2	-1.6	2.56						

$\dfrac{332.32}{41} \approx 8.105, \quad s = \sqrt{8.105} \approx 2.85$

46.

# of Child.	# of Presidents
0 - 1	8
2 - 3	15
4 - 5	10
6 - 7	6
8 - 9	1
10 - 11	1
12 - 13	0
14 - 15	1

47. and 48.

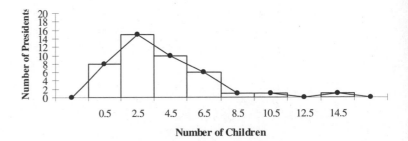

Number of Children of U.S. Presidents

49. No, it is skewed to the right.
50. No, some families have no children, more have one child, the greatest percent may have two children, fewer have three children, etc.
51. No, the number of children per family has decreased over the years.

Chapter Test

1. $\bar{x} = \dfrac{180}{5} = 36$

2. 37

3. 37

4. $\dfrac{21+46}{2} = 33.5$

5. $46 - 21 = 25$

6.

x	$x-\bar{x}$	$(x-\bar{x})^2$
21	-15	225
37	1	1
37	1	1
39	3	9
46	10	100
	0	336

$\dfrac{336}{4} = 84, \ s = \sqrt{84} \approx 9.17$

7.

Class	Frequency
25 - 30	7
31 - 36	5
37 - 42	1
43 - 48	7
49 - 54	5
55 - 60	3
61 - 66	2

8. and 9.

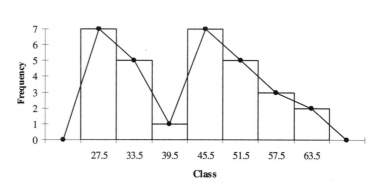

10. Mode = $695

11. Median = $670

12. 100% - 25% = 75%

13. 79%

14. 100(700) = $70,000

15. $700 + 1($40) = $740

16. $700 - 1.5($40) = $640

17. $z_{50,000} = \dfrac{50,000-75,000}{12,000} = \dfrac{-25,000}{12,000} \approx -2.08$

$z_{70,000} = \dfrac{70,000-75,000}{12,000} = \dfrac{-5000}{12,000} \approx -0.42$

$0.481 - 0.163 = 0.318 = 31.8\%$

18. $z_{60,000} = \dfrac{60,000-75,000}{12,000} = \dfrac{-15,000}{12,000} = -1.25$

$0.500 + 0.394 = 0.894 = 89.4\%$

19. $z_{90,000} = \dfrac{90,000-75,000}{12,000} = \dfrac{15,000}{12,000} = 1.25$

$0.500 - 0.394 = 0.106 = 10.6\%$

20. From #17 and #18,

$z_{60,000} = -1.25$ and $z_{70,000} \approx -0.42$

$0.394 - 0.163 = 0.231 = 23.1\%$

approximately $0.231(300) = 69.3 \approx 69$ cars

21. a)

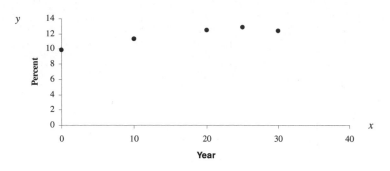

b) Yes

c)

x	y	x^2	y^2	xy
0	9.8	0	96.04	0
10	11.3	100	127.69	113
20	12.5	400	156.25	250
25	12.8	625	163.84	320
30	12.4	900	153.76	372
85	58.8	2025	697.58	1055

$$r = \frac{5(1055) - 85(58.8)}{\sqrt{5(2025) - 7225}\sqrt{5(697.58) - 3457.44}} = \frac{277}{\sqrt{2900}\sqrt{30.46}} \approx 0.932$$

d) Yes, $\mid 0.932 \mid > 0.878$

e) $m = \dfrac{5(1055) - 85(58.8)}{5(2025) - 7225} = \dfrac{277}{2900} \approx 0.1$

$b = \dfrac{58.8 - \dfrac{277}{2900}(85)}{5} \approx 10.1, \quad y = 0.1x + 10.1$

f) $y = 0.1(40) + 10.1 = 14.1\%$

CHAPTER FOURTEEN

GRAPH THEORY

Exercise Set 14.1

1. A **graph** is a finite set of points, called **vertices**, that are connected with line segments, called **edges**.

3.
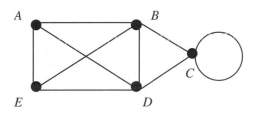

5. If the number of edges connected to the vertex is even, the vertex is **even**. If the number of edges connected to the vertex is odd, the vertex is **odd**.

7. a) A **path** is a sequence of adjacent vertices and the edges connecting them.
   b) A **circuit** is a path that begins and ends at the same vertex.
   c)
   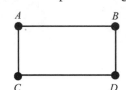

   The path *A, B, D, C* is a path that is not a circuit.
   The path *A, B, D, C, A* is a path that is also a circuit.

9.

   A, B, and *C* are all even.

11.
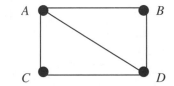

   B and *C* are even. *A* and *D* are odd.

301

13.

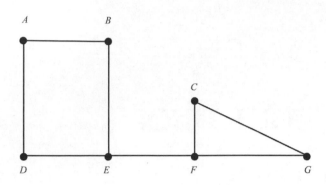

15. No. There is no edge connecting vertices B and C. Therefore, A, B, C, D, E is not a path.

17. Yes. One example is A, C, E, D, B.

19. Yes. One example is C, A, B, D, E, C, D.

21.

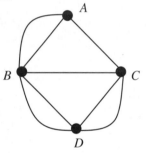

23.

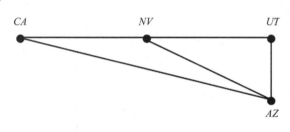

25.

27.

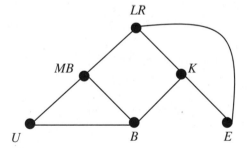

29.

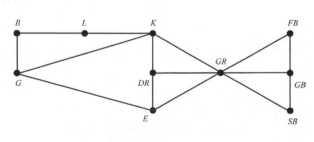

31.

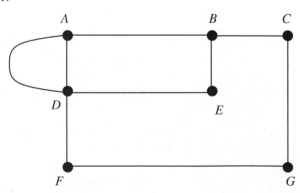

33. Disconnected. There is no path that connects *A* to *C*.

35.   Connected

37. Edge *AB*

39.   Edge *EF*

41.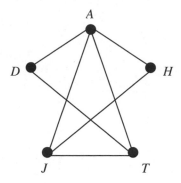

Other answers are possible.

43. It is impossible to have a graph with an odd number of odd vertices.

45. a) and b) Answers will vary.

Exercise Set 14.2

1. a) An **Euler path** is a path that must include each edge of a graph exactly one time.

   b) and c)

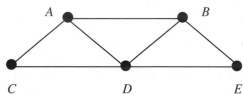

   b) The path *A*, *B*, *E*, *D*, *C*, *A*, *D*, *B* is an Euler path.
   c) The path *A*, *B*, *E*, *D*, *C* is a path that is not an Euler path.

3. a) Yes, according to Euler's Theorem.
   b) Yes, according to Euler's Theorem.
   c) No, according to Euler's Theorem.
5. If all of the vertices are even, the graph has an Euler circuit.
7. *A*, *B*, *C*, *D*, *E*, *B*, *E*, *D*, *A*, *C*; other answers are possible.
9. No. This graph has exactly two odd vertices. Each Euler path must begin with an odd vertex. *B* is an even vertex.
11. *A*, *B*, *A*, *C*, *B*, *E*, *C*, *D*, *A*, *D*, *E*; other answers are possible.
13. No. A graph with exactly two odd vertices has no Euler circuits.
15. *A*, *B*, *C*, *E*, *F*, *D*, *E*, *B*, *D*, *A*; other answers are possible.
17. *C*, *B*, *A*, *D*, *F*, *E*, *D*, *B*, *E*, *C*; other answers are possible.
19. *E*, *F*, *D*, *E*, *B*, *D*, *A*, *B*, *C*, *E*; other answers are possible.

21. a) Yes. There are zero odd vertices.   23.   a) No. There are more than two odd vertices.
    b) Yes. There are zero odd vertices.   b) No. There are more than zero odd vertices.

25. a) Yes. Each island would correspond to an odd vertex. According to item 2 of Euler's Theorem, a graph with exactly two odd vertices has at least one Euler path, but no Euler circuit.
    b) They could start on either island and finish at the other.

In Exercises 27-31, one graph is shown. Other graphs are possible.

27. a)

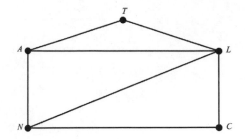

b) Vertices *A* and *N* are both odd. According to item 2 of Euler's Theorem, since there are exactly two odd vertices, at least one Euler path, but no Euler circuits exist.

Yes; *A, T, L, C, N, L, A, N*

c) No. (See part b) above.)

29. a)

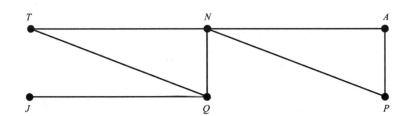

b) Vertices *J* and *Q* are both odd. According to item 2 of Euler's Theorem, since there are exactly two odd vertices, at least one Euler path, but no Euler circuits exist.

Yes; *J, Q, T, N, A, P, N, Q*

c) No. (See part b) above.)

31. a)

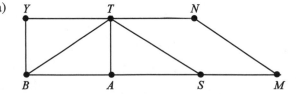

b) Vertices *A, S, T,* and *B* are odd. According to item 3 of Euler's Theorem there is neither an Euler path nor an Euler circuit.

c) No. (See part b) above.)

33. a) No. The graph representing the floor plan:

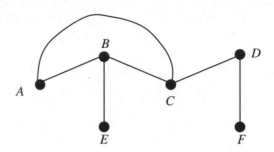

The wood carver is seeking an Euler path or an Euler circuit. Note that vertices *B, C, E,* and *F* are all odd. According to item 3 of Euler's Theorem, since there are more than two odd vertices, no Euler path or Euler circuit can exist.

b) No such path exists.

35. a) Yes. The graph representing the floor plan:

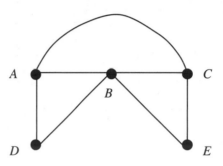

The wood carver is seeking an Euler path or an Euler circuit. Note that vertices *A* and *C* are both odd. According to item 2 of Euler's Theorem, since there are exactly two odd vertices, at least one Euler path, but no Euler circuits exist.

b) One path is *A, D, B, E, C, B, A, C.*

37. a) Yes. The graph representing the map:

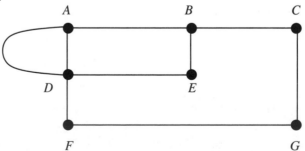

They are seeking an Euler path or an Euler circuit. Note that vertices *A* and *B* are both odd. According to item 2 of Euler's Theorem, since there are exactly two odd vertices, at least one Euler path, but no Euler circuits exist.

b) The residents would need to start at the intersection of Maple Cir., Walnut St., and Willow St. or at the intersection of Walnut St. and Oak St.

39. *F, G, E, F, D, E, B, D, A, B, C, E*; other answers are possible.
41. *H, I, F, C, B, D, G, H, E, D, A, B, E, F*; other answers are possible.
43. *A, B, E, F, J, I, E, D, H, G, C, D, A*; other answers are possible.
45. *A, E, B, F, C, G, D, K, G, J, F, I, E, H, A*; other answers are possible.
47. *A, B, C, E, B, D, E, F, I, E, H, D, G, H, I, J, F, C, A*; other answers are possible.
49. *F, C, J, M, P, H, F, M, P*; other answers are possible.
51. *B, E, I, F, B, C, F, J, G, G, C, D, K, J, I, H, E, A, B*; other answers are possibe.
53. *J, F, C, B, F, I, E, B, A, E, H, I, J, G, G, C, D, K, J*; other answers are possible.
55. It is not possible to draw a graph with an Euler circuit that has a bridge. Therefore, a graph with an Euler circuit has no bridge.
56. a)   b)   c)

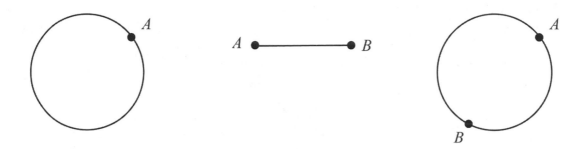

Exercise Set 14.3

1. a) A **Hamilton circuit** is a path that begins and ends with the same vertex and passes through all other vertices exactly one time.
   b) Both **Hamilton** and **Euler circuits** begin and end at the same vertex. A **Hamilton circuit** passes through all other *vertices* exactly once, while an **Euler circuit** passes through each *edge* exactly once.
3. a) A **weighted graph** is a graph with a number, or weight, assigned to each edge.
   b) A **complete graph** is a graph in which there is an edge between each pair of vertices.
   c) A **complete, weighted graph** is a graph in which there is an edge between each pair of vertices and each edge has a number, or weight, assigned to it.
5. a) The number of unique Hamilton circuits in a complete graph with n vertices is found by computing $(n-1)!$

   b) $n = 4; (n-1)! = (4-1)! = 3! = 3 \cdot 2 \cdot 1 = 6$

   c) $n = 9; (n-1)! = (9-1)! = 8! = 8 \cdot 7 \cdot 6 \cdot 5 \cdot 4 \cdot 3 \cdot 2 \cdot 1 = 40,320$

7. To find the optimal solution using the **Brute Force method**, write down all possible Hamilton circuits and then compute the cost or distance associated with each Hamilton circuit. The one with the lowest cost or shortest distance is the optimal solution to the traveling salesman problem.
9. *A, B, C, G, F, E, D* and *E, D, A, B, F, G, C*; other answers are possible.
11. *A, B, C, D, G, F, E, H* and *E, H, F, G, D, C, A, B*; other answers are possible.
13. *A, B, C, E, D, F, G, H* and *F, G, H, E, D, A, B, C*; other answers are possible.
15. *A, B, D, E, G, F, C, A* and *A, C, F, G, E, D, B, A*; other answers are possible.
17. *A, B, C, F, I, E, H, G, D, A* and *A, E, B, C, F, I, H, G, D, A*; other answers are possible.

19.

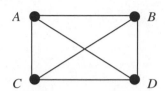

21. The number of unique Hamilton circuits within the complete graph with eight vertices representing this situation is $(8-1)! = 7! = 7 \cdot 6 \cdot 5 \cdot 4 \cdot 3 \cdot 2 \cdot 1 = 5040$ ways

23. The number of unique Hamilton circuits within the complete graph with eleven vertices representing this situation is $(11-1)! = 10! = 10 \cdot 9 \cdot 8 \cdot 7 \cdot 6 \cdot 5 \cdot 4 \cdot 3 \cdot 2 \cdot 1 = 3,628,800$ ways

(The vertices are the 10 different farms he has to visit and his starting point.)

In Exercises 25-31, other graphs are possible.

25. a)

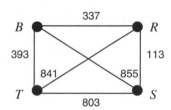

b)

Hamilton Circuit	First Leg/Cost	Second Leg/Cost	Third Leg/Cost	Fourth Leg/Cost	Total Cost
S, R, B, T, S	113	337	393	803	$1646
S, R, T, B, S	113	841	393	855	$2202
S, T, B, R, S	803	393	337	113	$1646
S, T, R, B, S	803	841	337	855	$2836
S, B, R, T, S	855	337	841	803	$2836
S, B, T, R, S	855	393	841	113	$2202

The least expensive route is S, R, B, T, S or S, T, B, R, S

c) $1646

27. a)

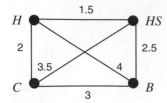

b)

Hamilton Circuit	First Leg/Distance	Second Leg/Distance	Third Leg/Distance	Fourth Leg/Distance	Total Distance
H, HS, B, C, H	1.5	2.5	3	2	9 miles
H, HS, C, B, H	1.5	3.5	3	4	12 miles
H, B, HS, C, H	4	2.5	3.5	2	12 miles
H, B, C, HS, H	4	3	3.5	1.5	12 miles
H, C, HS, B, H	2	3.5	2.5	4	12 miles
H, C, B, HS, H	2	3	2.5	1.5	9 miles

The shortest route *is H, HS, B, C, H* or *H, C, B, HS, H*

c) 9 miles

29. a)

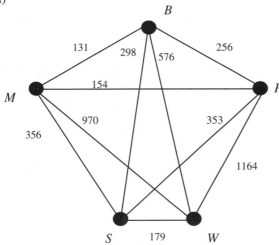

b) *B, M, P, S, W, B* for 131 + 154 + 353 + 179 + 576 = $1393

c) Answers will vary.

31. a)

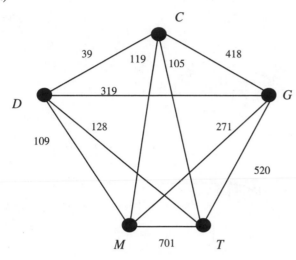

b) *C, D, M, G, T, C* for 39 + 109 + 271 + 520 + 105 = $1044
c) Answers will vary.

33. a) – d) Answers will vary.

35. *A, E, D, N, O, F, G, Q, P, T, M, L, C, B, J, K, S, R, I, H, A*; other answers are possible.

Exercise Set 14.4

1. A **tree** is a connected graph in which each edge is a bridge.
3. Yes, because removing the edge would create a disconnected graph.
5. A **minimum-cost spanning tree** is a spanning tree that has the lowest cost or shortest distance of all spanning trees for a given graph.

7.

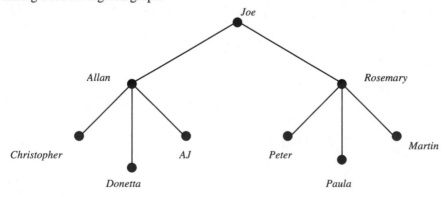

9.

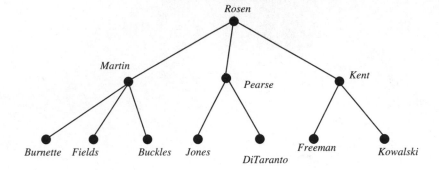

11.

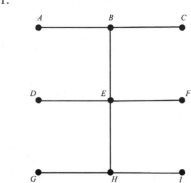

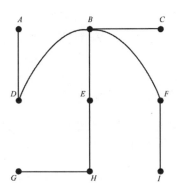

Other answers are possible.

13.

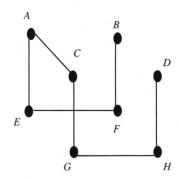

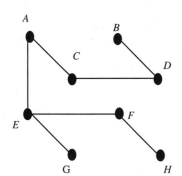

Other answers are possible.

15.

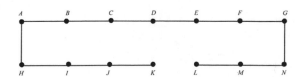

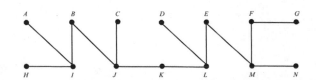

Other answers are possible.

17.

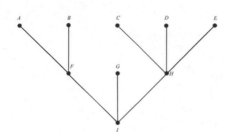 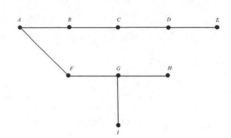

Other answers are possible.

19.

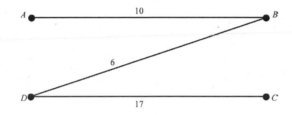

Choose edges in the following order:
DB, BA, DC

21.

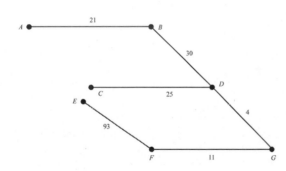

Choose edges in the following order:
DG, GF, AB, CD, BD, EF

23.

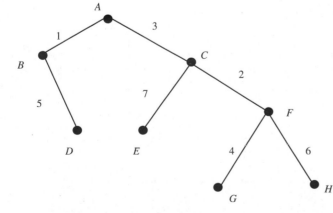

Choose edges in the following order:
AB, CF, AC, FG, BD, FH, EC

25.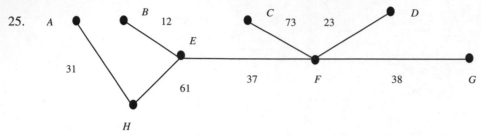

Choose edges in the following order: *BE*, FD, AH, EF, FG, HE, CF

27. a)

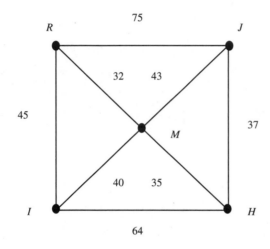

Other answers are possible.

b)

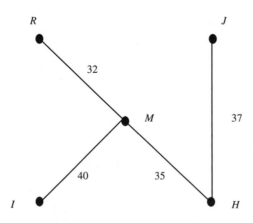

Choose edges in the following order:
RM, MH, JH, IM

c) 15(32 + 35 + 37 + 40) = 15(144) = $2160

29. a)

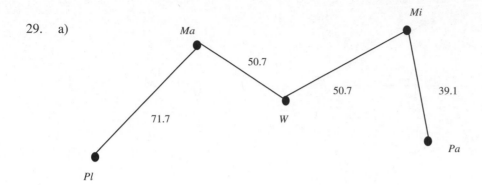

Choose edges in the following order: *Mi Pa, W Mi, Ma W, Ma Pl*

b) 895(39.1 + 50.7 + 50.7 + 71.7) = 895(212.2) = $189,919

31. a)

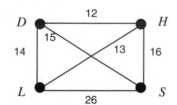

Other answers are possible.

b)

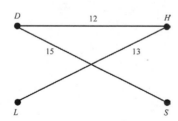

Choose edges in the following order:
DH, HL, DS

c) 3500 (12 + 13 + 15) = 3500 (40) = $140,000

33. a)

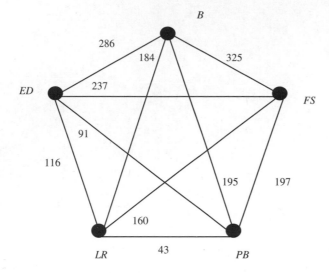

Other answers are possible.

b)

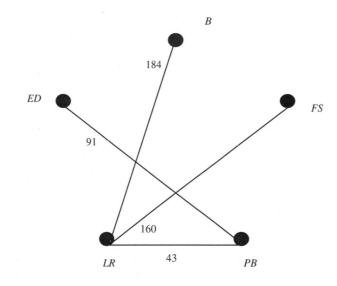

Choose edges in the following order: *LR PB, ED PB, LR FS, B LR*

c) 2500(43 + 91 + 160 + 184) = 2500(478) = $1,195,000

35. Answers will vary.
37. Answers will vary.
38. a) EULER
     b) FLEURY
     c) HAMILTON
     d) KRUSKAL

Review Exercises

1.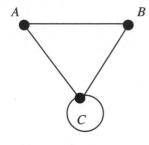

A, B, and C are all even. There is a
loop at vertex C.
Other answers are possible.

2.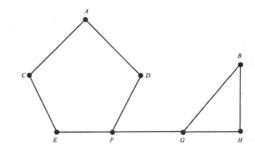

Edge FG is a bridge.
Other answers are possible.

3. A, B, C, A, D, C, E, D; other answers are possible.

4. No. The graph has two odd vertices, so a path that includes each edge exactly once must start at one odd vertex and end at the other; B is an even vertex.

5.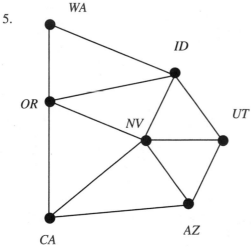

6.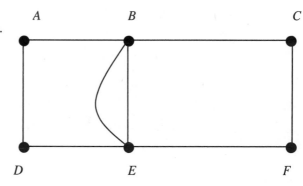

7. Connected

8. Disconnected. There is no path that connects A to C.

9. Edge CD

10. C, B, A, F, E, D, C, G, B, A, G, E, D, G, F; other answers are possible.

11. F, E, G, F, A, G, D, E, D, C, B, A, B, G, C; other answers are possible.

12. B, C, A, D, F, E, C, D, E, B; other answers are possible.

13. E, F, D, E, C, D, A, C, B, E; other answers are possible.

14. a) No. The graph representing the map:

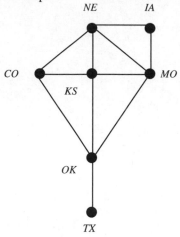

b) Vertices *CO* and *TX* are both odd. According to item 2 of Euler's Theorem, since there are exactly two odd vertices, at least one Euler path, but no Euler circuits exist.
Yes; *CO, NE, IA, MO, NE, KS, MO, OK, CO, KS, OK, TX*; other answers are possible.
c) No. (See part b) above.)

15. a) Yes. The graph representing the floor plan:

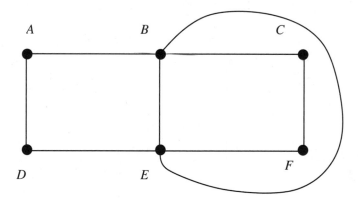

We are seeking an Euler path or an Euler circuit. Note that there are no odd vertices. According to item 1 of Euler's Theorem, since there are no odd vertices, at least one Euler path (which is also an Euler circuit) must exist.

b) The person may start in any room and will finish in the room where he or she started.

16. a) Yes. The graph representing the map:

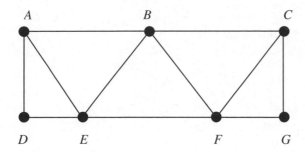

The officer is seeking an Euler path or an Euler circuit. Note that vertices A and C are both odd. According to item 2 of Euler's Theorem, since there are exactly two odd vertices, at least one Euler path but no Euler circuits exist.

b) The officer would have to start at either the upper left-hand corner or the upper right-hand corner. If the officer started in the upper left-hand corner, he or she would finish in the upper right-hand corner, and vice versa.

17. $A, B, F, A, E, F, G, C, D, G, H, D$; other answers are possible.
18. $A, B, C, D, H, G, C, F, G, B, F, E, A$; other answers are possible.
19. A, C, B, F, E, D, G and A, C, D, G, F, B, E; other answers are possible.
20. A, B, C, D, F, E, A and A, E, F, B, C, D, A; other answers are possible.

21.

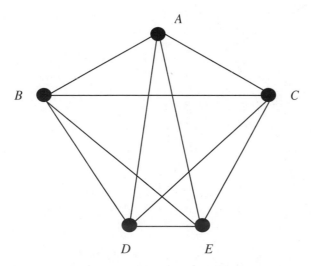

22. The number of unique Hamilton circuits within the complete graph with 5 vertices representing this situation is $(5-1)! = 4! = 4 \cdot 3 \cdot 2 \cdot 1 = 24$ ways

23. a)

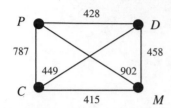

b)

Hamilton Circuit	First Leg/Cost	Second Leg/Cost	Third Leg/Cost	Fourth Leg/Cost	Total Cost
P, D, C, M, P	428	449	415	902	$2194
P, D, M, C, P	428	458	415	787	$2088
P, C, M, D, P	787	415	458	428	$2088
P, C, D, M, P	787	449	458	902	$2596
P, M, D, C, P	902	458	449	787	$2596
P, M, C, D, P	902	415	449	428	$2194

The least expensive route is *P, D, M, C, P* or *P, C, M, D, P*

c) $2088

24. a)

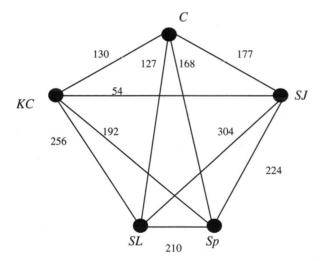

b) *SJ, KC, C, SL, Sp, SJ* traveling a total of 54 + 130 + 127 + 210 + 224 = 745 miles

c) *Sp, C, SL, KC, SJ, Sp* traveling a total of 168 + 127 + 256 + 54 + 224 = 829 miles

25.

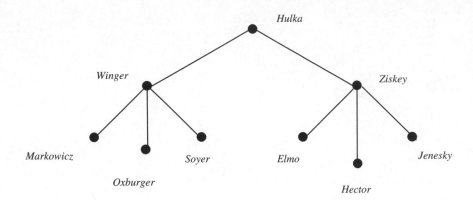

26.

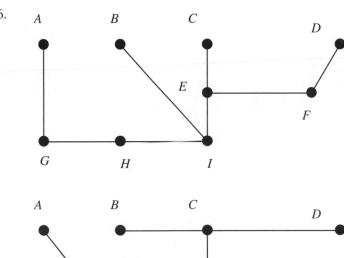

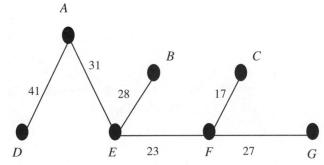

Other answers are possible.

27.

Choose edges in the following order:
CF, EF, FG, BE, AE, AD

28.   a)

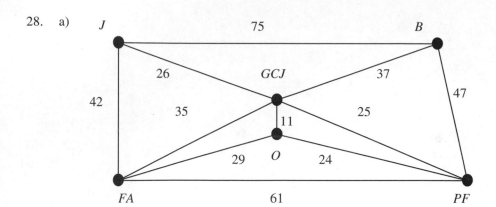

b)

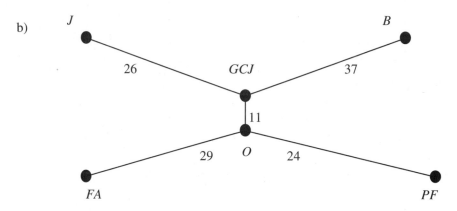

Choose edges in the following order:
O GCJ, O PF, J GCJ, FA O, GCJ B

c)   2.50 (11 + 24 + 26 + 29 + 37) = 2.50 (127) = $317.50

Chapter Test

1.

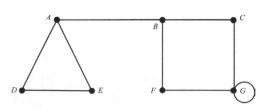

Edge *AB* is a bridge.   There is a loop at vertex *G*.
Other answers are possible.

2.

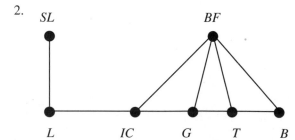

3. One example:

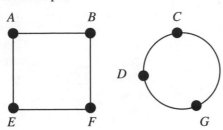

4. *D, A, B, C, E, B, D, E*;
   other answers are possible.

5. Yes. The graph representing the floor plan:

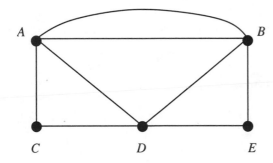

We are seeking an Euler path or an Euler circuit. Note that there are no odd vertices. According to item 1 of Euler's Theorem, since there are no odd vertices, at least one Euler path (which is also an Euler circuit) must exist.

The person may start in any room and will finish in the room where he or she started.

6. *A, D, E, A, F, E, H, F, I, G, F, B, G, C, B, A*; other answers are possible.
7. *A, B, C, D, H, I, L, K, J, G, F, E, A*; other answers are possible.
8. The number of unique Hamilton circuits within the complete graph with 8 vertices representing

   this situation is $(8-1)! = 7! = 7 \cdot 6 \cdot 5 \cdot 4 \cdot 3 \cdot 2 \cdot 1 = 5040$ ways

9. a)

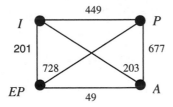

9. b)

Hamilton Circuit	First Leg/Cost	Second Leg/Cost	Third Leg/Cost	Fourth Leg/Cost	Total Cost
I, P, EP, A, I	449	728	49	203	$1429
I, P, A, EP, I	449	677	49	201	$1376
I, A, P, EP, I	203	677	728	201	$1809
I, A, EP, P, I	203	49	728	449	$1429
I, EP, A, P, I	201	49	677	449	$1376
I, EP, P, A, I	201	728	677	203	$1809

The least expensive route is *I, P, A, EP, I* or *I, EP, A, P, I* for $1376.

c) *I, EP, A, P, I* for $1376

10.

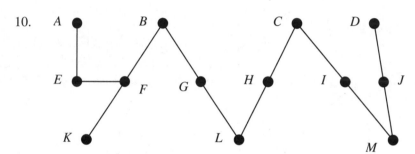

Other answers are possible.

11.

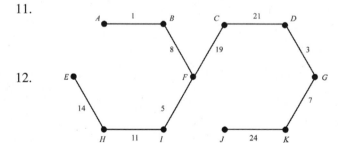

Choose edges in the following order:
AB, DG, FI, KG, BF, HI, EH, FC, CD, JK

12.

Choose edges in the following order:

V2 V4, V3 V4, V4 V5, V1 V2

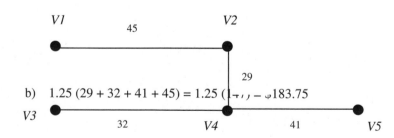

b)   1.25 (29 + 32 + 41 + 45) = 1.25 (147) = $183.75

CHAPTER FIFTEEN

VOTING AND APPORTIONMENT

Exercise Set 15.1

1. When a candidate receives more than 50% of the votes.

3. Voters rank candidates from most favorable to least favorable. Each last place vote is awarded one point, each next to last place vote is awarded two points, each third from last place vote is awarded three points, etc. The candidate receiving the most points is the winner.

5. Voters rank the candidates. A series of comparisons in which each candidate is compared to each of the other candidates follows. If candidate A is preferred to candidate B, then A receives one point. If candidate B is preferred to candidate A, then B receives one point. If the candidates tie, each receives ½ point. The candidate receiving the most points is declared the winner.

7. A preference table summarizes the results of an election.

9. a) Jeter is the winner; he received the most votes using the plurality method.

   b) No. $\dfrac{265128}{192827+210361+265128} = \dfrac{265128}{668316} \approx 0.40$ is not a majority.   Majority is > 334,158 votes.

11.

Number of votes	3	1	2	2	1
First	B	A	C	C	A
Second	A	B	B	A	C
Third	C	C	A	B	B

13. $9 + 5 + 3 + 2 = 19$ employees

15. Votes – (M): 9, (V): 5+3 = 8, (W): 2. Mop wins with the most votes.

17. A majority out of 19 votes is 10 or more votes.
    First choice votes: (M) 9, (V) 8, (B) 0
    None receives a majority, thus B with the least votes is eliminated.
    Second round: (M) 9, (V) 5+3+2 = 10
    Vacuum wins with a majority of 10 votes.

19. B: 4 1st place votes = (4)(3) = 12
       2 2nd place votes = (2)(2) = 4
       3 3rd place votes = (3)(1) = 3
  G: 2 1st place votes = (2)(3) = 6
       4 2nd place votes = (4)(2) = 8
       3 3rd place votes = (3)(1) = 3
  M: 3 1st place votes = (3)(3) = 9
       3 2nd place votes = (3)(2) = 6
       3 3rd place votes = (3)(1) = 3
  B = 19 points; G = 17 points; M = 18 points
  Beach wins with 19 points.

21. B vs. G: M = 3+2+1 = 6   G = 2+1 = 3
  B gets 1 pt.
  B vs. M: B = 3+1 = 4   M = 2+2+1 = 5
  M gets 1 pt.
  G vs. M: G = 3+2 = 5   M = 2+1+1 = 4
  G gets 1 pt.
  All get 1 point, which indicates no winner.

23. Votes: (S) 8+3+2 = 13   (L) 6+3 = 9
  (H) 4+3+2 = 9   (T) 1
  San Antonio wins with the most votes.

25. A majority out of 32 votes is 17 or more votes.
  First choice votes: (S) 13, (L) 9, (H) 9, (T) 1
  None receives a majority, thus T with the least
  votes is eliminated.
  Second round: (S) 13, (L) 9, (H) 10
  No majority, thus eliminate L.
  Third round: (S) 16, (H) 16
  Since S and H tied, there is no winner.

27. W: 5 1st place votes = (5)(3) = 15
       4 2nd place votes = (4)(2) = 8
       3 3rd place votes = (3)(1) = 3
  D: 1 1st place votes = (1)(3) = 3
       7 2nd place votes = (7)(2) = 14
       4 3rd place votes = (4)(1) = 4
  J: 6 1st place votes = (6)(3) = 18
       1 2nd place votes = (1)(2) = 2
       5 3rd place votes = (5)(1) = 5
  W = 26 points; D = 21 points; J = 25 points
  Williams wins with 26 points.

29. W vs. D: W = 5+4 = 9   D = 1+2 = 3
            W gets 1 pt.
  W vs. J: W = 5   J = 1+4+2 = 7
          J gets 1 pt.
  D vs. J: D = 5+1 = 6   J = 4+2 = 6
         D and J get 0.5 pt.
  W = 1 pt.   D = 0.5 pt.   J = 1.5 pts.
  Johnson wins with 1.5 points.

31. A majority out of 12 votes is 7 or more votes.
  Most last place votes: (W) 3, (D) 4, (J) 5
  Thus J with the most last place votes is eliminated.
  Second round using the most last place votes:
  (W) 1+2 = 3, (D) 5+4 = 9
  Williams wins with the least last place votes.

33. L: 5 1st place votes = (5)(3) = 15
     6 3rd place votes = (6)(1) = 6
  E: 2 1st place votes = (2)(3) = 6
     9 2nd place votes = (9)(2) = 18
  O: 4 1st place votes = (4)(3) = 12
     2 2nd place votes = (2)(2) = 4
     5 3rd place votes = (5)(1) = 5
  L = 21 points; E = 24 points; O = 21 points
  Erie Road wins with 24 points.

35. L vs. E: L = 5   E = 2+4 = 6   E gets 1 pt.
 L vs. O: L = 5   O = 2+4 = 6   O gets 1 pt.
 E vs. O: E = 5+2 = 6   O = 4   E gets 1 pt.
 Erie Road wins with 2 points.

37. a) Votes: (TI): 10, (C): 3, (HP): 2
 Texas Instruments wins with the most votes.
  b) TI: 10 1st place votes = (10)(4) = 40
   5 2nd place votes = (5)(3) = 15
  C: 3 1st place votes = (3)(4) = 12
   6 2nd place votes = (6)(3) = 18
   6 3rd place votes = (6)(2) = 12
  S: 9 3rd place votes = (9)(2) = 18
   6 4th place votes = (6)(1) = 6
   9 4th place votes = (9)(1) = 9
  HP: 2 1st place votes = (2)(4) = 8
   4 2nd place votes = (4)(3) = 12
   9 4th place votes = (9)(1) = 9
 TI= 55 points; C = 42 points; S = 24 points,
 HP = 29 points   TI wins with 55 points.

37. c) A majority out of 15 votes is 8 or more votes.
 First choice votes: (TI) 10,   (C) 3
   (S) 0,   (HP) = 2
 Because TI already has a majority, TI wins.
  d) TI vs. C: TI = 6+4+2 = 12   C = 3
 TI gets 1 pt.
 TI vs. S: TI = 6+4+3+2 = 15   TI gets 1 pt.
 TI vs. HP: TI = 6+4+3 = 14   HP = 2
 TI gets 1 pt.
 C vs. S: C = 6+4+3+2 = 15   C gets 1 pt.
 C vs. HP: C = 6+3 = 9   HP = 4+3 = 7
 C = gets 1 pt.
 S vs. HP: S = 6+3 = 9   HP = 4+2 = 6
 S gets 1 pt.
 TI wins with 3 points.

39. a) A: 6 1st place votes = (6)(4) = 24
   1 2nd place vote = (1)(3) = 3
   2 3rd place votes = (2)(2) = 4
   5 4th place votes = (5)(1) = 5
  B: 1 1st place vote = (1)(4) = 4
   4 2nd place vote = (4)(3) = 12
   9 3rd place votes = (9)(2) = 18
  C: 5 1st place votes = (5)(4) = 20
   6 2nd place vote = (6)(3) = 18
   1 3rd place vote = (1)(2) = 2
   2 4th place votes = (2)(1) = 2
  D: 2 1st place votes = (2)(4) = 8
   3 2nd place vote = (3)(3) = 9
   2 3rd place votes = (2)(2) = 4
   7 4th place votes = (7)(1) = 7
 A = 36 points; B = 34 points; C = 42 points;
 D = 28 points
 C wins with 42 points.

39. b) Votes: (A): 6,   (B): 1,   (C): 5,   (D): 2
 A wins with the most votes.
  c) A majority out of 14 votes is 8 or more votes.
 First choice votes: (A) 6,   (B) 1
   (C) 5,   (D) = 2
 None receives a majority, thus B with the least
 votes is eliminated.
 Second round: (A) 7, (C) 5, (D) 2
 No majority, thus eliminate D.
 Third round: (A) 9, (C) 5
 A wins with 9 votes.
  d) A vs. B: A = 6   B = 8   B gets 1 pt.
 A vs. C: A = 9   C = 5   A gets 1 pt.
 A vs. D: A = 7   D = 7   A / D get 0.5 pt.
 B vs. C: B = 34   C = 11   C gets 1 pt.
 B vs. D: B = 9   D = 5   B gets 1 pt.
 C vs. D: C = 12   D = 2   C gets 1 pt.
 B and C tie with 2 points.

41. a) If there were only two columns then only two of the candidates were the first choice of the voters. If each of the 15 voters cast a ballot, then one of the voters must have received a majority of votes because 15 cannot be split evenly.

    b) An odd number cannot be divided evenly so one of the two first choice candidates must receive more than half of the votes.

43. a) C: $4 + 1 + 1 = 6$ R: $4 + 4 + 3 = 11$
       W: $3 + 3 + 2 + 2 + 1 + 1 = 12$
       T: $4 + 3 + 2 + 2 = 11$
       The Warriors finished 1st, the Rams and the Tigers tied for 2nd, and the Comets were 4th.

    b) C: $5 + 0 = 5$ R: $5 + 5 + 3 = 13$
       W: $3 + 3 + 1 + 1 + 0 + 0 = 8$
       T: $5 + 3 + 1 \ 1 = 10$
       Rams - 1st, Tigers - 2nd, Warriors - 3rd, and Comets - 4th.

45. a) Each voter casts $4+3+2+1 = 10$ votes.
       $(15)(10) = 150$ votes

    b) $150 - (35+40+25) = 150 - 100 = 50$ votes

    c) Yes. Candidate D has more votes than each of the other 3 candidates.

47. Answers will vary.

Exercise Set 15.2

1. If a candidate receives a majority of first place votes, then that candidate should be declared the winner.

3. If a candidate is favored when compared individually with every other candidate, then that candidate should be declared the winner.

5. A candidate that is preferred to all others will win each pairwise comparison and be selected with the pairwise comparison method.

7. If a candidate receives a majority of first place votes, then that candidate should be declared the winner. Plurality counts only the 1st place votes.

9. The plurality method yields Tacos are the winner with a majority of 8 1st place votes. However, if the Borda count method is used:
   Tacos $(8)(3) + (3)(2) + (4)(1) = 24 + 6 + 4 = 34$
   Pizza $(4+3)(3) + (8)(2) = 21 + 16 = 37$
   Burgers $(4)(2) + (8+3)(1) = 8 + 11 = 19$
   The winner is Pizza using the Borda count method, thus violating the majority criterion.

11. Total votes $= 3+2+1+1 = 7$ Candidates A is the candidate of choice with a majority of 4 votes.
    A: 4 1st place votes $= (4)(4) = 16$
       3 4th place votes $= (3)(1) = 3$
    B: 3 1st place vote $= (3)(4) = 12$
       4 2nd place vote $= (4)(3) = 12$
    C: 2 2nd place vote $= (2)(3) = 6$
       4 3rd place vote $= (4)(2) = 8$
       1 4th place vote $= (1)(1) = 1$
    D: 1 2nd place vote $= (1)(3) = 3$
       3 3rd place votes $= (3)(2) = 6$
       3 4th place votes $= (3)(1) = 3$
     G $= 300$ points;   A $= 333$ points;
    A $= 19$ votes;   B $= 24$ votes;   C $= 15$ votes;
    D $= 12$ votes
    Candidate B is chosen with 24 votes, therefore the majority criterion is not satisfied.

13. P: 4 1st place votes = (4)(3) = 12
2 2nd place votes = (2)(2) = 4
3 3rd place votes = (3)(1) = 3
L: 3 1st place vote = (3)(3) = 9
5 2nd place vote = (5)(2) = 10
1 3rd place vote = (1)(1) = 1
S: 2 1st place votes = (2)(3) = 6
2 2nd place vote = (2)(2) = 4
5 3rd place vote = (5)(1) = 5
P = 19 votes;   L = 20 votes;   S = 15 votes
P vs. L: P = 4+1 = 5   L = 4   P gets 1 pt.
P vs. S: P = 4+1 = 5   S = 4   P gets 1 pt.
L vs. S: L = 4+1+2 = 7   S = 2   L gets 1 pt.
Because Parking wins its head-to-head comparisons
and the Lounge Areas win by Borda count method,
the head-to-head criterion is not satisfied.

15. A majority out of 25 votes is 13 or more votes.
First choice votes: A=7, B=15, C=3
Since B has > 13 votes, B wins by plurality with
elimination.

A vs. B: A = 7+3 = 10   B = 15   B gets 1 pt.
A vs. C: A = 7   C = 25-7 = 18   C gets 1 pt.
B vs. C: B = 15+7 = 22   C = 3   B gets 1 pt.

Yes, because B wins by both methods, the
head-to-head criterion is satisfied.

17. Votes: A: 8,   B: 4,   C: 5; thus, A wins.
If B drops out, we get the following:
Votes: A: 8,   C: 4 + 5 = 9, thus C would win.
The irrelevant alternatives criterion is not satisfied.

19. A receives 53 points, B receives 56 points, and
C receives 53 points. Thus, B wins using the
Borda count method. If A drops out, we get the
following: B receives 37 points, and C receives
44 points. Thus, C wins the second vote. The
irrelevant alternatives criterion is not satisfied.

21. A majority out of 32 voters is 17 or more votes.
Votes: A: 8 + 3 = 11, B: 9, C: 12; none has a
majority, thus eliminate B.
Votes: A:8 + 3 = 11,   C: 9 +12 = 21, thus C
wins. If the three voters who voted for A,C,B
change to C,A,B, the new set of votes becomes:
Votes: A: 8, B: 9, C: 15; none has a
majority, thus eliminate A.
Votes: B: 17, C:15, B wins.
Thus, the monotonicity criterion is not satisfied.

23. A majority out of 23 voters is 12 votes.
Votes: A: 10,   B: 8,   C: 5; none has a majority,
thus eliminate C.
Votes: B: 10,   B: 8 + 5 = 13; thus B wins.
After A drops out, the new set of votes is
B: 8,   C: 10 + 5 = 15; thus C wins.

The irrelevant alternatives criterion is not satisfied.

25. A receives 2 points, B receives 3 point, C receives
2 points, D receives 1 point, and E receives 2 pts.
B wins by pairwise comparison.
After A, C and E drop out, the new set of votes is
B: 2   D: 3, thus D wins. The irrelevant
alternatives criterion is not satisfied..

27. Total votes = 7   A wins with a majority of
4 votes.
A: 4 1st place votes = (4)(3) = 12
3 3rd place votes = (3)(1) = 3
B: 2 1st place vote = (2)(3) = 6
5 2nd place vote = (5)(2) = 10
C: 1 1st place votes = (1)(3) = 3
2 2nd place votes = (2)(2) = 4
4 3rd place vote = (4)(1) = 4
A = 15 points;   B = 16 points;   C = 11 points
B wins with 16 points. No. The majority
criterion is not satisfied.

29. Total votes = 31   Majority = 16 or more
    a) Museum of Natural History
    b) Museum of Natural History
    c) Museum of Natural History
    d) Museum of Natural History
    e) Museum of Natural History
    f) None of them

31. a) A majority out of 82 votes is 42 or more votes.
    First choice votes: (A) 28, (C) 30, (D) 24
    None receives a majority, thus D with the least votes is eliminated.
    Second round: (A) 52, (C) 30
    Thus, Jennifer Aniston is selected..

    b) No majority on the 1ˢᵗ vote; C is eliminated with the fewest votes.
    Second round: (A) 38, (D) 44
    Denzel Washington is chosen.

    c) Yes.

33. A candidate who holds a plurality will only gain strength and hold and even larger lead if more favorable votes are added.

35. AWV   37. AWV   39. AWV

Exercise Set 15.3

1. If we divide the total population by the number of items to be apportioned we obtain a number called the standard divisor.

3. The standard quota rounded down to the nearest whole number.

5. An apportionment should always be either the upper quota or the lower quota.

7. Jefferson's method, Webster's method, Adams's method

9. a) Webster's method   b) Adams's method   c) Jefferson's method

11. a) $\dfrac{7500000}{150} = 50,000$ = standard divisor

    b) and c)

State	A	B	C	D	Total
Population	1,220,000	2,730,000	857,000	2,693,000	7,500,000
Standard Quota	24.40	54.60	17.14	53.86	
Lower Quota	24	54	17	53	148
Hamilton's Apportionment	24	55	17	54	150

13. a) and b)

State	A	B	C	D	Total
Population	1,220,000	2,730,000	857,000	2,693,000	7,500,000
Modified Quota	24.70	55.26	17.35	54.51	
Jefferson's Apportionment (round down)	24	55	17	54	150

15. a) and b)

State	A	B	C	D	Total
Population	1,220,000	2,730,000	857,000	2,693,000	7,500,000
Modified Quota	24.06	53.85	16.90	53.12	
Adams's Apportionment (round up)	25	54	17	54	150

17. a) and b)

State	A	B	C	D	Total
Population	1,220,000	2,730,000	857,000	2,693,000	7,500,000
Modified Quota	24.38	54.55	17.12	53.81	
Webster's Apportionment	24	55	17	54	150

19. a) and b)

Hotel	A	B	C	Total
Amount	306	214	155	675
Modified Quota	11.86	8.29	6.01	
Jefferson's Apportionment (rounded down)	11	8	6	25

21. a) and b)

Hotel	A	B	C	Total
Amount	306	214	155	675
Modified Quota	10.55	7.38	5.34	
Adams's Apportionment (rounded up)	11	8	6	25

23. a) and b)

Store	A	B	C	Total
Amount	306	214	155	675
Standard Quota	11.33	7.93	5.74	
Webster's Apportionment (standard rounding)	11	8	6	25

25. a) A standard divisor $= \dfrac{\text{total}}{30} = \dfrac{540}{30} = 18$

b)

Store	A	B	C	D	Total
Population	75	97	140	228	540
Standard Quota	4.18	5.39	7.78	12.67	

27. A divisor of 17 was used.

Store	A	B	C	D	Total
Population	75	97	140	228	540
Modified Quota	4.41	5.71	8.24	13.41	
Jefferson's Apportionment (round down)	4	5	8	13	30

29. The standard divisor was used.

Store	A	B	C	D	Total
Population	75	97	140	228	540
Standard Quota	4.18	5.39	7.78	12.67	
Webster's Apportionment	4	5	8	13	30

31.

School	LA	Sci.	Eng.	Bus.	Hum	Total
Enrollment	1746	7095	2131	937	1091	13000
Standard Quota	33.58	136.44	40.98	18.02	20.98	
Lower Quota	33	136	40	18	20	247
Hamilton's Apportionment	34	136	41	18	21	250

33. A divisor of 52.5 was used.

School	LA	Sci.	Eng.	Bus.	Hum	Total
Enrollment	1746	7095	2131	937	1091	13000
Modified Quota	33.26	135.14	40.59	17.85	20.78	
Adams's Apportionment (round up)	34	136	41	18	21	250

35. a) A standard divisor $= \dfrac{\text{total}}{150} = \dfrac{13500}{150} = 90$

Dealership	A	B	C	D	Total
Annual Sales	4800	3608	2990	2102	13500
Standard Quota	53.33	40.09	33.22	23.36	150.00

37. A divisor of 88.5 was used.

Dealership	A	B	C	D	Total
Annual Sales	4800	3608	2990	2102	13500
Modified Quota	54.24	40.77	33.79	23.75	
Jefferson's Apportionment	54	40	33	23	150

39. A divisor of 89.5 was used.

Dealership	A	B	C	D	Total
Annual Sales	4800	3608	2990	2102	13500
Modified Quota	53.63	40.31	33.41	23.47	
Webster's Apportionment	54	40	33	23	150

41.

Precinct	A	B	C	D	E	F	Total
Crimes	743	367	432	491	519	388	2940
Standard Quota	53.07	26.21	30.86	35.07	37.07	27.71	
Lower Quota	53	26	30	35	37	27	208
Hamilton's Apportionment	53	26	31	35	37	28	210

43. The divisor 14.2 was used.

Precinct	A	B	C	D	E	F	Total
Crimes	743	367	432	491	519	388	2940
Modified Quota	52.32	25.85	30.42	34.58	36.55	27.32	
Adams's Apportionment (round up)	53	26	31	35	37	28	210

45. a) Standard divisor = $\dfrac{\text{total}}{200} = \dfrac{2400}{200} = 12$

b)

Shift	A	B	C	D	Total
Room calls	751	980	503	166	2400
Standard Quota	62.58	81.67	41.92	13.83	

47. The divisor 11.9 was used.

Shift	A	B	C	D	Total
Room calls	751	980	503	166	2400
Modified Quota	63.11	82.35	42.27	13.95	
Jefferson's Apportionment (round down)	63	82	42	13	200

49. The divisor 12.02 was used.

Shift	A	B	C	D	Total
Room calls	751	980	503	166	2400
Modified Quota	62.48	81.53	41.85	13.81	
Webster's Apportionment (standard rounding)	62	82	42	14	200

51. a) DIVISOR b) ADAMS c) WEBSTER d) MODIFIED e) HAMILTON

Exercise Set 15.4

1. The Alabama paradox occurs when an increase in the total # of items results in a loss of items for a group.

3. The population paradox occurs when group A loses items to group B, although group A's population grew at a higher rate than group B's.

5. Hamilton's, Jefferson's

7. New divisor = $\dfrac{900}{51} = 17.65$

School	A	B	C	D	E	Total
Population	210	165	160	175	190	900
Standard Quota	11.90	9.35	9.07	9.92	10.76	
Lower Quota	11	9	9	9	10	48
Hamilton's Apportionment	12	9	9	10	11	51

No. No school suffers a loss so the Alabama paradox does not occur.

9. a) Standard divisor $= \dfrac{900}{30} = 30$

State	A	B	C	Total
Population	161	250	489	900
Standard Quota	5.37	8.33	16.30	
Hamilton's Apportionment	6	8	16	30

b) New divisor $= \dfrac{900}{31} = 29.03$

State	A	B	C	Total
Population	161	250	489	900
Standard Quota	5.55	8.61	16.84	
Hamilton's Apportionment	5	9	17	31

Yes, state A loses 1 seat and states B and C each gain 1 seat.

11. a) Standard divisor $= \dfrac{25000}{200} = 125$

City	A	B	C	Total
Population	8130	4030	12,840	25,000
Standard Quota	65.04	32.24	102.72	
Hamilton's Apportionment	65	32	103	200

b) New divisor $= \dfrac{25125}{200} = 125.625$

City	A	B	C	Total
New Population	8150	4030	12,945	25,125
Standard Quota	64.88	32.08	103.04	
Hamilton's Apportionment	65	32	103	200

No. None of the Cities loses a bonus.

13. a) Standard divisor $= \dfrac{5400}{54} = 100$

Division	A	B	C	D	E	Total
Population	733	1538	933	1133	1063	5400
Standard Quota	7.33	15.38	9.33	11.33	10.63	
Lower Quota	7	15	9	11	10	52
Hamilton's Apportionment	7	16	9	11	11	54

b) New divisor $= \dfrac{5454}{54} = 101$

Division	A	B	C	D	E	Total
Population	733	1539	933	1133	1116	
Standard Quota	7.26	15.24	9.24	11.22	11.05	
Lower Quota	7	15	9	11	11	53
Hamilton's Apportionment	8	15	9	11	11	54

Yes. Division B loses an internship to Division A even though the population of division B grew faster than the population of division A.

15. a) Standard divisor = $\dfrac{4800}{48} = 100$

Tech. Data	A	B	Total
Employees	844	3956	4800
Standard Quota	8.44	39.56	
Lower Quota	8	39	47
Hamilton's Apportionment	8	40	48

b) New divisor = $\dfrac{5524}{55} = 100.44$

Tech. Data	A	B	C	Total
Employees	844	3956	724	5524
Standard Quota	8.40	39.39	7.21	
Lower Quota	8	39	7	54
Hamilton's Apportionment	9	39	7	55

Yes. Group B loses a manager.

17. a) Standard divisor = $\dfrac{990000}{66} = 15,000$

State	A	B	C	Total
Population	68970	253770	667260	990000
Standard Quota	4.60	16.92	44.48	
Hamilton's Apportionment	5	17	44	66

b) New divisor = $\dfrac{1075800}{71} = 15,152.11$

State	A	B	C	D	Total
Population	68970	253770	667260	85800	1075800
Standard Quota	4.55	16.75	44.04	5.66	
Hamilton's Apportionment	4	17	44	6	71

Yes. State A loses a seat.

Review Exercises

1. a) Robert Rivera wins with the most votes (13).

   b) A majority out of 27 voters is 14 or more votes. Robert Rivera does not have a majority.

2. a) Michelle MacDougal wins with the most votes (224).

   b) Yes. A majority out of 421 voters is 211 or more votes.

3.

# of votes	3	2	1	3	1
First	B	A	D	C	D
Second	A	C	C	B	A
Third	C	D	A	A	B
Fourth	D	B	B	D	C

4.

# of votes	2	2	2	1
First	C	A	B	C
Second	A	B	C	B
Third	B	C	A	A

5. Number of votes = 6 + 4 + 3 +2 + 1 + 1 = 17

6. Park City wins with a plurality of 7 votes.

7. P: 50 points,   V: 47 points,   S: 35 points,
    A: 38 points. Park City wins with 50 points.

8. A majority out 17 voters is 9 or more votes.
    Votes: P: 6+1 = 7, V: 4, S: 3+2 = 5, A:1.
    None has a majority, thus eliminate A.
    Votes: P: 6+1 = 7, V: 4, S: 3+2+1 = 6
    None has a majority, thus eliminate V.
    Votes: P: 6+4+1 = 11, S: 3+2+1 = 6.
    Park City wins.

9. P: 3 pts.,   V: 2 pts.,   S: 0 pts.,   A: 1 pt.
    Park City wins with 3 points.

10. Votes: P: 7,   V: 4,   S: 5,   A: 1   None has a
    majority, thus eliminate S with most last place
    votes. Votes: P: 10, V: 4, A: 3; Park City wins.

11. 38+30+25+7+10 = 110 students voted

12. Volleyball wins with a plurality of 40 votes.

13. S: 223 pts., V: 215 pts., B: 222 pts.
    Soccer wins.

14. A majority out of 110 voters is 56 or more votes.
    Votes: S: 38, V: 40, B: 32; None has a majority,
    thus eliminate B. Votes: S: 45, V: 65
    Volleyball wins.

15. S: 1 pt.,   V: 1 pt.,   B: 1 pt.   A 3-way tie

16. Votes: S: 38,   V: 40,   B: 32   None has a
    majority, thus eliminate V with the most last place
    votes. Votes: S: 68, B: 42.   Soccer wins.

17. a) Votes: A: 161+134 = 295,   F: 45,   M: 12,
        P: 0   AARP wins.
    b) Yes. A majority out of 372 voters is 187 or more
        votes. AARP receives a majority.
    c) A: 1387 pts.,   F: 740 pts.,   M: 741 pts.,
        P: 852 pts.   AARP wins.
    d) 187 or more votes is needed for a majority.
        Votes: A: 295,   F: 45,   M: 12,   P: 0
        AARP wins.
    e) A: 3 pts.,   F: 1 pt.,   M: 1 pt.,   P: 1 pt.
        AARP wins.

18. Votes: (NO): 70, (LV): 55, (C): 30, (SD): 45
    a) A majority out of 200 voters is 101 or more
        votes. None of the cities has a majority.
    b) New Orleans win a plurality of 70 votes.
    c) (NO):410 pts.,   (LV): 600 pts.,   (C): 495 pts.,
        (SD): 495 pts.   Las Vegas wins.
    d) Las Vegas wins with 130 pts. to 70 pts. for NO.
    e) NO: 0 pts., LV: 3 pts., C: 1 pt., SD: 2 pt.
        Las Vegas wins with points.

19. a) A majority out of 16 voters is 9 or more votes.
        Votes: (EB): 4+3+ = 7,   (FW): 1+1 = 2,
        (G): 0,   (WB): 6+1 = 7   None has a majority,
        thus eliminate G. Votes: (EB): 4+3 = 7,
        (FW): 1+1 = 2,   (WB): 6 + 1 = 7   None has a
        majority, thus eliminate FW
        Votes: (EB): 4+3+1 = 8,   (WB): 6+1+1 = 8.
        Thus, EB and WB tie.
    b) Use the Borda count method to break the tie.
        (EB) = 46 points,   (WB) = 50 points;
        World Book wins.

19. c) (EB) vs. (WB): EB: 4+3+1 = 8 points,
        (WB): 6+1+1 = 8 points.
        EB and WB tie again.

20. A: 33 pts.,   B: 39 pts,   C: 28 pts.,   D: 20 pts.
    Using the Borda count, method B wins.
    However, B only has 3 first place votes, thus the
    majority criterion is not satisfied.

21. A wins all its head-to-head comparisons but B
    wins using the Borda count method.
    The head-to-head criterion is not satisfied.

22. a) A majority out of 42 voters is 21 or more votes.
    Votes: A: 12, B: 10+6 = 16, C: 14
    None has the majority, thus eliminate A.
    Votes: B :10+6 = 16, C: 14+12 = 26 C wins.

b) The new preference table is

Number of votes	10	14	6	12
First	B	C	C	A
Second	A	B	B	C
Third	C	A	A	B

Votes: A: 12, B: 10, C: 20; None has a
majority, thus eliminate B.
Votes: A: 22, C: 20 A wins. When the
order is changed A wins. Therefore, the
monotonicity criterion is not satisfied.

22. c) If B drops out the new table is

Number of votes	10	14	6	12
First	A	C	C	A
Second	C	A	A	C

Votes: A: 10+12 = 22, C: 14+6 = 20 A wins.
Since C won the first election and then after B
dropped out A won, the irrelevant criterion is
not satisfied.

23. a) M has 0 pts., S has 3 pts., F has 2 pts., and
    E has 1 pt. Thus, Starbucks wins.
    b) Maxwell House wins w/a plurality of 33 votes.
    c) M = 228 pts., S = 277 pts., F = 293 pts., and
    E = 292 pts. Thus, Folgers wins.
    d) Eight O'clock wins over Maxwell House with
    76 points.
    e) Same results as in a), thus, Starbucks wins.
    f) The plurality, plurality with elimination, and
    Borda count methods all violate the
    head-to-head criterion.

25. The Borda count method
26. Plurality and plurality w/elimination methods
27. Pairwise comparison and Borda count methods

24. a) Yes. Fleetwood Mac is favored when
    compared to each of the other bands.
    b) Votes: A: 15, B: 34, C: 9+4 = 13,
    F: 25 Boston wins.
    c) A: 217 points, B: 198 points, C: 206 points,
    F: 249 points Fleetwood Mac wins.
    d) A majority out of 87 voters is 44 or more votes.
    Votes: A: 15, B: 34, C: 13, F:25
    None has a majority, thus eliminate C.
    Votes: A: 15+9+4 = 28, B: 34, F: 25
    None has a majority, thus climate F.
    Votes: A: 28+25 = 53, B: 34 Abba wins.
    e) A = 2 pts., B = 0 pts., C = 1 pt., F = 3 pts.
    Thus, Fleetwood Mac wins.
    f) Plurality and plurality w/elimination methods

28. Standard divisor = $\dfrac{6000}{10}$ = 600

Region	A	B	C	Total
Number of Houses	2592	1428	1980	6000
Standard Quota	4.32	2.38	3.30	
Lower Quota	4	2	3	9
Hamilton's Apportionment	4	3	3	10

29. Using the modified divisor 500.

Region	A	B	C	Total
Number of Houses	2592	1428	1980	6000
Modified Quota	5.18	2.86	3.96	
Jefferson's Apportionment (rounded down)	5	2	3	10

30. Using the modified divisor 700.

Region	A	B	C	Total
Number of Houses	2592	1428	1980	6000
Modified Quota	3.70	2.04	2.83	
Adams's Apportionment (rounded up)	4	3	3	10

31. Using the modified divisor 575.

Region	A	B	C	Total
Number of Houses	2592	1428	1980	6000
Modified Quota	4.51	2.48	3.4	
Webster's Apportionment (normal rounding)	5	2	3	10

32. Yes. Hamilton's Apportionment becomes 5, 2, 4.   Region B loses one truck.

33. Standard divisor = $\dfrac{690}{23} = 30$

Course	A	B	C	Total
Number of Students	311	219	160	690
Standard Quota	10.37	7.30	5.33	
Lower Quota	10	7	5	22
Hamilton's Apportionment	11	7	5	23

34. Use the modified divisor 28

Course	A	B	C	Total
Number of Students	311	219	160	690
Modified Quota	11.11	7.82	5.71	
Jefferson's Apportionment (round down)	11	7	5	23

35. Use the modified divisor 31.5

Course	A	B	C	Total
Number of Students	311	219	160	690
Modified Quota	9.87	6.95	5.08	
Adams's Apportionment (round up)	10	7	6	23

36. Use the modified divisor 29.5

Course	A	B	C	Total
Number of Students	311	219	160	690
Modified Quota	10.54	7.42	5.42	
Webster's Apportionment (standard rounding)	11	7	5	23

37. The new divisor is $\dfrac{698}{23} = 30.35$

Course	A	B	C	Total
Number of Students	317	219	162	698
Standard Quota	10.44	7.22	5.34	
Lower Quota	10	7	5	22
Hamilton's Apportionment	11	7	5	23

No. The apportionment remains the same.

38. The Standard divisor $= \dfrac{55000}{55} = 1000$

State	A	B	Total
Population	4862	50138	55,000
Standard Quota	4.86	50.14	
Hamilton's Apportionment	5	50	55

39. The apportionment is 4, 51.

40. The apportionment is 5, 50.

41. The apportionment is 5, 50.

42. The new divisor is $\dfrac{60940}{60} = 1015.67$

State	A	B	C	Total
Population	4862	50138	5940	60940
Standard Quota	4.79	49.36	5.85	
Hamilton's Apportionment	5	49	6	60

Yes. State B loses a seat.

Chapter Test

1. 6+5+5+4 = 20 members voted.

2. No candidate has a majority of ≥ 11 votes.

3. Chris wins with a plurality of 9 votes.

4. D = 41 pts., C = 44 pts., S = 35 pts. Chris wins.

5. Donyall wins with 11 pts.

6. D = 1.5 pts., C = 1 pt., S = 0.5 pt. Donyall wins.

7. a) Votes: H: 26+14 = 40, I: 29, L: 30, S: 43
   Thus, the snail wins.

   b) (H) 1st $(40)(4) = 160$
           2nd $(59)(3) = 177$
           3rd $(0)(2) = 0$
           4th $(43)(1) = 43$ H receives 380 points.
     (I) 1st $(29)(4) = 116$
          2nd $(40)(3) = 120$
          3rd $(73)(2) = 146$
          4th $(0)(1) = 0$ I receives 382 points
    (L) 1st $(30)(4) = 120$
          2nd $(43)(3) = 129$
          3rd $(43)(2) = 86$
          4th $(26)(1) = 26$ L receives 361 points

7. b) (S) 1st $(43)(4) = 172$
         2nd $(0)(3) = 0$
         3rd $(26)(2) = 52$
         4th $(73)(1) = 73$ S receives 297 points.
   The iguana (I) wins with the most points.

   c) A majority out of 142 voters is 72 or more votes.
      Votes: H: 40, I: 29, L: 30, S: 43; None has a
      majority, thus eliminate I.   Votes: H: 69,
      L: 30, S: 43   None has a majority, thus
      eliminate L.   Votes: H: 99, S: 43
      The hamster wins.

   d) H vs. I: I gets 1 pt.   H vs. L: L gets 1 pt.
      H vs. S: H gets 1 pt.   I vs. L: L gets 1 pt.
      I vs. S: I gets 1 pt.   L vs. S: L gets 1 pt.
      Ladybug wins with 3 points.

8. Plurality: Votes: W: 86,   X: 52+28 = 80,   Y: 60,
   Z: 58   W wins.

   Borda count: W gets 594 points, X gets 760 points,
   Y gets 722 points, Z gets 764 points   Z wins

   Plurality with elimination: A majority out of 284
   voters is 143 or more votes.

   Votes: W: 86,   X: 80,   Y: 60,   Z: 58

   None has a majority, thus eliminate Z.

   Votes: W: 86,   X: 80+58 = 138,   Y: 60

   None has a majority, thus eliminate Y.

   Votes: W: 86,   X: 138+60 = 198   X wins.

8. Head-to-Head: When Y is compared to each of the
   others, Y is favored. Thus Y wins the
   head-to-head comparison.

   Plurality, Borda count and Plurality with elimination
   each violate the head-to-head criterion. The pairwise
   method never violates the head-to-head criterion.

9. A majority out of 35 voters is 18 or more votes.
   Louisiana (L) has a majority.

   However, Mississippi (M) wins using the Borda
   count method with 115 points. Thus the majority
   criterion is violated.

10. a) The standard divisor $= \dfrac{33000}{30} = 1100$

State	A	B	C	Total
Population	6933	9533	16534	33,000
Standard Quota	6.30	8.67	15.03	
Hamilton's Apportionment	6	9	15	30

b) The divisor 1040 was used.

State	A	B	C	Total
Population	6933	9533	16534	33,000
Modified Quota	6.67	9.17	15.90	
Jefferson's Apportionment (round down)	6	9	15	30

c) The new divisor 1064.52

State	A	B	C	Total
Population	6933	9533	16534	33,000
Standard Quota	6.51	8.96	15.53	
Hamilton's Apportionment	6	9	16	31

The Alabama paradox does not occur, since none of the states loses a seat.

d) The divisor $= \dfrac{33826}{31} = 1091.16$

State	A	B	C	Total
Population	7072	9724	17030	33,826
Standard Quota	6.48	8.91	15.61	
Hamilton's Apportionment	6	9	16	31

The Alabama paradox does not occur, since none of the states loses a seat.

e) The new divisor is $\dfrac{38100}{36} = 1058.33$

State	A	B	C	D	Total
Population	6933	9533	16534	5100	38100
Standard Quota	6.55	9.01	15.62	4.82	
Hamilton's Apportionment	6	9	16	5	36

The new states paradox does not occur, since none of the existing states loses a seat.

APPENDIX

GRAPH THEORY

Exercise Set

1. A **vertex** is a designated point.

3. To determine whether a vertex is odd or even, count the number of edges attached to the vertex.
   If the number of edges is odd, the vertex is **odd**. If the number of edges is even, the vertex is **even**.

5. 5 vertices, 7 edges

7. 7 vertices, 11 edges

9. Each graph has the same number of edges from the corresponding vertices.

11. Odd vertices: *C, D*
    Even vertices: *A, B*

13. Yes. The figure has exactly two odd vertices, namely *C* and *D*. Therefore, the figure is traversable. You may start at *C* and end at *D*, or start at *D* and end at *C*.

15. Yes. The figure has no odd vertices. Therefore, the figure is traversable. You may start at any point and end where you started.

17. No. The figure has four odd vertices, namely *A, B, E,* and *F*. There are more than two odd vertices. Therefore, the figure is not traversable.

19. Yes. The figure has exactly two odd vertices, namely *A* and *C*. Therefore, the figure is traversable. You may start at *A* and end at *C*, or start at *C* and end at *A*.

21. a) 0 rooms have an odd number of doors.
    5 rooms have an even number of doors.
    b) Yes because the figure would have no odd vertices.
    c) Start in any room and end where you began. For example: *A* to *D* to *B* to *C* to *E* to *A*.

23. a) 2 rooms have an odd number of doors.
    4 rooms have an even number of doors.
    b) Yes because the figure would have exactly two odd vertices.
    c) Start at *B* and end at *F*, or start at *F* and end at *B*.
    For example: *B* to *C* to *F* to *E* to *D* to *A* to *B* to *E* to *F*

25. a) 4 rooms have an odd number of doors.
    1 room has an even number of doors.
    b) No because the figure would have more than two odd vertices.

27. a) 3 rooms have an odd number of doors.
    2 rooms have an even number of doors.
    b) No because the figure would have more than two odd vertices.

339

29. The door must be placed in room *D*. Adding a door to any other room would create two rooms with an odd number of vertices. You would then be unable to enter the building through the door marked "enter" and exit through the new door without going through a door at least twice.

31. Yes because the figure would have exactly two odd vertices. Begin at either the island on the left or on the right and end at the other island.

33.

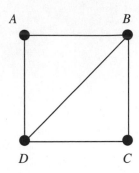

35. a) Kentucky, Virginia, North Carolina, Georgia, Alabama, Mississippi, Arkansas, Missouri
    b) Illinois, Arkansas, Tennessee

37. a) 4
    b) 4
    c) 11

39.

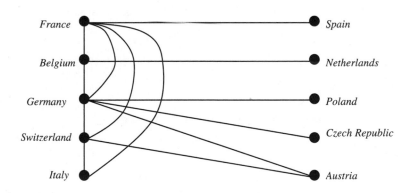

41.   a) Yes, the graph has exactly two odd vertices, namely *C* and *G*.
       b) *C, A, B, E, F, D, G, C*